WHEN
WE
DIE

WHEN
WE
DIE

The Science, Culture, and Rituals of Death

Cedric Mims

St. Martin's Griffin
New York

Photos: Simon Marsden, The Marsden Archive

www.stmartins.com

Library of Congress Cataloging-in-Publication Data

Mims, Cedric A.
    When we die : the science, culture, and rituals of death / Cedric Mims
        p.   cm.
    Includes bibliographical references and index.
    ISBN 0-312-20050-1 (hc)
    ISBN 0-312-26411-9 (pbk)
    1. Death.   I. Title.
R726.8.M56   1999
306.9—dc21   98-49816
                                                                  CIP

First published in the United Kingdom as *When We Die: What Becomes of the Body After Death*, by Robinson Publishing Ltd.

First St. Martin's Griffin Edition: July 2000

10 9 8 7 6 5 4 3 2 1

Contents

Preface   ix
Acknowledgements   xii
Sources   x

Part I: Death and its Causes   1

1 What is Death?   3
The Death of Cells   3
The Necessity of Death: Nature's Strategy   4
Why Immortality would Raise Problems   5
Death of a Whole Species: Extinction   7

2 The Main Causes of Death   9
The Broad Picture   9
How Long Do We Live?   10
The Quality of Life   11
What Do We Die Of?   12
Measuring the Impact of Death and Suffering   14
Causes of Death: The Big Killers   16

3 Suicide, Euthanasia, Homicide   30
Suicide   30
Euthanasia   45
Homicide   52

4 Ageing and Death   81
Life Expectancy and Life-span   81
Ageing in Animals and Humans   83
Ageing in the Test Tube   87
What Happens to us during Normal Ageing   89
What Causes Normal Ageing   93
How to Cheat Death   102
The Burden of Longevity   105

Part II: What Happens to Corpses? 109

5 The Body after Death 111
What Time of Day Do We Die? 112
Death as a Physical Event 112
Death and the Law 114
When Has Death Occurred? 115
From Rigor Mortis to Putrefaction 119
Consumption of the Corpse 121
How the Corpse Decomposes 122
Fossils and Petrification 124
Are Corpses Dangerous? 125
Nature's Need to Recycle the Corpse 125
Preliminary Treatment of the Corpse 126

6 Burial 132
Religion Makes a Difference 132
Is Burial Natural? 135
Simple Pit Burial 136
Mass Graves 137
Shrouds and Coffins 138
The Fear of Burial Alive 144
Delayed Burial and Reburial 145
Funerals 149
Burials at Sea 153
Tombs and Monuments 155
Cemeteries, Catacombs and the Columbarium 159
Green Burials 168

7 Exposure and Cremation 169
Exposure to the Elements 169
Cremation 171

8 Unusual Methods of Disposal 184
Cannibalism 184
The Acid Bath 188
The Compostorium 188
Eccentric Exits 189

9 Embalming and Mummification 190
Embalming 191
Mummification 195

10 Freezing and Other Methods of Preservation   210
Natural Freezing   210
Cryopreservation: Suspended Animation   213
The Bog Men   215
Alcohol, Formaldehyde and Other Preservatives   217
Eccentric Methods of Preservation   218
Preservation of DNA   220
Cloning   221

Part III: The Use and Abuse of Corpses   225

11 The Body and the Laboratory: Dissection   227
Who Owns a Dead Body?   227
Post-mortem Examination   228
Using the Corpse for Divination   231
The Study of Human History and Evolution   232
Ethical Problems with Human Remains   238
Dissection and the Training of Doctors   241
Alternatives to Dissection   249
Further Uses for Whole Corpses   250

12 Using Parts of the Body: Transplantation   252
Bones, Teeth, Placentas, Hair   252
Spare Part Surgery: Organ and Tissue Transplants   255
Where Do Transplanted Organs Come From?   259
Using Dead or Renewable Human Material   275
Medicinal Uses of Corpses   280
The Amazing World of Human Relics   281

13 The Abuse of Corpses   286
Punishing the Corpse   286
Cannibalism Revisited   288
Understanding Human Sacrifice   289
Using Corpses for Fertilizer, Fat, Leather   292
Necrophilia: Macabre Eroticism   294

14 Identifying Bodies and Parts of Bodies   295
Whose Body?   296
Murder Investigations and the Corpse   300
Whose Cells?   308
Who Owns DNA Sequences?   309
Removing Parts of the Body   310

Part IV: Death and Afterlife   315

15 Death and the Corpse: The Emotional Impact   317
The Prospect of Death   317
Seeing the Body   320
Corpses and Death Avoidance   322
The Comparison of Death with Sleep   323
Belief in an Afterlife and Anxiety about Death   326
Death as it Seems to Children   329
Old People and Death: Hospices and the Care of the Dying   330
Mourning and Grieving   333
What to Do when Someone Dies   337

16 The Afterlife and the Future of Corpse Disposal   338
The Afterlife and the Soul   338
Religion and the Afterlife   341
Reincarnation   349
Near-death Experiences   349
Ghosts, Spectres, Hauntings   353
The Transit to the Afterlife   355
The Future of Corpse Disposal   356

References   359
Index   364

Acknowledgements

What started me on this trail were my fascinating lunchtime conversations with the late Professor Keith Simpson at Guy's Hospital, London. Ruth Richardson's pioneering book *Death, Dissection, and the Destitute* has been a great inspiration. I cannot match its careful scholarship, but I owe much to her infectious enthusiasm.

I am also indebted to the following for their interest, encouragement and help, although they are in no way responsible for what I have written: Professor Matteo Adinolfi, Dr John Cairns, Vicky Mims, Dr Terry Gibson, Professor A. K. Mant, Jennifer Williams, Norman and Heike Mims, Dr Toshido Inada, Dr Bryan Furnass, Mr L. R. Squires and Mr J. R. Saunders.

Picture Credits

Sources

Lines from Julian Litten's *The English Way of Death. The Common Funeral since 1450*. Reprinted with permission from *The English Way of Death*, by Julian Litten, first published by Robert Hale Ltd in 1991.

Lines from *Ulysses* by James Joyce. Copyright © the Estate of James Joyce.

Extract from *Autobiography and Other Essays* by G.M. Trevelyan, reprinted with permission from *Autobiography and Other Essays*, by G.M. Trevelyan published by Longman, 1949.

Lines from *The Fatal Shore* by Robert Hughes. First published in Great Britain by Collins Harvill, 1987; © Robert Hughes 1986. Reproduced by permission of the Harvill Press.

Ruth Richardson for lines from *Death, Dissection and the Destitute*, Penguin Books, 1989.

Preface

Most of us would guess that the dead vastly outnumber the living. Three hundred years ago it seemed clear enough to the English poet Edward Young (1683–1765): 'Life is the desert, life the solitude; Death joins us to the great majority.' Today, though, it is not immediately obvious. The recent dramatic increase in human numbers from about 2,300 million in 1950 to their present level of 5,800 million, with a projected further rise to 8,300 million by the year 2025, makes one wonder whether at some stage there could be more of us alive than all of our ancestors put together. But it is not so. About 1 per cent of the population dies each year and altogether about 130,000 million humans have lived and died since the first emergence of our species. But most of them have lived in the past century, and it has been calculated that each of us today has no more than about twenty ancestors standing behind us. To hold all their corpses you would need a gargantuan cubical coffin measuring about 3 miles (5 km) along each side. They could then be disposed of *en masse* with no great impact on the landscape.

In his poem 'The Grave', Robert Blair (1699–1746) refers to the steady accumulation of corpses:

> What is this world?
> What but a spacious burial field unwalled,
> Strewed with death's spoils, the spoils of animals
> Savage and tame, and full of dead men's bones?
> The very turf on which we tread, once lived:
> And we that live must lend our carcases
> To cover our own offspring: in their turns
> They too must cover theirs. 'Tis here all meet!

Luckily, corpses do not stay around. They disappear, decompose, are degraded, and their component molecules are reutilized in nature's elemental cycles. In neolithic (pre-urban) times, with an estimated total human population of about 5 million and about one living person per 10 square km, disposal of the dead presented few problems. But when humans congregated into towns and cities, special burial sites became necessary. Cities, moreover, were until recently unhealthy places in which deaths exceeded births, and were often referred to as 'graveyards of mankind'. Human numbers have increased dramatically in the past 200 years and in the USA and western Europe there are now an average of 50–200 humans per square km. The weight in human flesh of those alive at the moment (5.8 billion) is about 300 billion tons. The never-ending stream of corpses has become a torrent: the 627,636 who died in the UK in 1994 would, if laid side by side, extend for 178 miles.

Increasingly efficient methods were needed to deal with such massive increases in the rate of production of dead bodies. Medieval churchyards overflowed long ago, being replaced first by public cemeteries and later by modern crematoria. Most people now die in hospitals or other institutions rather than at home, and their bodies are discreetly removed and cremated. France and England already have funeral supermarkets (the latter in Catford, London, but run by a French firm) where you can load coffin, urn or granite angel into your trolley. We have banished death from everyday life and many of us have never seen a corpse. Yet the funeral itself, which could in the end turn into a mere refuse disposal event, is still something that matters and still retains reminders of its ancient ritual character.

This book is a light-hearted but wide-ranging survey of death, the causes of death, and the disposal of corpses. It tells why we die and how we die, and what happens to the dead body and its bits and pieces. It describes the ways corpses are dealt with in different religions and in different parts of the world (burning, burying, exposing); the methods for preserving bodies (embalming, mummification, freezing); and the ways – fascinating in their diversity – in which corpses or parts of corpses are used and abused (dissection, organ transplants, cannibalism, human sacrifice).

I have also given consideration to our attitudes to death and to the afterlife. We are the only living things that know we will die. Evolution has endowed the human brain with the consciousness that is the core of humanity: 'I think, therefore I am.' We can contemplate the future, think about it and worry about it. Surely it is unacceptable that each one of us should be snuffed out for good, for ever, as soon as our nerve cells have died? Not unexpectedly, many believe that there is something other than the physical body, a soul or a spirit that will survive putrefaction or cremation. Belief in some sort of afterlife has been an almost universal human preoccupation, the driving force behind some of the great religions and the source of much of the world's art, architecture and literature.

This overview of death has led me into strange territories, and I have noticed that the more you think about the subject, the less gruesome it becomes. A variety of fascinating topics had to be explored, most of them with a brevity bordering on the scandalous, and then put in place so that the theme was pursued systematically. I hope the reader will be able to share my interest in the use of body parts (transplantation) and the subject of ageing. In many areas I have necessarily relied largely on secondary sources, and there must be errors, but the medical sections at least lay a claim to accuracy.

I have read widely and used material from other books, as listed in the References. Leonardo da Vinci (1452–1519) once said that 'anyone who invokes authors in discussion is not using his intelligence but his memory'. But in the last years of the twentieth century it is not easy to say things that have never been said before, and I take comfort from the advice of the French novelist and poet Anatole France (1884–1924): 'when a thing has been said and said well, have no scruple. Take it and copy it.'

What can we learn from this odd assemblage of data? Is it more than an entertaining collection of facts and anecdotes? Are there any take-home messages? The only conclusion I would suggest is that we should take a more matter-of-fact view of death, that we should accept it and talk about it more than we do – as we have done with the once taboo subject of sex. As we enter the third millennium, in an era of hitherto undreamed of possibilities for communication, and whether the old religions fade away, change or persist in their power, we should take on board death and its trappings.

PART I

Death and its Causes

1

What is Death?

When I consider the short duration of my life, swallowed up in the eternity before and after, the little space which I fill, and even can see, engulfed in the infinite immensity of spaces of which I am ignorant, and which know me not, I am frightened, and am astonished at being here rather than there, why now rather than then . . . the eternal silence of these infinite spaces frightens me.

From 'Pensées', by Blaise Pascal, French
mathematician and moralist (1623–62)

The Death of Cells

In the body cells are always dying. Blood cells, cells in the skin, those lining the intestines: all are either shed like leaves or degenerate and die. A white blood cell lives for only a few days, and each day we lose millions of cells from our skin and intestines. The fine white dust that you pick up on your finger when you run it over a shelf or other surface consists mostly of dead skin cells, and during a lifetime each of us sheds about 18 kg of skin. The live, naked cells lining the intestines suffer continual physical damage, and are shed after a few days. To replace all these lost cells and keep the body intact, other cells are constantly dividing. Cell death is therefore the natural state of affairs, and 'In the midst of life we are in death.' That passage from the the service for the burial of the dead in the Book of Common Prayer takes on a biological meaning.

You could say that these cells don't have to die, and that nature could have arranged for them to live much longer. But at the skin surface and in the intestine the inevitable mechanical damage is best met by continually shedding cells and replacing them with new ones. And the white blood cells, armed with powerful chemical weapons for use against invading microbes, need to be regularly dismantled and replaced by new ones. It is a common belief that cells in the test tube go on multiplying for ever. But the fact is that no cell can manage more than about sixty divisions (see chapter 4). After that, the cell ages and dies. A few types of cell, like nerve cells or heart muscle cells, stay as they are throughout life, all the time recreating themselves as old molecules are replaced by new ones, but not actually dividing.

The reasons why cells have to die becomes more obvious when we consider the development of the embryo. During this period the growing organs are always having to be remodelled and reshaped. Certain structures have to be demolished and cells destroyed. Our tail and our gill slits, for instance, present in the early embryo as we recapitulate our origin from primitive vertebrates, must be altered and diminished at later stages in development. In the same way the tadpole's tail is disposed of as the tadpole turns into a frog. As these events unfold, cells have to be killed off. A great deal of destruction accompanies the process of construction. Accordingly, to take care of these needs, all cells have a special 'autodestruct' or suicide program built into them. It can be switched on as required, and is an essential resource during development, and in certain infections when cell suicide is the best strategy for defeating the attack. It is called 'apoptosis' and is described in chapter 4.

The Necessity of Death: Nature's Strategy

Life is a process of constant change. All living things must reproduce and multiply, and when they die their offspring take their place. But the places are limited. The opportunities, or in modern jargon the ecological niches, are not infinite. There is a limited amount of room on the earth. This means competition, and the best fitted ones are going to survive and out-reproduce the others. This is how evolution works. Without death, the world would soon fill up with whatever creatures were present at the time, and there would be no more change, no more evolution. Without death, a single cell, after dividing each day for many weeks, would have produced hundreds of tons of cells and these would rapidly cover the surface of the earth. Nature is so prolific that death has to take a hand, even at the level of the elephant. If a female elephant bore six young during her lifetime and they all survived and reproduced at the same rate, then after 700 years the descendants of a single pair would number about 18 million. Hence the struggle for existence. Death is necessary. To die and leave the stage is the way of nature. It was put simply by the French moralist Montaigne (1533–92) in his essay 'To Study Philosophy is to Learn to Die': 'Give place to others, as others have given place to you!'

Death takes other things besides the individual at the end of life and the cells in the developing embryo. Certain objects are needed at one time but can be discarded when they have served their purpose. The placenta is doomed to die after the birth of the offspring, and is usually eaten by the mother (except in human beings). The umbilical cord soon dries up and dies, leaving its mark. Adam, strictly speaking, should be represented without a navel.

Why, we might ask, has nature chosen the strategy of death, the strategy of billions of short lifetimes, rather than some other basis for life? It is because this is the only way to ensure change, which, together with competition, is the driving force of evolution. The ultimate reason for sex is that it is a method for mixing together the genes of different individuals. It increases the variety of gene mixtures, and gives evolution something to work on.

The best way of looking at it is to remember that your germ cells (eggs, sperm) are fundamentally different from the rest of you. These cells, or a few of them at least, will outlive you, surviving after the egg has been fertilized and then dividing to form a new individual, your offspring. All the rest of you, all your somatic (bodily) organs and cells, die when you die. It is your DNA, your genes in the eggs and the sperm, that survive in your descendants. This is the pathway for changes in genes being handed down through the generations and allowing evolution. You and your body are no more than a device by which the germ cells ensure their immortality. To put it the other way round, your body sacrifices itself so that the germ cells can live on. Nature cares about the survival of your DNA rather than the survival of you yourself. The distinction between the immortal line of germ cells (the *germa*) and the mortal rest of the body (*soma*) was made more than a hundred years ago and is a useful one.

Why Immortality would Raise Problems

One other possibility would have been for nature to have produced superorganisms that never aged or died. But agelessness would have serious drawbacks. First, it would have made it impossible to achieve the drastic transformations in living creatures over millions of years that were needed to adapt to changing conditions. Animals and plants have had to undergo fundamental changes in response to alterations in climate, food, predators and so on. They did it by producing new individuals (offspring) at regular intervals, each generation showing slightly altered features. This allowed for change. The penalty for failure to change and adapt was death. Second, it would mean that as the old individuals accumulated there would soon have been no room for future generations. Third, there are daunting biological problems in designing an immortal.

One of these biological problems concerns DNA. Our genes have to put up with constant low-grade bombardment and irradiation damage which comes fom rocks and from outer space. All cells make occasional mistakes when they are making a second copy of their own DNA in preparation for division. These changes in DNA are called mutations, and they are nearly all harmful. Initially, most of the changes are corrected or repaired, but as cells get older they are less able to carry out the repairs. The DNA abnormalities accumulate and cell functions are interfered with. This has a lot to do with ageing and cell senescence, and is dealt with in more detail in chapter 4. How would the immortals get round this problem? An immortal species, if ever it arose, would be staking its existence on a single solution to the problem of survival: its solution. Assuming that thousands of other species were still around, these would still be undergoing the relentless change and adaptation that has been the stuff of life ever since it began. The immortals would have to watch out and keep these other species in their place, not permitting any developments that might threaten their supremacy. They would also have to control their own numbers at an optimal level for the environment. They would have to look after their own evolution, and in doing so would be distorting the archetypal rules of the game. They would have replaced nature.

Am I giving a description of the human species at some distant future time? Once all the secrets of DNA are solved and we know exactly how to make whatever we want, there will – theoretically – be no limits on what we can do to manipulate human development. We have probably already exempted ourselves from the ancient rules of nature, because the 'unfit' now survive; and little is known about what changes this is making in our gene pool. Once we had the capacity to modify, change, even improve our genes in the laboratory, we could take over our evolution – as we have already taken over the evolution of dogs, cats, cattle and other domestic animals. It seems unlikely that our present characteristics, the ones that evolution selected out as best for a hunter-gatherer life 100,000 years ago, for a life of uncertainty, famine and disease, would be appropriate for human life in the distant future.

Contemplation of such possibilities arouses a host of ancient fears. What opportunities for madmen, for despotic rulers, for mad or at least unethical, scientists! Words like 'cloning' (see chapter 12) add to the foreboding, and we regard this picture of the future with dismay. But it will almost certainly come to pass, and will probably not be so bad as many fear. For the apprehensive, there is the reassuring thought that checks and brakes will still exist in the form of the old-fashioned, uniquely human qualities like wisdom, common sense and, perhaps most old-fashioned of all, lovingkindness.

Death of a Whole Species: Extinction

So far, we have been looking at the individual cell or the individual living creature, but the idea of death applies also to the species. It is a fact that 99.9 per cent of all the species that ever existed are now extinct. Those alive today are the tips of the tiniest twigs of the tree of evolution. Only about one in every ten thousand species that ever lived are still around. Throughout life's history there have been one or two extinctions every week. In terms of its emotional impact the death of a species may be less disturbing than the death of an individual creature; but it is still death, on a larger and equally irreversible scale.

An extinction can take place quite quickly. When people think of an extinct species they often think of the dodo, a flightless bird about the size of a turkey that used to live on the island of Mauritius. Within a hundred years of their arrival (by about 1681), human settlers on the island had exterminated it. In 1810 another bird, the American passenger pigeon, existed in its millions. Migrating flocks darkened the skies and the weight of their numbers broke great branches from trees. As late as 1871, 136 million pigeons were concentrated in a single nesting area of 850 square miles in Wisconsin. Yet by the end of the century the species was rare, and the last individual died in Cincinnati Zoo in 1914. The International Council for Bird Preservation lists 108 species of birds, worldwide, as having become extinct since the year 1600 (the total number of bird species is about 9,000). Most of these extinctions have been caused by human activities. However, deaths of whole species have been a regular feature of evolution, essential for change and progress. At times, too, there have been mass extinctions, in which vast numbers of species have been wiped out. One such episode occurred at the end of the Cretaceous period, 66 million years ago, when the dinosaurs and much of marine life perished. One explanation is that an asteroid hit the earth, forming a global cloud of dust particles which blocked out light and heat from the sun, causing radical climatic change with which many species could not cope.

We appear now to be in the midst of another mass extinction, caused by our own actions. Thousands of species are dying out each year as humans destroy their habitats or kill them directly. Many species have been obliterated by human hunting. Maybe this is acceptable when it is done for food, as in the case of the giant birds called moas in New Zealand, hunted to extinction within a few hundred years by the Maoris; but doing it for fun is another matter. Elector John George II of Saxony, who reigned between 1656 and 1680, was crazy about hunting, and he shot an unbelievable total of 42,649 red deer. Luckily he did not extinguish the population. Today, we tend to make a fuss when larger, more familiar species are threatened with extinction; but count-less less well-known ones are disappearing all the time. It is hard to absorb the fact that living species are now becoming extinct at 100–1,000 times the average rate in the geological past. On the other hand, very many are still with us. Around 1.5 million species are named and recorded (a disproportionate number of them beetles), and there are probably a few million more still not recognized and classified.

The answer to the threat of accelerating extinction rates, as humans swarm over the surface of the globe, their houses and food-growing areas displacing other creatures, is to maintain selected species in zoos. The last wild specimen of the European bison (also known as the wisent) died in 1925, but the species has been saved by breeding it in zoos. Maintaining species in zoos is not just preservation for the sake of it, or to satisfy mere curiosity: it enables study and provides for the education and delight of future generations.

2

The Main Causes of Death

Anyone can stop a man's life, but no one his death; a thousand doors open on to it.

Marcus Seneca (4 BC–AD 65), Roman philosopher and poet

Until the nineteenth century the world's population stayed well below a thousand million, but since then it has increased at an alarming rate. By mid-1996 the total stood at 5,800 million, and it is increasing at 86 million a year (about 167 born every minute). By the year 2000, it is estimated, there will be 6,158 million of us; by 2025 8,300 million. Presumably the figure will soon be double what it is now. At present at least half of the world's population live in cities of more than a million people. There are 280 cities of this size, three times as many as in 1950, and all the new ones are in developing countries. Birth vastly outnumber deaths.

These are disturbing figures; but the future estimates are not uncontested. Some groups of experts say that the world population is unlkely to double in the twenty-first century, but instead will reach a peak around 2070–80 and then decline. They say, for instance, that fertility is already beginning to fall in many developing countries. Indeed, there are so many factors to take into account that all long-term predictions should be suspect. Most of the people who will be alive in 2020 have already been born, so we can take them into account and make reasonable calculations up to that time. But later in the next century? Will famines continue, will AIDS have been controlled, could new devastating infections appear, and, most important of all, are the elderly going to increase in number until we get near to the 'maximum' natural human life-span?

The Broad Picture

One of the attractions of studying birth and death is that, unlike health and happiness, they can easily be measured, counted. It is true that not everything that counts can be counted, but there is something to be said for knowing about numbers, knowing how often something occurs. For instance, being stung to death by a scorpion is an alarming thing to think about, but can be put into perspective by discovering that only 1,000 people a year worldwide die in this way, compared to 4 million a year who die as a result of of accidents and violence (a total in which automobile-related deaths loom large).

Death rates in developed countries (Europe, North America, Australia, New Zealand, Japan) have declined more or less continuously through the twentieth century. Before 1930 this was due for the most part to fewer deaths during infancy and childhood. By the early 1950s more than 94 per cent of newborn infants could expect to survive to adult life, and by the late 1980s 98–99 per cent. Now, nearly all of the 11 million deaths a year in developed countries are of adults. And the number of deaths is catching up with the number of births. In the UK deaths will exceed births from about the year 2024, at which time the population will have levelled off at about 60 million. The picture is different in developing, poorer countries in parts of Africa, South America and Asia. Here, infant and child mortality is still high but lower than in developed countries in earlier centuries. Most of these deaths are due to infectious and parasitic diseases, often coupled with malnutrition. But we have to make a distinction between the death rate, meaning the deaths per 100,000 population, and the total deaths in the population. For instance, nearly half of the total number of people dying with heart disease and stroke are in developing countries, but in the relation to population the rate per 100,000 people is higher in developed countries such as Finland, UK, New Zealand, Sweden and the USA.

How Long Do We Live?

The 'natural' span of a life used to be said to be three score years and ten—seventy—but the latest available figures (1996) shows that in England and Wales a baby boy can expect to live for seventy-four years, and a girl for eighty years. From a statistical point of view this is now the normal human life expectancy. What it means is that in future years more and more of the population will be elderly. In the UK in 1995, 7 per cent of the population were already over seventy-five; this proportion will increase to about 10 per cent by the year 2020. When you get old you are more likely to suffer from conditions like heart disease, stroke or cancer, and these determine when you die – that is, the natural life expectancy. Death before this 'natural' limit can be regarded as premature. We can divide deaths into those that are 'expected', on account of age, and those that are premature, occurring earlier than normal. For practical purposes, death is not generally expected when it takes places between adolescence and senescence, say between the ages of fifteen and sixty-five. Death below the age of sixty-five can therefore be counted as premature.

In chapter 4 we will consider something quite different, the 'maximum' life-span of humans: in other words, how long they would live if none of the usual causes of death (heart disease, stroke, cancer, etc.) killed them, and death was due to nothing but 'old age'. We shall see that the maximum human life span goes up to about 115 years.

The Quality of Life

When and why we die are not, of course, the only things that matter. The quality of life is of just as much significance as its length, though less easy to measure. Among the most common medically recognized conditions affecting the quality of life are what are called mood or 'affective' disorders (meaning depression or other disturbances bad enough to require medical help), mental retardation, blindness (which affects 27 million people), and parasitic worm infections such as filariasis. An astonishing 665 million people suffer from goitre (swelling of the thyroid gland in the neck).

When the World Health Organization was established fifty years ago it defined health as 'a state of complete physical, mental and social wellbeing and not merely the absence of disease or infirmity'. This was an inspiring idea, but it had its disadvantages. A state of complete physical, mental and social well-being is more like a definition of happiness than of health. The two are interconnected, but should be distinguished. Health is perhaps a universal human right, and to a large extent it can be delivered by society. Happiness, on the other hand, is more subjective and may involve love, friendships, religion and many other factors. It is harder to cater for, and although it partly depends on good health it cannot be delivered in the same way. The word 'happiness' itself originally (before about AD 1500) meant something that came by chance or good fortune. The American Declaration of Independence (1776) includes as inalienable rights life and liberty, but no more than the right to *pursue* happiness. Moreover, there can be conflicts between the two goals. If a man gives up alcohol or cigars he may be healthier; he will not necessarily be happier. The most important reason for distinguishing between the two is that if you try to guarantee the unattainable (happiness for everyone) it will divert resources from pursuit of the attainable goal of good health for all.

What Do We Die Of?

There were an estimated 52 million deaths across the globe in 1996, 40 million of them in the developing world. If 52 million is too large a number to comprehend, consider that it works out at about 100 deaths each minute. In the UK in 1994 there were 627,636 deaths (about one a minute), more than a half of them of people aged over sixty-five. The main causes of death, whether expected or premature, are listed in table 1, using figures given by the World Health Organization for 1996. They are as reliable as any figures can be, but it has to be remembered that death certificates are made out for only about a quarter of all deaths. They do not include the deaths due to great famines, massacres, wars or natural disasters. We will deal with these causes below when we come to consider premature deaths more specifically.

Table 1   The top ten causes of death in 1996 (all ages)

Cause of death	No. (millions)	% of total
Infectious and parasitic disease	17.3	33
Diseases of circulation (stroke, heart attack, etc.)	15.3	29
Cancer	6.3	12
Death at birth or shortly afterwards (prematurity, asphyxia, trauma, infection)	3.7	7
Chronic lung disease (chronic bronchitis, emphysema)	2.9	6
External causes (accidents, suicide)	1.0	2
Malnutrition, diabetes and related conditions	0.9	2
Death of mother as a result of pregnancy, childbirth	0.6	1
Mental disorders (e.g. dementia, alcohol dependence)	0.3	<1
Miscellaneous or unknown	3.7	7
Total	52	100

Source: World Health Organization figures.

When we die, and of what, depends on many factors. The very young and the old are especially vulnerable. Obviously, where you are in the world, and the health and wealth of your country, are also relevant. Infectious and parasitic diseases cause just 1 per cent of deaths in the developed world, but 43 per cent in the developing world. Stroke and heart disease, on the other hand, account for 46 per cent of deaths in the developed world but only 24 per cent in the developing world, although the actual number of deaths per year in the latter is nearly double that in the former.

One important influence is poverty and, in the UK at least, employment. Unemployment is now a fact of life in developed countries, and death rates are higher among the unemployed than among the employed. The exact reasons for this are not clear; to complicate matters, there is the possibility that unemployed people as a group differ in other ways from the employed, and that these differences make them more likely to die earlier. Socio-economic status influences the death rate, and this too is not well understood. Unskilled workers tend to smoke more and perhaps eat less healthy diets, but this is only part of the story.

In the UK the likelihood of your dying early depends a great deal on where you live. In parts of northern England you are almost twice as likely to die prematurely as in parts of the south. Death rates have fallen steadily since the 1950s, including deaths during the first year of life, but the 'north – south' difference applies also to these deaths. A girl born in Leeds is more than twice as likely to die during her first year than a girl born in a Dorset town. The reasons for the gap are not clear, though it looks as if poverty has a lot to do with it. But the socio-economic factor in disease and death has been with us for hundreds of years, as was shown recently in a study of Glasgow graveyards. The environment in childhood seems to matter more for some diseases, and the environment in adult life for others. In the English Civil Service, men in the lower grades were at three times the risk of dying from heart disease than men in higher grades. The level of risk depended also on how much control they had over their job, whether they had a say in planning and deciding exactly how the job was done and how fast it was done.

As a single cause of illness and death you cannot beat cigarette smoking. In the USA in 1995 no fewer than 2.25 million deaths, a quarter of all deaths, were due to diseases attributable to smoking. In Europe in the same year there were 1.2 million deaths attributable to smoking (14% per cent of all deaths), and in the UK five women and eight men die of smoking *every hour*. The smoking tragedy is now shifting to developing countries where anti-smoking forces are weaker. Nearly a third of the cigarettes made in the USA are now exported, and China already has more tobacco-related deaths than any other country. The argument that tobacco brings in tax revenues and is needed to keep tobacco farmers alive is fallacious. The Chinese government calculated that in 1993, while the tax revenues from cigarettes were 41 billion yuan (£3.5 billion), the economic losses due to tobacco illnesses and death amounted to 65 billion yuan.

Measuring the Impact of Death and Suffering

Death awaits us all, and is the only certain thing about life. Also, deaths are easy to count. It is harder to measure the great load of human suffering that represented by premature deaths and the extra burden caused by serious diseases and disabilities. But it can be done: one method is by calculating what has been called 'disability adjusted life years' (DALYs). These are the total years of life lost in a population because of early death, plus the years spent suffering from diseases and disabilities (the numbers being adjusted according to the severity of the conditions). A pioneering team at the the Harvard School of Public Health and the WHO, led by C.J.L. Murray and A.D. Lopez, have focused on DALYs in their 'Global Burden of Disease' study. By using the DALY to estimate the total burden of suffering and life-shortening conditions, interesting comparisons can be made. The developed countries suffer only 11.6 per cent of the global burden but account for 90 per cent of the total spending on health. This we might have predicted, but it is useful to set a figure on it.

The worldwide picture is changing all the time, and there are major differences between established market economies and the developing world. Worldwide the top ten causes of life-shortening and chronic suffering in 1990, in order of importance, were:

1 Lung infections (especially in children)
2 Diarrhoeal diseases and dysentery
3 Perinatal disease (occurring around the time of birth, including prematurity, pneumonia, birth asphyxia and trauma)
4 Major mental depression
5 Heart disease
6 Diseases of blood vessels (e.g. stroke)
7 Cancer
8 Measles

9   Road traffic accidents
10 Congenital abnormalities (birth defects)

Malaria, homicide, suicide, falls and other accidents come lower down the list. Taking into account changes since 1990, AIDS would now be in the top ten, but measles probably not. Calculating DALYs is a complicated business, with many variables, and not everyone agrees with the details, but it gives an idea of the main causes of suffering and life-shortening.

If we go further, and try to assess to what extent these conditions are attributable to environmental risks, an interesting picture emerges. This is worth doing because the environmental causes are theoretically preventable, and therefore the exercise is relevant to setting priorities in health programmes. The top ten risk factors come out as follows:

1	*Malnutrition*	Major impact in young children, results in increased susceptibility to measles, diarrhoea, etc.
2	*Poor sanitation*	Unclean water, poor sewage disposal and inadequate hygiene account for the burden of intestinal infections and parasites
3	*Unsafe sex*	Accounts for AIDS, a dozen other sexually transmitted diseases, and cancer of the cervix (due to genital warts)
4	*Alcohol*	Alcohol lurks behind road accidents, murders and domestic violence, as well as causing liver disease (the figure is corrected for the beneficial effect of moderate alcohol intake on heart disease)
5	*Occupation*	Exposure at work to physical and chemical injury
6	*Tobacco*	Causes most lung cancer and plays a part in several other diseases; the burden is increasing as smoking spreads across the developing world.
7	*High blood pressure*	Causes stroke
8	*Physical inactivity*	A factor in heart disease; overeating may also play a role here

9	*Illicit drugs*	Spread of infection via needles; also a cause of violent crime
10	*Air pollution*	Impact of particles and sulphur dioxide on chronic lung disease

Doing something about malnutrition and sanitation may cost money. But it looks as if *human behaviour* is responsible for most of the world's suffering and premature death.

Russia provides an example of the effect of behaviour. Since the collapse of the Soviet Union in 1991 there has been a huge rise in death rates, mainly in young adults and middle-aged people. The biggest rise is in deaths connected with alcohol, accidents and violence, although heart disease and infectious diseases also contribute. High on the list of causative factors are alcohol, smoking and diet. By 1993 the average alcohol consumption per head in Russia was 40 grams a day (28 units a week, compared to current recommended limits of 21 units a week for men, 14 for women), and diets tend to be low in vegetables and fruits.

Causes of Deaths: The Big Killers

Let us look more closely at the main causes of death, the things that kill us either at or before our allotted time.

Man-made Death: War, Massacre, Starvation

One category that does not appear in the WHO figures (see Table 1) is that of people killed directly (in war, massacres) or indirectly (by starvation, imprisonment, labour camps) *by other humans*. Man-made deaths in the twentieth-century have been estimated at about 110 million. About 20 million died in the First World War; about 35 million in the Second World War. These figures dwarf the numbers killed in earlier centuries – not because we are becoming inherently more evil, but because populations are so much larger. There are so many more of us to suffer and die in wars, labour camps, famines, etc. If anything, men were more evil in past centuries but the consequences were on a smaller scale.

The man-made deaths described here are caused by the actions of governments and nations. This, and the vast numbers involved, put them in a different category from homicide, which is discussed in chapter 3.

Killings that are systematically inflicted on an identifiable group of people, often over a period of years, merit the term 'genocide' rather than 'massacre'. Genocide has sometimes been carried out, or attempted, by indirect methods such as starvation and neglect. The Spaniards killed off more than a million Caribbean Indians between 1492 and 1600, and another million Central and South American Indians between 1498 and 1824. European settlers in North America, beginning in 1620, disposed of more than a million Native Americans. Genocide, as noted by Jared Diamond in *The Rise and Fall of the Third Chimpanzee*, has continued well into the twentieth century. Early settlers in Australia were responsible for the death of more than 100,000 Aborigines between 1788 and 1928, and between 1950 and 1990 genocidal killings took place in at least fifteen countries in Africa, South America and South-East Asia.

Our century has seen enormous and praiseworthy efforts into protecting and saving human life (by, for example, medical advances, better supplies of food), yet at the same time enormous ability and willingness to destroy life. In *The Twentieth-Century Book of the Dead* (1972) the writer Gil Elliot made some interesting although very approximate estimates of man-made deaths in the twentieth-century. (It was Elliot who came up with the figure of 110 million quoted above.) He made a distinction between deaths caused by guns and bombs (50 million) and deaths caused by privation – defined as deprivation of the necessities of life: water, food, shelter and clothing (60 million). He further divided this latter category into privation in cities (16 million), in labour camps, mostly in the Soviet Union (20 million) and in conditions of famine, blockade and war (24 million). Slightly more than half of these 110 million were males. Particular appalling episodes produced 'cities of the dead': In the First World War, Verdun (1 million); in the Second World War, Leningrad (1 million) and Auschwitz (1.5 million). In all, he estimates that at least 4.5 million Jews died in the Second World War: 2 million from hunger, disease and privation, 1.5 million by gas, 1 million by being shot. Elliot noted that figures of this magnitude are beyond our comprehension, and suggests that further research be undertaken so that such estimates can be put on a sounder statistical footing.

In the half-century since the end of the Second World War more than 20 million people have been killed in about 150 local wars in Africa, Asia, Europe and Latin America. Millions of them were children. Many more were disabled, especially by land-mines. Modern weapons are manufactured in industrialized countries, and about $35 billion worth a year are sold or given to developing countries.

It would be unthinkable to write a book about death without

mentioning some of the most terrifying tools of war manufactured by humans: chemical weapons. They have the capacity to kill people by the million, and unlike nuclear weapons they do not destroy cities and property but leave things (once the corpses have been disposed of) as good as new.

The first chemical weapon, the gas chlorine, was developed in 1915 by the German Nobel Prize winner Fritz Haber. It was discharged from cylinders in the direction of enemy positions at Ypres in the First World War. In a single month in the spring of 1915 no less than 500 tons were released from 20,000 cylinders. But it needed to be combined with a large-scale infantry attack, was too dependent on the weather and was not very effective. The British used it later in the year with no greater effect.

The Germans then set to work on the next possibility, which was phosgene. It was first used in December 1915, and, unlike chlorine, which formed a visible cloud, it was practically colourless and odourless. After inhaling it the victim might feel nothing except some irritation of the eyes and throat, but the lungs then filled up with fluid and death followed. Phosgene is about twenty times as powerful as chlorine. The Allies used it, together with chlorine, at the Battle of the Somme in 1916, and over the next nine months 1,500 tons were discharged. The meteorological problems were taken care of by firing shells which exploded and released the gas on reaching the enemy up to a mile away. By the end of the First World War 95 per cent of all gas used was delivered by artillery.

Next, in 1917, the Germans began to use a new weapon: mustard gas (dichlorethyl sulphide). It smelt a bit like mustard and caused only slight irritation after being inhaled, but then devastated the lungs. Death occurred over the next few days, as the lungs solidified. It was used in liquid form, and stayed around on the ground after use, polluting water. By the end of the war the British had lost about 2,000 men from mustard gas; another 125,000 were casualties and had to be taken to hospital. The military impact was considerable. The British began to produce it, but the war ended before they were able to make much use of it. One survivor of a British mustard gas attack in October 1918 was a twenty-nine-year-old corporal in the Bavarian Reserve Infantry called Adolf Hitler.

The next major development in chemical warfare was the emergence of nerve gases. In about 1936 the German scientist Dr Gerhard Schrader discovered a series of poisons of unprecedented power: nerve agents, which block the action of an essential substance in nerves called cholinesterase, causing the body to lose control over muscles of the limbs and essential functions such as respiration. Death is due to asphyxiation. The first of these substances was tabun. The next, more powerful still, was discovered in 1938 and called sarin, after the German scientists who had developed it (Schrader, Ambros, Rudriger, and van der Linde). One drop on the skin is lethal. Whole cities could be wiped out.

During the Second World War the Germans produced immense quantities of these poisons. By the end of 1944 Germany had about 12,000 tons of tabun, ready for use in bombs and shells, stored in secret munitions dumps. But it was never used, because the German air force was so depleted and because of the fear of retaliation. If it had been used the war would have been prolonged, because the British scientists were well behind, still trying to develop new forms of mustard gas.

In the Cold War, chemical weapons continued to be stockpiled and biological (germ) weapons developed, although the latter were less potent and less reliable. Finally, in 1969, US President Nixon called a halt to the chemical and biological arms race. Nevertheless, new methods of delivery were devised, the most important being the 'binary weapon', which was a shell containing two chemicals, harmless in themselves, which would mix when the shell exploded on its target and form a lethal nerve gas.

Today, since the collapse of the Soviet Union the threat of a major war has receded, and a chemical disarmament treaty has become a reality. However, the fact that humans manufactured enough chemical weapons to wipe out most of the world's population without making the planet uninhabitable should not be forgotten. The threat is still there, as events on the Tokyo subway in 1995 reminded us, when a Japanese doomsday sect released sarin gas, killing twelve and injuring nearly 1,000. The ensuing panic was far greater than these figures suggest. Fortunately, most terrorist organizations are too sensitive about their public image to use weapons that would alienate their sympathizers.

Infectious and Parasitic Diseases

These conditions killed over 17 million people in 1996, mostly children and mostly in the least developed countries. This is because it is in these countries that poor hygiene, malnutrition, lack of antibiotics and lack of good medical services make infections commoner and more lethal. The underlying cause of sickness and death here is poverty.

The top three killers (worldwide) are:

- Respiratory (lung) infections, accounting for the death of 4 million each year, most of them children under the age of five. The WHO is making great progress with vaccines, notably against whooping cough, diphtheria and measles, but it is only in wealthy, developed countries that these diseases have been properly controlled.
- Tuberculosis, which has now moved up to second place in the list of the great infectious killers, accounting for 3 million deaths in 1996: more than any other single infection. Increasing numbers of those with TB also have AIDS, and AIDS plus tuberculosis forms a lethal combination. We have the vaccine and the drugs that could largely overcome tuberculosis; the problem is that many poor countries cannot afford them and lack the means for delivering them to people.
- Diarrhoeal infections, claiming another 2.5 million children each year (about 8,000 a day). In parts of Africa, Asia and Latin America the average child has diarrhoea for a total of two to three months a year. Some die; others suffer stunted growth. The diarrhoea is due to a great variety of microbes and other parasites, but when mothers give children a low-tech treatment (drinks containing salt and sugar) the death rate is greatly reduced.

The enormous amount of diarrhoea is mostly due to unclean water supplies and inadequate sanitation. The United Nations Childrens Fund (UNICEF) points out that about half the world's population cannot safely dispose of their bodily waste. Intestinal infections spread easily through their communities because they live in medieval squalor. And yet $6 million spent over ten years could solve the problem – a mere 1 per cent of the global military budget.

In earlier centuries diarrhoea was a common cause of death in European children. An epitaph in a churchyard in Cheltenham, England refers to an infant dying at the age of three weeks:

> It is so soon that I am done for,
> I wonder what I was begun for

The next most common causes of death in this category, after the top three, are:

- Malaria (over 2 million deaths in 1996). There is now no reliable preventative drug: the only solution is to avoid mosquito bites.
- HIV/AIDS (1.5 million deaths in 1996). The infection is still spreading rapidly, in the Indian subcontinent as well as in sub-Saharan Africa.

- Hepatitis B (1.1 million deaths in 1996). This virus also causes liver cancer.
- Measles (1 million deaths in 1996) and whooping cough (0.35 million deaths in 1996) are still great killers of children in developing countries, though preventable by vaccines.
- Tetanus of the newborn (0.31 million deaths in 1996). This occurs when bacteria get into the cut surface of the umbilical cord
- Sleeping sickness (0.15 million deaths in 1996). This infection is spread by tsetse flies in Africa.

Note that AIDS is *not* one of the top three infectious killers. However, by the end of 1996, altogether (worldwide, since the disease first appeared), 29.4 million children and adults had been infected with HIV, 8.4 million had developed AIDS, and 6.4 million had died. By the end of the century, just twenty years after being discovered, AIDS will have killed about 20 million people: one million a year. To appreciate its total impact you have to remember that it kills young adults, leaving orphans behind, and it makes people much more susceptible to tuberculosis, so that in Africa one in three AIDS patients dies of tuberculosis. It is a deadly partnership. Nevertheless, AIDS deaths are fewer than those due to motor cars, and although in 1996 1.4 million died of AIDS, an equal number committed suicide or were murdered. Three-quarters of the HIV infections have been transmitted through unprotected sexual intercourse, and three-quarters of these involved heterosexual relations. Infants and children are infected in the womb or in early infancy.

I have not included in this section Lassa fever, Ebola fever, schistosomiasis, cholera, typhus, rabies, poliomyelitis or a host of other well-known and often lethal infections. This is because numerically they are less important than the ones I have mentioned. They can be dramatic and newsworthy, but are not major causes of death. Ebola fever, for instance, kills most people who are infected, but it is restricted to certain parts of Africa and the total number of deaths so far is less than a thousand.

In the past, infectious diseases had a dramatic impact, killing millions. The great influenza pandemic of 1918 killed 20 million in just two years – more than died in the whole of the First World War. For other examples see the boxes in this section on plagues, sweating sickness and smallpox.

Plagues

Great pestilences have been a feature of recorded history, and have occasionally influenced the actual course of events.

In 701 BC Sennacherib, king of the Assyrians, was forced to withdraw from Judah without conquering Jerusalem when 185,000 of his army died of what must have been an infectious disease. The episode is immortalized in Byron's poem 'The Destruction of Sennacherib'.

Plague used to be a big killer. The bacteria that cause it are carried from rats to humans by rat fleas, and the lymph nodes in the armpit or groin (Greek *bubo* = groin), which drain the area of the flea bite, swell up and turn into great bags of pus. In bubonic plague the powerful toxins from the bacteria kill half the sufferers. If the bacteria invade the lung (pneumonic plague) the death rate climbs to 100 per cent, and the infection can now spread directly from person to person.

The plague arrived in Europe from the Far East in 1348, and the effect was devastating. In England the disease, called the Black Death, killed about a third of the population over the course of two and a half years. An epidemic in 1665, the year before the Great Fire of London, was graphically described by Daniel Defoe (who was only five years old at the time) in his 'Journal of a Plague Year in London': 'They had dug several pits in another ground, when the distemper began to spread in our parish, and especially when the dead carts began to go about . . . Into these pits they had put perhaps fifty or sixty bodies each; then they made larger holes, wherein they buried all that the cart brought in a week, which . . . came to from 200 to 400 a week; and they could not well dig them larger, because of the order of the magistrates confining them to leave no bodies within six feet of the surface.'

The medieval mind attributed the plague variously to earthquakes, to the movement of the planets, or to a Jewish or Arab plot (350 massacres of Jews took place during the plague years in Europe), or most commonly to God's punishment for human wickedness.

Accidents and Natural Disasters

Each year about 4 million people die as a result of violence and accidents, which include falls, fires, floods, drowning, homicide and road deaths. In 1996 accidents ranked as the sixth most numerous cause of death, worldwide. Developing countries have nearly four times as many deaths from these causes as developed countries. Interesting details include the fact that in India women are more than twice as likely as males to die from a burn, presumably because they are the ones who have to look after the fires for cooking.

The English sweating sickness: a mysterious pestilence

A strange and terrifying sickness struck England in the summer of 1485. It began three weeks after the entry of the Earl of Richmond's army (which included mercenaries from France) into London. The Lord Mayor, his successor and six aldermen died within a week. Affected people developed sudden headache, muscle pains, fever, difficulty in breathing, and profuse sweating. As decribed by a contemporary physician, in the florid language of that period: 'it is on account of the fetid, corrupt, putrid, and loathsome vapours close to the region of the heart and of the lungs whereby the panting of the breath magnifies and increases and restricts itself.'

Death occurred often within twenty-four hours of the onset of the illness. We do not know how many died, but the disease seems to have been more severe in young healthy men than in women, children or older people. It was called the Sudor Anglicus, later the English Sweating Sickness, and four more epidemics came in the summers of 1508, 1517, 1528 and 1551. The disease then disappeared as mysteriously as it had come, and now we can only guess at cause. It sounds like a virus affecting the lungs, rather than a disease such as the plague, typhus or malaria – possibly an especially virulent type of influenza, were it not for the fact that influenza generally strikes in midwinter. Perhaps it was a virus carried by rodents, which were plentiful in those days: in London there was at least one family of black rats per household! On the other hand, the epidemics were explosive and widespread, and for this the infection would probably have needed to spread directly from person to person by coughs and sneezes.

Road deaths, regrettably, are almost an accepted feature of our lives (see box). If the same number who die each month on the roads died in aeroplanes, trains or passenger ships, there would be a public outcry and demands for action. In the USA, the millionth person to die in a road accident did so in 1973. But the road carnage continues and seems an inevitable result of our hunger for personal mobility. Some vehicles are safer than others. In England and Wales in 1989, 343 male motorcyclists aged sixteen to twenty-four died, whereas only 323 car drivers of the same age died – in spite of the fact that there are so few motorcyclists compared with car drivers.

Natural disasters attract more attention than road deaths: they hit the headlines, whereas the latter, though much more numerous, are a continuous phenomenon accepted by society. Some disasters attract worldwide attention because they are particularly dramatic or because they involve children – like the 1966 catastrophe in Aberfan, south Wales, when semi-liquid coal slurry from a collapsed mine-tip swept over a school and smothered 140 people, nearly all children.

Goodbye to smallpox

From the beginning of recorded history, smallpox has been a killer. Kings died of it, and when, between the fifteenth and eighteenth centuries, explorers and colonists introduced it into Africa, the Americas and Australia, the impact was comparable to that of guns and the Bible. In the Aztec and Inca kingdoms an estimated 3.5 million died within a few years, and the fact that the Spanish conquistadores were resistant made them seem even more invincible.

Smallpox was a regular feature of life in the towns and cities of Europe 200 years ago. For children it was like the measles, a hurdle to be got over, and it caused about a third of all childhood deaths. People learnt that you could protect against smallpox by scratching into skin the material from the blisters (pocks) of sufferers, and Lady Mary Wortley Montagu, wife of the British ambassador to Turkey, brought this method back with her to England, where it was widely used. In 1721 two royal princes were protected in this way, after preliminary tests on six condemned criminals from Newgate prison.

Then a country doctor of enquiring mind, Edward Jenner, who lived in Gloucestershire, England, discovered the modern vaccine for smallpox. On 14 May 1796 he took material from a cowpox blister on the hand of a dairymaid, Sarah Nelmes (who had caught it from a cow called Blossom), and rubbed it into scratches on the arm of eight-year-old James Phipps. James suffered a very mild indisposition, and six weeks later Jenner inoculated him with virulent smallpox material. The boy remained well. It was a risky thing to do, but the boy had been protected.

Jenner's method was called vaccination (*vacca*, Latin = cow), and by the first half of the twentieth century smallpox had been almost eradicated from Europe and North America. But the disease continued in Asia and Africa and in 1974 there were nearly a quarter of a million cases. In 1969 the WHO began a campaign to wipe out smallpox, using Jenner's vaccine. There were daunting difficulties, such as cultural barriers, warfare and the problems of transport to remote areas. How do you get to villages in the highlands of Ethiopia that are more than 20 miles from the nearest tracks negotiable by land rover? But the campaign was eventually successful, and the last case of smallpox was recorded in October 1977 in a small town in Somalia, Africa. Smallpox has gone for ever; it is an extinct species.

The total cost of the campaign, arguably the greatest achievement of medical science in the twentieth century, was US $150 million. For it to be successful the disease had to be an exclusively human one (if it infected animals or insects, it could come back again) and one that did not stay around in the body, (which would mean there could be healthy carriers). Also, there had to be a good and cheap vaccine (Jenner's vaccine, by the way, would not even be licensed by today's strict standards), and there had to be the resources and organization to carry the programme through on a global scale. These were provided by the WHO and its devoted band of scientists, administrators, vaccinators and epidemiologists.

The infernal combustion engine

The car is an unprecedented gateway to personal mobility and a visible token of income and status. None of us needs to be reminded of its central role in our lives. Let us consider the negative aspects of the car, with its internal combustion engine.

Ever since the first recorded road traffic death, that of a pedestrian in New York in 1895, cars have been killing us. About 40,000 people a year die on the roads in the USA. Road deaths in England and Wales totalled 3,600 in 1995 (about ten deaths each day). But they have been falling from their peak of 4,968 in 1990 – in spite of the fact that more miles are driven each year. In 1995 the average expenditure on cars per head per week was £13.67, as opposed to £1.40 on railways and buses.

To put this in perspective, the figure of 3,600 road deaths in 1995 can be compared with the 3,547 suicides in the same year. The UK, moreover, has one of the lowest road accident death rates per person in the European Union, the highest rate (on 1994 figures) being in Portugal.

Cars kill the innocent unborn as well as the innocent passenger or pedestrian. A US researcher calculated that in the USA in 1968 about 5,000 unborn children died with their pregnant mothers in car crashes – a loss possibly balancing out the number of children conceived in cars during the same period ...

Cars not only injure and kill us, but also have other unpleasant effects. They make a massive contribution to atmospheric pollution, producing substances injurious to health such as carbon monoxide, lead and nitrous oxide. They produce carbon dioxide, a contributor to global warming; they are associated with physical unfitness and thus heart attacks; they make a noise; the roads on which they run destroy the environment. Society pays these largely hidden costs, as well as the injuries and deaths, for the convenience of having the car.

The trouble is, the way we live right now, we need those cars. There is no doubt at all that as the fossil fuel crisis draws nearer and as the developing nations get their cars too, something will be done about the threat of the motor car. Either we drastically curtail our constant travelling to and fro, or technology will have to provide a viable alternative.

The following list gives some examples of the numbers of deaths caused by great natural disasters in the past. It includes famines, although as often as not famine is a direct result of human action – or inaction.

20,000,000 dead – famine in North China, 1969–1971
6,000,000 dead – the 'great famine' in China, 1333–7
1,500,000 dead – famine in Bengal, India, 1943–4
1,000,000 dead – flood in Henan Province, China, 1939
900,000 dead – flood in Henan province, China, 1887

830,000 dead – earthquake in Shanxi province, China, 1556

242,000 dead – earthquake in Tientsin, China, 1976

100,000 dead – flood in Friesland, Holland, 1228

99,000 dead – earthquake in Tokyo, 1923: many of the deaths occurred in the firestorms that raged through the city after the quake, caused by damaged power cables and gas mains or by cooking stoves that were flung to the ground

36,000 dead – violent eruption in Krakatoa, a volcanic island between Java and Sumatra, on 27 August 1883: the sound was heard 3,000 miles away, and the shock was felt in California, 9,000 miles away; most of the deaths were due to the giant waves (tsunamis) formed by the eruption

28,000 dead – volcanic eruption at St Pierre, Martinique in the Caribbean, 1902: a hot cloud of volcanic gas swept through the city and killed all but three people in a few minutes

20,000 dead – eruption of Vesuvius obliterates Pompei, 24 August AD 79 (see chapter 7).

Rare lethal accidents – such as snakebite, being struck by lightning, bitten by scorpions or killed by sharks or crocodiles – also tend to attract publicity. They are given more space here than the major disasters, but they are numerically trivial in comparison.

Snakebite has always been common in India and Pakistan, and in 1889 accounted for 22,480 deaths. Today, more than a thousand die each year in Maharashtra state alone. Encounters with venomous snakes are also frequent in hunter-gatherer tribes of Papua New Guinea, Tanzania and Ecuador, where 2–4 per cent of all adult deaths are due to snakebite. Brazil suffers about 2,000 deaths a year from snakebite; and there are parts of Burma where snakebite is one of the commonest causes of death in men. Between the rice fields are earthen ridges containing holes inhabited by rats, which eat the rice. The holes also house venomous snakes (mainly Russell's viper), which flourish on a diet of rats. There have been moves to develop a vaccine (the word generally refers to something you are given before encountering an infection) against the venom of these snakes, to protect the workers in the rice fields. By comparison, snakebite death figures in Western countries are very low: in the USA there are 50–100 snakebite deaths (mostly from rattlesnakes) each year, and in the UK adders have killed just fourteen people over the course of the last 100 years.

Lightning is an unusual cause of death. In the UK it kills about a dozen people a year, in the USA 400–500, and more in the tropics where thunderstorms are commoner. There was a well-recorded lightning strike at Ascot racecourse, England, on 14 July 1955. The lightning struck a tea-stall which had unearthed, metal-topped props, and many people standing around in wet clothes were thrown to the ground. Some were dazed, some unconscious. Two died and forty-six others were taken to hospital. Nine recovered by the time they arrived there and the rest suffered only minor effects – burns, nervous pain or pins and needles. The only long-term injury caused was one or two cases of hearing loss because of damage to the eardrum. The lightning strike also hit a nearby stand, but this had lightning conductors and no one was hurt.

Stings cause significant numbers of deaths. Scorpions are highly venomous, and their stings kill between 100 and 200 people a year in Mexico – up to ten times as many as die of snakebite. Bee and wasp venoms are not normally powerful enough to kill unless hundreds of stings are received, but those who are hypersensitive to the venom are very vulnerable. A single sting can then be lethal, but an early injection of adrenalin is life-saving.

Choking to death. The classic accident is when a piece of food or an object like a peanut goes down the wrong way. It lodges in the windpipe and the victim clutches at his chest, his face a picture of distress because he cannot breathe. Other people present may think he has had a heart attack. His pulse quickens and his blood pressure rises, as the oxygen leaches out from the blood and the carbon dioxide accumulates. As he strives frantically to draw air into the lungs this merely wedges the foreign object more securely into position or sucks it down into the lungs. If he is lucky, someone in the vicinity will know how to perform the Heimlich manoeuvre. They will stand behind him, put their arms round his abdomen at the belt level below the ribcage, one hand a fist and the other grasping it, and give a sudden forceful pull in the upward direction. This will empty his lungs and with any luck the sudden blast of air will expel the object. A piece of steak is a frequent offender, because it does not dissolve or disintegrate like a piece of bread.

The impact of teeth. Worldwide there are about 100 shark attacks each year, half of them fatal. They attract attention because of the sinister circumstances, and because such an attack could happen to any swimmer in the right part of the world. Crocodiles take a heavier toll, killing about 1,000 a year in Africa, and the occasional unlucky person in Australia. A few deaths are due to attacks by carnivorous animals. A man-eating tiger in Champawat, India, killed 436 people before it was shot by the hunter and writer, Jim Corbett, more than fifty years ago.

Radioactivity. It is worth pointing out here that, despite the considerable anxiety it causes, radiation is a very uncommon cause of illness or death. Nearly all (85 per cent) of the radiation dose we receive each year comes from natural sources: in other words, from cosmic rays arriving from outer space, and from inhaled radon gas whose source is the uranium present in many earth materials. The other 15 per cent comes from medical exposure, and although X-rays now involve lower doses, this is probably outweighed by the increased use of newer procedures such as CT scans. In a CT or CAT (computerized axial tomography) scan, radiological data are processed by a computer to display slices (cross-sections) through the human body. The International Commission on Radiological Protection reported in 1990 that natural sources are likely to cause a fatal illness (e.g. cancer) in one or two people per 100,000 per year. We know that the whole world is radioactive, and even low doses carry health risks. Risks can be higher in certain places and in certain occupations, and we obviously need to decide what level of risk is tolerable and what is unacceptable.

Cardiovascular Disease

This covers heart disease (more than 5 million deaths a year worldwide) and strokes (4 million). These account for 48 per cent of all deaths in developed countries, but only 11 per cent in developing countries. However, nearly half of the total number of those deaths occur in developing countries. Many of them, of course, are in older people and not 'premature'.

The vast majority of *sudden* natural deaths – that is, death within an hour of being taken ill – are due to heart disorders. Once the heart stops beating consciousness is lost, and the person dies in minutes. A half of all heart disease deaths are sudden.

Disease of the blood vessels of the heart (coronary arteries) is often thought of as a disease of middle-aged men. The standard picture is of a stressed executive who is overweight, smokes and takes little exercise. But it is also common in women, though it comes on about ten years later in life, and in the USA it kills more women than men.

Cancer

This disease, too, affects mostly older people (about 2.4 million a year in developed countries). In 1996 6,346,000 died of it, worldwide, and by the year 2000, the WHO estimates, about 7 million (4 million males and 3.2 million females) will die from cancer each year. Although most of these deaths are in the developing world, they form a smaller proportion of total deaths in these countries. Unfortunately the developed world is exporting to the developing world some of its bad habits (cigarette-smoking, and a low-fibre, high-fat diet) with their associated higher risk of cancer.

The commonest types of cancer are:

1   Lung cancer, which since the Second World War has accounted for 35 per cent of male deaths at age 35–69, and about a quarter of all cancer deaths worldwide. It killed 989,000 in 1996 and is mostly caused by smoking. In the USA in the late 1970s 30–35 per cent of all cancer deaths were due to smoking. In the UK it now exceeds breast cancer as a cause of death in women, and accounts for one in 6.5 deaths in women and one in three deaths in men.

2   Cancer of the stomach. This killed about 776,000 worldwide in 1996.

3   Cancer of the intestines (colon and rectum): about 495,000 deaths in 1996.

4   Cancer of the liver: 386,000 deaths in 1996.

5   Cancer of the female breast: 376,000 deaths in 1996, accounting for about 15 per cent of female deaths. It is a rarity in men.

6   Cancer of the oesophagous: 358,000 deaths in 1996.

7   Cancer of the mouth and throat: 324,000 deaths in 1996.

8   Cancer of the cervix: 247,000 deaths in 1996.

Cancer of the prostate comes only twelfth in the worldwide list, but is the second commonest cancer killer in men in England and Wales, causing 9,000 deaths in 1994. It is often a slow-growing type of cancer, occurs in older men, and tends to be neglected and not diagnosed.

A great deal of research effort is devoted to cancer, and the results are of immense interest. We are gradually learning how the body controls and coordinates its hundreds of millions of cells, what makes them divide, how they are formed in the embryo, and how they age and die. Nevertheless, most cancer deaths could be prevented by the application of a few well-known principles. For instance, 20 per cent of all cancer deaths could be prevented if tobacco smoking were miraculously eliminated, and deaths from liver and cervical cancer would be greatly reduced by global vaccination against the hepatitis B virus (the main cause of liver cancer) and the widespread introduction of cervical smear tests. Again, it seems likely that diet (fibre content, certain foods, certain cooking practices) is significant in causing and preventing cancers of the stomach and intestines, and if so these cancers could be reduced by changes in eating habits.

3

Suicide, Euthanasia, Homicide

Poor man, what art! a Tennis ball of Error,
A Ship of Glass, toss'd in a sea of terror,
Issuing in blood and sorrow from the womb,
Crawling in tears and mourning to the tomb,
How slippery are thy paths, how sure thy fall,
How art thou Nothing when th'art most of all!
John Hall (1627–56)

This chapter considers certain 'unnatural' deaths perpetrated by individuals. The distinction between these deaths and those caused by wars, massacres and infections as described in chapter 2 is that these are usually on a smaller, more personal scale. It is precisely this that makes them so fascinating. The section on homicide covers human sacrifice, abortion and infanticide as well as murder and manslaughter.

Suicide

Suicide, or self-murder, is a perennial source of fascination. Who does it, and why and how? It doesn't happen in non-human animals, and many people have found it hard to accept or condone it in humans because it seems to violate a natural law of nature. All animals have an instinctive urge to preserve their lives, and suicide goes aginst this natural law. It therefore merits consideration at some length.

Interest in the phenomenon of suicide is by no means new. Since antiquity, it has been studied by physicians, theologians, philosophers and jurists, some of whom approved it as a rational act. The Roman philosopher and poet Seneca (4 BC–AD 65) argued: 'If I can choose between a death of torture and one that is simple and easy, why should I not select the latter? As I choose the ship in which I sail and the house which I inhabit, so will I choose the death by which I leave life.' In 1971 it was estimated that about 5,000 learned papers and books had already been written on the subject. Many artists, too, have illustrated suicide. Delacroix and Millais both portrayed Ophelia, who drowned herself in Shakespeare's *Hamlet* after her father's death. Other painters who have portrayed suicide include Edvard Munch (*The Dead Couple*) and Toulouse-Lautrec (*La Pendue*). Suicide figures in many novels, and some writers seemed especially interested in it. A heartrending passage in Thomas Hardy's *Jude the Obscure* describes how Jude's eldest child kills the younger children and then himself, leaving a note that reads: 'Done because we are too menny.' There are at least seven suicides in Ibsen's plays, and fourteen in Shakespeare's.

Worldwide, more than three-quarters of a million people kill themselves each year, about 135,000 of them in Europe. This works out at about 2,000 a day or 80–100 each hour. If you assume that five people (close family, friends) are associated with each case, then you have 4 million a year suffering the emotional consequences of those suicides.

Unsuccessful attempts at suicide are at least ten to fifteen times as common as successful ones, which means that there are about between 10 million and 20 million suicides or attempted suicides in the world each year.

How Do They Do It?

The method of suicide depends, of course, on where it happens and what is available. Where there are many firearms in the community (the USA), coal gas in most houses (the UK until the 1970s), sedatives in the bathroom (most Western countries) or agrochemical poisons on many people's shelves (Sri Lanka, China), it is easiest to use these methods. Some methods, such as jumping and slashing, are less reliable, and others verge on the ridiculous. For instance, the English comic opera and verse writer W. S. Gilbert (1836–1911) noted that 'Self-decapitation is an extremely difficult, not to say dangerous, thing to attempt.'

A drink gives one more courage, and about a half of suicides have some alcohol in the blood. Considerate suicides do it away from home, in an impersonal place such as a train or a hotel bedroom, saving loved ones from the trauma of finding the body. One of the worst effects on surviving friends and relatives is the feeling of responsibility, or of having failed to prevent the death.

Suicide by firearms or hanging is commoner in males. Hanging has been a fairly common method in modern Europe, but there was a powerful taboo against it in ancient Rome. Women are more likely to commit suicide by poisoning or drowning, and rarely shoot themselves. A poem by Emily Dickinson (1830–86) describes a suicide who

> Groped up, to see if God was there –
> Groped backward at Himself
> Caressed a Trigger absently
> And wandered out of Life.

On 4 January 1704 George Edwards, a prosperous Essex man, rigged up three guns to fire simultaneously at himself, and set them off when his wife entered the house. For some time he had been questioning Christian beliefs, wondering, for instance, how if all men were descended from Adam some of them could be black. These views had made him unpopular with his neighbours and with his wife, who had refused to sleep with him. Shooting, of course, is not necessarily by gun: in 1622 the Earl of Berkshire shot himself with his crossbow.

Drugs and poisoning are commoner in female suicides, especially given the ready availability today of antidepressants and sedatives. In Sri Lanka there are more than a thousand deaths a year fom poisoning by agrochemicals, about three-quarters of them suicidal. The substances used are herbicides (e.g. Paraquat) or organophosphorous compounds such as parathion. The method is also common in Chinese women.

Suicide by coal gas (putting one's head in the oven) was the commonest method in England and Wales in the 1960s, but virtually disappeared when carbon monoxide in domestic gas was reduced to low levels. But this seems to have had little impact on the overall suicide rate; and in spite of catalytic converters there is still enough carbon monoxide in exhaust fumes to be lethal. In Australia in 1995, 509 of 2,367 suicides, most of them in young to middle-aged men, were achieved with car exhaust fumes, making this method the second most common after hanging, ahead of firearms. Carbon monoxide works by combining in the blood with haemoglobin and preventing the haemoglobin from carrying oxygen round the body: thus the person inhaling it soon dies. The carbon monoxide makes the haemoglobin stay red, and so the corpse has a healthy pinkish tinge in spite of the fact that death was due to oxygen shortage.

The writer Virginia Woolf drowned herself in a river near her house in Rodmell, Sussex, England. It is often difficult to decide whether a death by drowning was a suicide or accidental (see chapter 15).

Some commit suicide by jumping from a great height or throwing themselves under a train or car. Certain places are particularly popular for the jumper from on high, such as the 500-foot chalk cliffs at Beachy Head, Sussex, England, and the Golden Gate Bridge in San Francisco; but any tall building or high point will do. In 1600 William Doddington, a rich puritan merchant, flung himself to his death from the steeple of St Sepulchre's church in London. He left a suicide note, blaming his financial ruin on a business rival. Having an 'accident' on the road is another possibility, for instance by driving into a wall or tree; the advantage of this method is that it need not look like suicide. Contrariwise, a murderer can engineer a 'fake suicide', but he must be careful. One man strangled his wife, placed the body in a car and pushed it over the edge of a mountain road. At first it looked like suicide, but unfortunately for him he had left the ignition in an 'off' position.

Slashing or cutting, usually of the throat or wrist, is often unsuccessful because the wound will stop bleeding unless it is very deep. People who slash their wrists, especially, often start off with a series of shallow, tentative incisions before making deeper ones. Stabbing or falling on a sword are more reliable.

The very careful, practical suicide may use multiple methods. An extreme example would be blowing one's brains out as one drives a car (with the exhaust directed into the car) over a cliff, having taken an overdose beforehand. Then there is the impulsive suicide, and the nonchalant suicide. The poet and translator Thomas Creech (1659–1701), working on his translation of Lucretius, wrote in the margin: 'NB I must remember to hang myself when I have finished.'

The expression *felo de se*, formerly used of suicide, means committing a felony (serious crime, murder) on oneself, in other words self-murder. It can be argued that not all self-killing is self-murder. Are the martyrs who died rather than recant their faith true suicides? And what about the soldiers who rush to certain death in battle, or even the people who risk their lives in dangerous sports? Do we include all types of risky, self-destructive behaviour? Some psychologists recognize an additional type of suicide, called chronic suicidal behaviour. People who inflict on themselves chronic drug abuse, alcoholism, obesity or cigarette smoking, knowing that these may well kill them, are in a sense deliberately destroying themselves. Saudi Arabia has an unacceptably high number of deaths in road traffic accidents. Religious authorities in the country warned the young in July 1997 that they could go to hell if they died because of careless driving. Such a death would be considered suicide, which is a serious crime.

Suicide in Different Countries

The statistics are interesting, but care needs to be taken in making comparisons because of different practices for certifying death as suicide and differences in willingness to accept and record a death as suicide. For these reasons, too, suicide rates in history are difficult to study.

Currently, Greenland has the highest suicide rate in Europe: 127 per 100,000 population per year. Hungary, Finland and Denmark also have high rates (up to 60 per 100,000 per year); the UK, Sweden and Norway have moderate rates, while Italy, Spain and Portugal have low rates.

In the eighteenth century, many Europeans thought that the English were lax about suicide, failing to treat it seriously as a crime, and were in any case more likely to have recourse to it because, as the French philosopher and historian Montesquieu (1689–1755) argued, England had such a dismal climate, and its natives were predisposed to gloominess and suffered from an impaired ability of the bodily machinery to filter nervous juices properly. In fact the suicide rate was no higher in England than elsewhere at this time.

Predominantly Roman Catholic countries tend to have low rates. Most female suicides occur in China and nearly half of all male suicides are in sub-Saharan Africa. Between 1973 and 1983 there was an increase in all European countries (especially in Ireland, Northern Ireland, Norway and Belgium), mainly in young people (under twenty years old). In the UK the rate in 15–24-year-olds rose by 80 per cent between 1980 and 1992, and at this age four in every five suicides are in males. The reasons for this are not clear.

Who Commits Suicide?

The effect of age. Children under fifteen rarely kill themselves nowadays, but in sixteenth- and seventeenth-century England child suicide was not all that uncommon. In those days children aged ten to fourteen went out to work or were apprenticed, and they were often beaten or terrorized by elders at work as well as at home. At present the highest rates are in the 15–25 age group. In the USA the commonest cause of death at this age is accident (especially in automobiles), followed by homicide, then suicide. In 1986 there were an average of fourteen suicides a day in this age group, and for some reason the rate was highest in Alaska and lowest in New Jersey. The rate is also high in those over fifty years of age. In 1980 in the USA this age group made up 26 per cent of the population yet accounted for 39 per cent of the suicides. The rate rises in those over sixty-five and is higher still in those over eighty. Obvious reasons for suicide in old people are retirement, death of a spouse or partner, illness, depression and poverty.

The effect of sex. In Western countries suicide rates are two to three times as high in males as in females, but as differences in gender roles diminish, the disparity in suicide rates is diminishing. It is intriguing that in Japan there are approximately equal numbers of male and female suicides, and it is one of the commonest causes of death in young Japanese women. In China suicide is commoner in women than in men, and is responsible for nearly one in four of deaths in women, most of them by drinking pesticides. Fifty-six per cent of all female suicides worldwide are in China. The reasons are not clear.

The effect of marital status. Suicide is less common in married people, especially when there are children. In the USA the rate is twice as high in single men. Presumably a person is less likely to become suicidal when there is someone to talk things over with, and/or the responsibility for dependants (spouse, children) to consider. Conversely, people who do not marry are more likely to be isolated and lonely, and so more susceptible to suicidal feelings.

The effect of occupation. Doctors, dentists, vets and pharmacists have higher than normal rates, probably because they have easier access to drugs and know what to do (poisoning is the commonest method in these occupations). Among doctors, rates are higher in pathologists and psychiatrists; it could be argued that those with abnormal personalities tend to be attracted to these particular specialities. Rates are also relatively high among sailors. Is this because of isolation from family and society, or do suicide-prone individuals take to the sea? There are also high rates in farmers, forestry workers and others in isolated rural areas, suggesting isolation as a factor. Farmers tend to use firearms. Student suicides are quite common, with the stress of exams and family expectations applying pressure to young people. It is the third commonest cause of death in US students.

The effect of mental condition. Some would say that no one who commits suicide can be in their right mind. A third or more of all suicides have a history of mental illness. It is certainly commoner in those actually suffering from depression (unfortunately, many antidepressants can be used for suicide) and from schizophrenia. Fewer schizophrenics are in hospital these days, and may not always be taking their drugs regularly; it could also be that contact with 'normal' society makes suicide more likely.

The effect of long-term stress. The unemployed immigrants, refugees, people in prison – all these groups have stressful, often disappointing lives, with major disruption of normal activity, and may succumb to depression and despair.

The effect of addiction. For alcoholics and drug addicts, life depends on a supply of expensive substances and most of their days are spent in what might be called altered states of consciousness. Their attitudes to death are abnormal.

There are, however, popular misconceptions about suicide. Some of the most common – and most dangerous – are:

- People who talk about it don't do it: wrong.
- Those that do it unsuccessfully will succeed in the end: wrong – 89–90 per cent do not try it again and later are pleased that the attempt failed.
- It comes without warning: wrong – in most cases there have been direct or indirect warning signs.
- People who kill themselves are insane: wrong – most are not insane by ordinary standards.
- Suicide is prevented by good social circumstances: wrong.

Why Do They Do It?

Sometimes the imagination has to fill in the answer to this question. An anonymous man who drowned himself in England in 1719 had written on a slate that was discovered in his pocket: ' 'Tis better to die than be exposed.'

If you exclude those suffering from mental illness, there usually is a reason, although it may seem to others too trivial to justify such an irrevocable response. For example, the English author and letter-writer Horace Walpole (1717–97) wrote to a friend that John Crowley, a man who had inherited £400,000 a year earlier, had shot himself during a bad attack of gout. The pain of gout, in the days before we learned how to treat it, could be intolerable. Somewhere there is the story of a sufferer lying in bed, looking up at the ceiling from which a spider hangs, and praying to heaven that the spider does not fall on his swollen and agonizingly painful toe.

In some societies in the past widows, servants or slaves killed themselves or allowed others to kill them after the master had died. An example in British India was suttee, the practice in which the widow allowed herself to be burned on the funeral pyre of her dead husband. The word 'suttee' comes from a Sanskrit word meaning a virtuous wife.

Suttee was made illegal in 1892 but continued nevertheless. When the ancient city of Ur on the Euphrates river was excavated a royal tomb was found, containing the bodies of sixty-two people besides the dead monarch. They included women (perhaps wives, other ladies of the court) and men (possibly soldiers, priests), who had been ritually killed and buried at the same time, nearly 5,000 years ago.

At times the old and the sick have been expected to sacrifice themselves for the good of the others during famines and other hard times. The Japanese and the Inuit used to follow this practice. Elderly people went out into the freezing cold and soon died of hypothermia. Starving oneself to death is a slower, less pleasant method. Self-sacrifice has also occurred in other circumstances. On Scott's last expedition to the Antarctic, Captain Oates was taken ill and feared that he would be an encumbrance to his companions. On 17 March 1912, as the explorers were sheltering during a blizzard, he famously said: 'I am just going outside; I may be some time'. Then, knowing that he was walking to his death, he went out into the driving snow and was never seen again. In retrospect, all members of the expedition were probably destined to die as a result of vitamin C deficiency (scurvy), as a consequence of their failure to take the appropriate foods or eat their dogs, as Amundsen (a medical man) had done. By the time Oates died, Amundsen had already arrived at the pole, reaching it on 14 December 1911.

Suicide as a personal choice when confronted by incurable disease, unbearable pain, or merely by old age and loneliness, is discussed below under euthanasia. If you are old and ill and want to end your life, it is best to make sure no one finds you before you are dead. It is not enough merely to say you want to do it. In May 1997 a retired headmistress in Birmingham, England, aged eighty-one and in deteriorating health, decided to end her life. She took an overdose of sleeping pills and placed a bag over her head. She had always said she would do this if her health failed, and she hung a note round her neck threatening to sue anyone who revived her. But this did not prevent a paramedic battling (in vain) to save her life.

Although it could be argued that most people who commit suicide are mentally disturbed, true madness accounts for only a small proportion of suicides. However, about 10 per cent of those with severe mental illness commit suicide.

Poverty and actual or threatened financial ruin are powerful causes of suicide, and not only among high-stakes gamblers or the once very wealthy. Many ordinary people, ground down by the hopelessness of poverty and privation, have taken their lives. For instance, the suicide rate in England rose in 1574, 1587 and 1597–1600 – years that had been blighted by failures of the harvest and low wages. Again, when the South Sea Bubble, an infamous period of financial speculation, 'burst' in 1720, there was widespread panic and despair. Shares in the South Sea Company had risen in price from £110 to £1,000 and then fallen. Many people were ruined. During the next year the number of suicides in London, as recorded in the Bills of Mortality, almost doubled, from twenty-seven to fifty-two. In 1731 a fashionable society lady of Bath, Fanny Braddock, killed herself. She made a noose, stood on a chair and hanged herself from the door. The noose was made from two of her girdles, one silver and one gold, after a red girdle had broken on the first attempt. She had inherited a fortune of £12,000 but had gambled it all away.

Humiliation has driven many to suicide, for example after defeat in war, or rape. Long ago in China a general who had been defeated was sent a silk scarf by the Emperor, with which to hang himself. The great Carthaginian general, Hannibal, at the end of his extraordinary career, took poison rather than surrender to the hated Romans. Lucretia, the innocent young Roman woman, stabbed herself to death after being raped by Sextus Tarquinius. Other Lucretias have killed themselves when chastity was threatened or lost, deeming death preferable to living under the shadow of shame and dishonour. The shame of being publicly punished was behind the suicide in 1654 of Simon Bourne, who poisoned himself in Worcester Castle, England, rather than suffer the shame of being drawn and quartered in public.

The Japanese practice of hara-kiri, or ritual suicide, originated over a thousand years ago as an honourable form of death to avoid capture. It could also be imposed on a nobleman who had committed a criminal act, or undertaken voluntarily as a protest about someone or something. A fixed ceremony was prescribed, with a special knife or sword, and an assistant. After a short thrust into the left of the abdomen the knife was pulled across to the right side and then removed. It was reinserted into the midriff and the cut continued upwards. Finally the throat was cut. This was the full ritual. Obviously there was profuse bleeding and pain. It needed bravery. One day in 1663 a certain George Gibbs, deciding that things had gone too far and blaming Satan, got up from bed, slashed open his abdomen, pulled out his bowels, and lay dying for eight hours. Although hara-kiri was forbidden by English law in 1868 it was still practised.

A later Japanese version of 'official' suicide was that practised by the kamikaze pilots of the Second World War who bombed American warships in Pearl Harbor, knowing they would die in the operation.

The midget, one-man Japanese submarines were also manned by suicides. In a letter to his family one kamikaze pilot wrote: 'Please congratulate me. I have been given a splendid opportunity to die. This is my last day. The destiny of our homeland hinges on the decisive battle in the seas to the south, where I shall fall like a blossom from a radiant cherry tree.'

The romantic suicide has been celebrated widely in literature. The best-known and classic example is Shakespeare's *Romeo and Juliet*; he also dramatized the tale of *Antony and Cleopatra*. Then there is young Werther, the sentimental hero of the romance by Goethe (1749–1822), whose unrequited love for Lotte drove him to kill himself. Legend and history provide other examples. Dido was the legendary daughter of the King of Tyre, sister to Pygmalion, and the founder of Carthage in 853 BC. She stabbed herself in public rather than marry the King of Libya. The story that she did it after being deserted by her loved one Aeneas, the hero of Troy, is an invention. The tragic deaths of the lovers Pyramus and Thisbe are described in a story by Ovid, the poet of love. In reality, although so fascinating to writers and dramatists, whose public adore heart-rending stories of lovelorn youths and maidens killing themselves, love suicide is a rarity. But there have been occasional couples who, faced with insuperable barriers to union, have chosen to die rather than be parted.

Suicide pacts are commonest in older people. In England and Wales sixty-two couples commited suicide together over a five-year period up to 1993. But it is rare, and the 124 who died in this way accounted for less than one in a hundred of all suicides. The mean age was fifty-six years, and most of them used motor-vehicle exhaust fumes, with a smaller number preferring pills, such as barbiturates or analgesics. As might be expected, most of them were in stable relationships. There were four male–male and two female–female partnerships, but nearly all were married couples. Only four of the pairs were described as 'lovers'. These suicides are more likely to leave suicide notes, presumably to explain their decision to die together, and fifty-two of the sixty-two pairs did so.

Another type of romantic suicide is exemplified by that of the English poet Thomas Chatterton (1752–70), who, living in squalid poverty, poisoned himself with arsenic – not for love, but for want of recognition and fame. Chatterton became the prototype of the brilliant young suicide. Keats dedicated the poetic romance 'Endymion' (1818) to his memory, and he was eulogized by Coleridge and Wordsworth.

Suicide, and the threat of it, has also been used as a weapon. The pacifist Gandhi, in his fight against the British in India, went on fourteen hunger strikes lasting up to twenty-one days, with his death as the threat. He was at least partially successful. The suffragettes in England in the early twentieth-century refused food in prison for the same reason, and were force-fed. In the late 1980s Irish republicans in the Maze prison, Belfast, used the same method to fight for political justice. They died after between forty-five and sixty-one days, and were hailed as martyrs. Some other martyrs' deaths can be interpreted as suicidal. The death of Jesus Christ was really suicidal, and an effective 'weapon' for the spreading of Christianity. The English poet John Donne (1572–1631) wrote a prose work called *Biathanatos* in which he maintained that Jesus committed suicide. In 1963 several South Vietnamese monks committed suicide by pouring petrol over themselves and igniting it. Certain Buddhist sects allow suicide under special circumstances, and this was a dramatic way of protesting against the victimization of Buddhists by Catholics. In ancient China suicide was sometimes driven by revenge, when the individual who sacrificed himself hoped that his spirit would haunt and punish the cruel person who had driven him to it.

Suicide also occurs by imitation. Suicide rates increase when a celebrity commits suicide. During the month after Marilyn Monroe's death in 1962 there was a 40 per cent increase in the number of suicides in Los Angeles. The model may only be a soap opera hero or heroine, but it is someone the suicidal person identifies with, and imitates. In a German television series a nineteen-year-old student killed himself by jumping under a train, and during the next five weeks there were three times as many suicides as expected in the 15–29 age group. The subway in Vienna became a favourite place for suicides in 1978, and the rate increased after dramatic reporting of cases in local newspapers. A decision was made to keep the reporting at a more subdued level and the rate fell. The great authority on suicide Emile Durkheim (1858–1917) tells the story of the fifteen patients who hanged themselves from the same hook in a dark passage of a hospital. The hook was removed and there was an end to the epidemic. The most impressive imitation suicides, as far as sheer numbers go, were those who jumped into the mouth of the volcano on the island of Oshima, Japan in 1933. First was Kiyoko Matsumoto, a nineteen-year-old schoolgirl, and during the rest of the year another 143 did it. After two years, when the total had reached 1,208, a fence was put round the volcano, and buying a one-way ticket to the island was made illegal.

Mass suicides are partly imitation suicides. In November 1978 no fewer than 163 women, 138 men and 82 children killed themselves by drinking a fruit juice laced with cyanide. They were followers of the fanatical Reverend Jim Jones, who had organized the suicides and who afterwards shot himself and his family. A historical example of mass suicide is that of the besieged people on the Masada, described in the box below.

Masada, the almost impregnable fortress

The Masada (Hebrew = 'rock fortress') is a fortress on an isolated mountain plateau above the west shore of the Dead Sea. It has been inhabited for the past 3,000 years and in the second century BC a castle was built on its flat summit, towering 1,540 feet above the Dead Sea. In AD 73 it was the last stronghold of the Jewish Zealots and the Essenes in their battle against Rome.

The castle had been enlarged by Herod, King of the Jews, in AD 36–30, to become the largest fortress in the country. The reason for enlarging it was the constant threat from Rome. Cleopatra, Queen of Egypt, had been putting pressure on the Roman Mark Antony to depose Herod, and in AD 72 (by which time Herod had died) the Tenth Legion plus auxiliary troops began to besiege Masada. The defending garrison had plenty of food, so instead of starving them out the Romans constructed a ramp of stone, wood and sand up the side of the rocky fortress, built a tower on it, and kept the fortress under fire. At the same time they started using a battering ram against its walls.

After nearly two years of this bombardment, the besieged inhabitants realized their position was hopeless, and resolved to kill themselves rather than surrender to the Romans – who would surely have killed them anyway. There were 960 men, women, and children in the fortress. They drew lots to decide who should die first and who should do the killing. When the bloody business was over and nearly everyone was dead, the survivors (the executioners) drew lots and selected ten men to kill the others. Each victim lay down, embracing his already dead family, and offered his throat to the executioners. Finally only one man was left, and when he had checked that all the others were dead, he thrust his sword through his body and collapsed beside his family. According to the Jewish historian Josephus Flavius, two women and five children who had hidden in a cistern survived this mass suicide.

Strictly speaking of course, the only suicide was the unfortunate last man. The ramp built by the Romans is still visible and Masada is now an Israeli national monument.

America's worst mass suicide hit the headlines on 26 April 1997, when no fewer than thirty-nine bodies were found in the Rancha Santa Fe mansion, outside San Diego, California. They ranged from twenty to seventy-two years old, were male and female in about equal numbers, and belonged to a cult led by Marshall H. Appelwhite, a sixty-five-year-old former music professor. They believed that the planet earth was about to be 'recycled' and that they had to leave it and rendezvous with a UFO that was said to be trailing behind the comet Hale-Bopp. Their only hope of survival was to leave their bodies behind so that they could enter a non-perishable, non-corruptible world on another planet. They would be resurrected on the other planet after three or four days. All had died willingly after a group decision. Suitcases were packed, and, equipped with money and passports for the journey, they consumed a lethal dose of pentobarbital mixed with pudding and apple sauce, drank some vodka, and lay down to die with plastic bags over their heads. Appelwhite was not one of those cult leaders who seek sexual favours from his followers: on the contrary, he not only enforced celibacy but had himself castrated. He did, however, demand their money, possessions, obedience and worship. The cult had a website (www.heavens gate.com) that had foretold the fate of the disciples and their leader by putting a red alert over the logo saying that 'closure' was coming soon.

When Do They Do It?

The suicide rate is lower in winter, in spite of the shorter days and long hours of darkness that often cause seasonal affective depression or SAD, a condition that can be helped by increasing the exposure to (artificial) light. The suicide rate is higher in spring and early summer, and has been so at least since the sixteenth century. Sexual activity and sexual crimes are also more frequent at this time of the year. Perhaps at this time the person is confronted with his problems and his isolation just when life is blossoming out for others.

Religions and Cultural Attitudes to Suicide

The early Christian martyrs – and, as noted above, Jesus himself – were in effect suicides. There are four suicides recorded in the Old Testament, including those of Samson and Saul. There were many suicides in the middle ages, although there are no figures to enable comparisons with later periods. In York, England, in 1190, 600 Jews killed themselves in protest against the pogroms and repression they were enduring under King Richard I (the Lionheart). In France in the twelfth century, 5,000 Albigensians, labelled as heretics and persecuted by the church, killed themselves. However, suicide came to be frowned upon in Christian society. The idea became current that a suicide would not find peace in the next world and might return as a ghost. The body was not allowed to be buried in consecrated soil (that is, in a churchyard): in England the northern sides of church-yards were reserved for the bodies of executed felons, those who had been excommunicated, unbaptized infants, and suicides. In many parts of Europe, up to the late eighteenth century, the suicide's corpse was dragged through the streets and buried at a crossroads with a stake through it and a stone placed on the face or mouth (perhaps to stop the malevolent spirit escaping and haunting people?).

In Tudor England suicide became a terrible crime in the eyes of the church and the law: the story is set out in scholarly detail in the book *Sleepless Souls* by M. MacDonald and T.R. Murphy. In the year 1600 George Abbot, a future Archbishop of Canterbury, made an interesting comparison with the Gadarene swine. He noted that 'it is worse than beastly to kill or drown or make away with ourselves; for . . . the very swine would not have run into the sea but that they were carried by the devil'. Between 1500 and about 1650 suicide was punished more severely than ever before or afterwards. Offenders were tried posthumously and, unless considered insane, if they were found guilty all their possessions were forefeited to the Crown and their heirs punished. Thus a man who committed suicide reduced his survivors to pauperism. Eventually atti-tudes began to change, and although in 1660 it was still the case that only 5 per cent of suicides were 'let off' (judged to be lunatics), the figure had risen to more than 90 per cent by 1710–20. It was because of the new liberal trend in thought that the word 'suicide' itself was invented. Until about 1636 it had been called self-murder, self-destruction or *felo de se*, and considered an abominable crime in which Satan had a hand. Now, men like Sir Thomas Browne (1605–82) were saying that the act of self-destruction had nothing to do with witchcraft and the devil, but was a response of free human beings to intolerable circumstances and should be regarded in that light. The new word suicide, which did not have those superstitious trappings, was more appropriate. By the eighteenth century views were softening further, and philosophers and writers, including Rousseau, Voltaire and David Hume, were arguing for a more liberal attitude. Some prominent individuals, however, continued to take a harsh view. The English preacher and founder of methodism John Wesley (1709–91) suggested that the corpses of suicides be gibbeted and left to rot in public. Gradually, suicides in Britain more or less ceased to be punished in practice, although suicide was not finally decriminalized until 1961. In spite of changing attitudes suicide retains a slight stigma for most people, and it is still something that tends to be disguised by both doctor and family in many cases.

Other religions have taken a variety of attitudes to suicide. Hinduism permits it; as noted above, it was formalized for widows in the practice of suttee. For Buddhists, too, it can be correct in certain circumstances, and in Japanese Shintoism it was allowed until recently. By contrast, Confucianism prohibits it, and in Islam it is serious crime, worse even than murder.

In ancient Greece suicide was generally considered a shameful act, life being a gift from the gods. Nevertheless some took this path. The founder of the Stoics, Zenca, had a good life until he was ninety-eight, then fell and injured his big toe, after which he felt so unsettled he went home and hanged himself. In ancient Rome, for most people attempted suicide was a punishable offence; however, the Roman Stoic Seneca, as noted above, defended it to put an end to suffering and physical decline. Heroic, sacrificial suicide was permissible for the good of the state, but in the case of a Roman soldier it was deemed equivalent to desertion: thus, although the death penalty was superfluous, it also brought disgrace.

Perhaps the most famous suicides of the imperial Roman period are Antony and Cleopatra. Cleopatra was an intelligent and ambitious woman with great seductive power. The daughter of Ptolemy II, she became joint ruler of Egypt at the age of seventeen, in 52 BC. Julius Caesar was fascinated by her, waged a war on her behalf and took her to Rome as his mistress until his assassination. After returning to Egypt she became mistress and ally of Mark Antony. Their relationship was unpopular in Rome, and the two of them were defeated in a sea battle by Augustus. Hearing a rumour of her death, Mark Antony stabbed himself. Cleopatra then tried to captivate Augustus but failed, and to avoid being led in triumph through Rome as a captive she killed herself by clasping an asp (cobra) to her bosom.

No culture is free of suicide, but the rate seems to increase when primitive societies disintegrate after encountering 'civilized' societies. The old social patterns and fabric of family life tends to weaken, and alcohol and other disruptive influences play their part in dislocating social structures. Greenland, which has the highest suicide rate in the world, has experienced a clash between native culture and that of the West (Denmark). The rate is also high in Native American reservations in the USA. As other developing societies tread the path to 'civilization' and development, suicide rates can be expected to rise.

Prevention of Suicide

The Humane Society, founded in the Coffee House in St Paul's Cathedral, London in 1774 by two doctors, was originally intended to rescue drowning victims and resuscitate them. Before long it found that its main function was to haul suicides out of the water. During its first twenty-five years the Society tried to save more than 500 suicides, and succeeded in more than 350 cases. More recently, organizations have been set up specifically to prevent suicide. In England the Salvation Army established an anti-suicide department as early as 1906, and the National Save-a-Life League was started in the same year in New York City. These bodies took the view that suicidal despair was often a social and spiritual problem rather than a medical one. The Samaritans, started in 1953 by a vicar in a London parish, now have 165 centres across the UK, staffed by more than 30,000 volunteers, taking millions of phone calls and offering care, compassion and friendship. There are about 200 similar centres in the USA and also, of more recent origin, in Germany, Switzerland, and Finland.

To what extent do these admirable organizations actually reduce the suicide rate? The answer is not clear, although they obviously at the least delay many attempts. The interventions that have proved their worth are:

- treatment of psychiatric patients who are at risk;
- control of guns in the community;
- reducing carbon monoxide levels in domestic gas and in car emissions;
- control of dangerous chemicals;
- toning down reports of suicide in the press (to minimize the imitation factor).

Euthanasia

The word 'euthanasia' means, in Greek, 'good death'. It is used now to mean actively intervening to bring about a person's death, and it usually refers to a physician helping a patient to die. In nearly all countries it is regarded as unethical, and it is also against the law. Nevertheless it is practised to a limited but unknown extent. There is increasing public support for the right to die, for euthanasia and for physician-assisted suicide. Faced with a terminally ill, incurable patient who is in pain, who wants to die and who asks for assistance, the caring physician is tempted to give that help. The relatives are distressed and there seems no moral reason for withholding it. He eases the dying process. It is an act of commission, an actively taken step, as distinct from an act of omission, when, for instance, antibiotics are not given to a dying patient, or heroic methods of resuscitation are not carried out, or when a newborn baby suffering from gross and untreatable malformations is allowed to die. This generally arouses less controversy. Nevertheless, some would argue that there is not much difference in principle between allowing doctors to stop treatment to hasten death, and allowing them to adminster a lethal injection. In a recent case an eighty-five-year-old woman who had had a series of strokes that left her weak, semi-conscious and with little movement was being kept alive by special high-nutrient feeding. Her relatives asked that feeding be withdrawn so that she could die without further suffering, and this was eventually done. By the time she died she weighed less than 25 kg. The doctor was arrested on suspicion of murder, but the case was later dropped. Surely it would have been better to give her a lethal injection than, in effect, to starve her to death?

The question of euthanasia arises not only in respect of the terminally ill, but also in the intensive care ward. Six of every ten Britons and eight of nine Americans now die in hospitals, with life-support technology at hand. By tradition, the doctor's mission is to save life wherever possible. The patient may end up hopelessly dependent on a life-support machine but without fulfilling all the criteria for brain death (as described in chapter 5). In this case the patient cannot ask to be allowed to die and the decision has to be taken by close relatives. Switching off the machine is surely a type of euthanasia, and there have been a few instances when a patient in such a condition has made a recovery. But it is becoming a widely held view that advancing medical knowledge has allowed life to be extended far beyond what many expect, or indeed want. The dying process has been unacceptably prolonged, and people should be permitted to die in dignity and peace.

The 'Right to Die'

In English law the patient is legally entitled to refuse life-prolonging treatment, as long as he or she is 'competent' and capable of understanding the treatment and the alternatives. If the patient is competent and refuses treatment it is irrelevant whether others find the decision unwise or foolish. This happens when, for instance, a Jehovah's Witness refuses a life-saving treatment. But most people agree that the parents' views should not be allowed to interfere with the treatment of a child.

It is possible to draw up a document called a 'living will', which consists of instructions in advance as to what kind of medical treatment you want, or don't want, should you be in a condition where you cannot express your wishes. This means that if you don't want treatment you needn't have it, even if this results in your death. It has legal force because a doctor refusing to carry out those wishes would be guilty of assault. Living wills, or Advanced Medical Directives, are popular in the USA and are becoming common in England. They can be a comfort, for instance, to those with progressive illnesses who are worried about their last days. The will should be carried as a card, and a copy lodged with the GP.

More and more ordinary people believe that euthanasia should at least be available. In Australia fourteen different public opinion polls conducted between 1962 and 1996 showed that the vast majority supported the availability of euthanasia. There is a 'right to die' library available on the Internet (http:www.islandnet.com/-deathnet/open.html). Physicians, on the other hand, aware of their professional ethical responsibilities, are more divided. In the USA, if a patient asks for help with suicide, their alternative to giving it is to put the patient in touch with the Hemlock Society, with one of the suicide prevention schemes, or with another physician. The Hemlock Society was founded by Derek Humphry, who wrote a book called *Final Exit*, advocating suicide in certain circumstances and giving methods and examples. American and English law still says euthanasia is murder, but juries often acquit the defendant – as, for example, in 1986 in the trial of a doctor who gave a massive injection of phenobarbitone to a patient suffering from inoperable lung cancer. The Supreme Court of the USA ruled as recently as August 1997 that the US Constitution gives the right to refuse life-saving treatment and nutrition, but does *not* give a general right to physician-assisted suicide, although the responsibility for banning the latter was left to individual states.

The Netherlands has been debating and researching the subject for more than twenty years and takes a pragmatic approach, as it does on other difficult ethical issues like drugs and abortion. Euthanasia is still a criminal act, but doctors are acquitted in the courts as long as they do it under strict guidelines. It must be carried out on a voluntary request from the patient, who must be suffering unbearably and hopelessly, and a second doctor must be consulted. For instance, the patient may be depressed and this depression can be treated – although one could answer that a patient dying of cancer might have something to be depressed about. The doctor's report was until recently sent to the local magistrate, who decided whether to prosecute, the theoretical maximum penalty being twelve years in prison. One Friesian doctor was temporarily jailed on remand, facing murder charges, because of an alleged failure to follow the strict euthanasia guidelines. He was accused of giving a lethal injection of insulin to a seventy-two-year-old cancer patient without proper consultation. Since 1997 the doctor's report has been sent instead to a review committee of doctors, lawyers and ethicists, which is more appropriate. A firm line is still drawn between acceptable medical care and murder, but the law needs to be clarified. It is estimated that in the Netherlands about a third of cases are not reported because of the requirement for a full report and the fear of criminal prosecution. If euthanasia is decriminalized, as seems possible, doctors will no longer need to report cases. The welfare state provides nursing care for the chronically ill and everyone's medical expenses are covered, so finances are not a factor in the request for euthanasia. The Netherlands are different from the USA and UK. People trust their doctors and hardly ever sue them. The need is there; in 1990 there were 3,200 cases of euthanasia, accounting for 2.4 per cent of all deaths. Even so, only about a third of explicit requests for euthanasia are actually carried out. On the other hand, in 1995 about 900 people were subjected to euthanasia without asking for it, although usually the matter had been discussed at an earlier stage, and they were in great pain and receiving morphine in the last days of life.

In 1988 the British Medical Association said that patients do not have the right to demand treatment that the doctor cannot, in conscience, provide, and this applies to euthanasia. But times are changing, and already in Switzerland, Germany, Japan and Poland, mercy killing of the terminally ill, where the victim knowingly requests death, receives only nominal penalties.

What Are the Alternatives?

Two things may reduce the pressure on doctors to carry out euthanasia. In the first place, better terminal care, with adequate pain relief, would mean there was less need for euthanasia. This sort of care is available in a well-run hospice and should be more widely available. Health workers should be better educated about the needs of the terminally ill patient. The wish to die, however, is not necessarily affected by pain control. The pressure for euthanasia might also diminish if attitudes to death changed (see chapter 16), if people were not so afraid of the dying process. The dying Empress Maria Theresa (1717–80) is said to have refused to take morphia, saying 'I want to meet God awake.' Secondly, physician-assisted suicide offers a way round the medical ethical dilemma. This includes giving advice about suicide, giving a prescription for a suicide drug, or actually setting up equipment for suicide. In the last case the doctor prepares a lethal injection, inserts the needle into the patient's vein, and then the patient himself, in full knowledge of what he is doing, activates an automatic injection process. In other words the patient kills himself, and the doctor is not responsible directly for destroying a life. It is not euthanasia, but the legal view is unclear, and varies from country to country.

Certain physicians in the USA, notably Dr Jack Kevorkian, Michigan's 'Doctor of Death', and Dr Phillip Nitschke in the Northern Territory of Australia have been pioneers in physician-assisted suicide. In the Northern Territory, the Rights of the Terminally Ill Act was passed in 1996, and four people were able to make use of it before it was overturned by the Australian federal parliament in 1997. The first was seventy-two-year-old Bob Dent, who had prostatic cancer and ended his life by pushing the commands on a computer joined to a machine that administered three drugs intravenously into his arm. The euthanasia issue has divided Australian doctors, and although more than half the population are in favour, most politicians and of course church leaders have vehemently opposed it. It has been described as 'treating suffering by eliminating the sufferer'. Opposition to the Act was fuelled by the knowledge that the Northern Territory has the highest proportion of indigenous people (Aborigines) of any state, and half of all deaths are in these people. The traditional Aboriginal viewpoint prohibits euthanasia: death is seen as something natural, so that intervention by an outside agent would be murder or sorcery. There would be a need to protect such a vulnerable group from abuse. Furthermore, there are not enough dedicated palliative care units or hospices in the Northern Territory to ease the suffering of the terminally ill and reduce the call for euthanasia. Nevertheless, the issue will not go away, and legalization of some way of carrying out the wishes of the terminally ill seems probable in the future. Dr Nitschke is reported to be building a 'coma machine' that would allow terminally ill people automatically to receive enough drugs to keep them comatose until they die. Should the patient begin to regain awareness, the machine boosts the dosage, keeping him to a previously specified level of consciousness.

Physician-assisted suicide, however is unlikely to be sanctioned in a Catholic or Muslim country. Recently Oregon became the only state in America where physicians can play a key role, in the sense that a person with less than six months to live can request a lethal dose of barbiturates from the doctor and then decide when and where to kill herself. But the US Supreme Court ruled in June 1977 that terminally ill patients do not have a constitutional right to have a doctor help them to die. In other words, physician-assisted suicide is unconstitutional. Yet there are no legal barriers to patients being given drugs to ease their suffering, even if the drugs hasten death.

One of the problems with this route is that up to 50 per cent of patients terminally ill with cancer suffer from depressive conditions which can often be treated. It can then be difficult to judge the 'competence' of the patient who asks for physician-assisted suicide: that is, does he really mean it? But if there is also irremediable mental suffering, the case for suicide is clearer. The question then arises, if such a terminally ill person is an acceptable candidates for physician-assisted suicide, what is the position on irremediable mental suffering on its own? This is where physicians would always draw the line.

Severely ill AIDS patients may wish to end their lives, and the doctors who look after them often help. In a 1996 survey of 118 San Francisco-based physicians, half of the respondents said that they had prescribed lethal doses of drugs to AIDS patients. But a different story comes from London, where only one of 600 patients admitted to an AIDS hospice over a three-year period made a request for euthanasia. Whatever the cause of the difference, it is likely that improved palliative care and increased attention to the psychological and spiritual needs of such patients will reduce such requests.

The doctor, of course, must be sure that the patient's life is irreversibly lost and that his suffering is unacceptable. Difficulties may arise here too. Supposing the patient is not actually dying, but maintains that his life is not being preserved at an acceptable level, at an acceptable quality. In 1994 a Dutch physician assisted the suicide of a physically well woman who wanted to die because her two sons and father had died during the preceding few years. Again, the patient with advanced Alzheimer's disease, if asked earlier in life, might well have opted for euthanasia in this condition. The Hippocratic oath, still taken by some doctors, is little help in confronting these problems, because it is set out in very general terms, with no clear guidance on such matters. At present any individual can carry a card expressing the wish that heroic steps should not be taken to maintain life in intensive care if in a hopeless, irreversible state; but this wouldn't apply in the case of Alzheimer's disease or a crippling stroke. Even if most members of the public might like to feel that their physician could help them make a peaceful exit if they asked for it, the physicians themselves, with their tradition of healing and saving life, are more divided, and at the moment the law says no. A 1994 survey of National Health Service doctors found that 46 per cent would consider granting a request to end a patient's life, if it was legal. At present a doctor risks prosecution for murder (which carries an automatic life sentence) or a charge under the Suicide Act (maximum fourteen years' prison sentence).

A way around current legal prohibitions is offered by what is called the 'doctrine of double effect'. In the UK euthanasia is against the law, but the doctor is allowed to give a drug with the intention of relieving pain, even when death is the probable or inevitable effect. In June 1997 a forty-seven-year-old patient with motor neurone disease won in the courts a guarantee that her physician would lawfully be able to give her enough morphia to relieve her distress, even though the dose might be fatal. Though she had a reasonable quality of life at that stage, in the final stages of the disease she knew she would be paralysed, incapable of speech, and likely to die of suffocation. (In the event she died peacefully, without needing the drug in this way, the following December.) The doctrine of double effect may sound like hair-splitting, but legally and ethically it makes a big difference. Giving morphia to relieve pain (although it also hastens death by depressing breathing) is quite different from injecting a substantial dose of potassium into a vein, which stops the heart beating but has no beneficial effect.

Of course, not all terminally ill patients are in such possession of their faculties as the woman with motor neurone disease mentioned above. About 80 per cent of people in the USA die without an attempt at resuscitation. The idea of withholding or withdrawing treatment, commonly initiated by the physician, usually comes up only hours or days before death, and by that time the patient is often mentally incapable of discussing the matter. Cancer patients are more likely to have made arrangements in advance, perhaps because they are more likely to be in hospices, although less than 5 per cent of them do so.

In July 1997 the British Medical Association urged police to look into the activities of two respected English doctors who have been helping terminally ill patients to die by means of the double effect. The doctors announced that, with a clear conscience, and after discussing it with relatives, they had been giving the patients lethal overdoses of pain-killers, hastening their death while relieving their suffering. One of them provided patients with a 'customized exit bag' containing a lethal dose of the hypnotic drug (sleeping pill) Temazepam. A member of the BMA Medical Ethics Committee said that this amounted to execution. But if the doctors were tried for murder the prosecution would have to prove that they intended to kill the patient rather than merely relieve pain. If a standard painkilling drug was used, the prosecution would not have much of a case.

A note of caution must, of course, be sounded. If the law ever permitted euthanasia it might move us on to the slippery slope towards *compulsory* euthanasia. As described in the following section on homicide, in Nazi Germany psychiatric patients were considered to be nothing more than a burden, and thousands were put to death. Even today, if one regards the unborn foetus as a person, then in the UK and US hundreds of thousands of unwanted people are put to death every year. We have already seen that in certain societies in the past it was accepted that the elderly, being a burden to the others, should be encouraged to kill themselves; sometimes, too, they were killed by other members of their group. It is not inconceivable that if euthanasia became too easily available we could end up with the legalized disposal of expensive, troublesome and unwanted elderly people.

Killing Old People

It has often been argued that old people should be at liberty to kill themselves. In bygone days they were often given a helping hand. Old or sick people were an impossible burden in harsh environments or when journeys had to be made. Father Paul LeJeune, in his acount of his travels as a Jesuit missionary in Canada 1610–1791, decribes the killing of an old mother:

'On the second day of January, I saw a number of savages trying to cross the great river St Lawrence in their canoes . . . I saw a savage dragging his mother behind him over the snow . . . being unable to take her down the common path of a mountain which borders the river along which he was going, he let her roll down the steepest place to the bottom . . . I could not bear this act of impiety and said so to some of the savages who were near me. They answered "What wouldst thou have him do with her? She is going to die anyway . . . she will not suffer so much . . . he is unable to cure her or to drag her after him." This is the way they take care of the sick that they think are going to die. They hasten death by a blow from a club or an axe when they have a long journey to make, and do this through compassion.'

Homicide
Is Killing Always Wrong?

The prohibition on killing is one of the oldest moral rules. But when we look at it closely we see that it is riddled with difficulties. In the first place there are exceptions.

- In most societies, many or all non-human animals are not covered by the prohibition, and can be killed. Not everyone accepts this. Jainism, an old Indian religion, insists, like Buddhism, that all life is sacred; but the Jainist monk goes further than the Buddhist. He must strain water before drinking it, to filter out living creatures. He wears a mask so that he doesn't inhale innocent insects, and he sweeps the floor before him as he walks to avoid crushing living things under his foot.

- In war, it is considered acceptable to kill enemy soldiers (though pacifists would disagree).
- In many countries the law still says that certain criminals may be executed.
- Killing may be permissible in extenuating circumstances – for example to protect an innocent person, or even oneself, from a homicidal attack.
- Euthanasia may be considered acceptable. This controversial topic is dealt with above.

Execution of Criminals

The putting to death of people convicted of crimes is one of the most ancient of practices, in almost all societies. It is called capital punishment because in Latin *caput* means a head: thus 'capital' means concerning (loss of) the head, hanging or decapitation being the favoured methods.

In England the number of offences that carried the death penalty increased hugely in the eighteenth and early nineteenth century to a peak of 220 in 1830. These included cattle theft, housebreaking, or stealing five shillings or more from a shop. On 22 July 1777, Robert Biggin was whipped through the streets of Cary, Somerset, by the hangman, before being hanged. His crime? He had stolen potatoes. It was considered a terrible crime to forge Bank of England notes. The first forger to be executed was R. W. Vaughan in 1758, and the last was Thomas Maynard in 1829. Between 1805 and 1818 there were 207 executions for bank note forgery (mostly of one pound notes). Between 1749 and 1758 in London and Middlesex alone 527 people were convicted of capital offences and 365 of them (about thirty-six each year) were hung. Later in the eighteenth century more lenient sentences became commoner, and in 1789–98, out of 770 convictions, only 191 were hanged. Generally the chances of arrest were small, because there was no proper police force until the mid-nineteenth century – except for the runners organized by the Fielding brothers from their house in Bow Street, London (the 'Bow Street Runners'). Once convicted, the passing of the death penalty depended on the judge, and many offenders were pardoned on condition they agreed to be transported to the American or Australian colonies. In 1837 the ridiculous number of capital offences was reduced to fifteen. Opinion began to turn against hangings as public spectacles, and after 1868 the deed was done in the privacy of prisons.

There was often an element of public vengeance in the death penalty for murderers. In London in 1811 John Williams committed suicide while awaiting trial for a series of brutal murders. To make an example of him and to satisfy public opinion they put his body in a cart, together with the heavy hammer and ripping chisel that had been his murder weapons. The cart was then taken in slow progression from the prison to the grave, led by important citizens and hundreds of police constables, and calling on the way at the houses of his victims. The grave had been made deliberately too small and the corpse was crammed into it and a stake driven through the heart using the murder weapons. As soon as this had been done the crowd of ten thousand shouted their approval. The grave-digger sold bits of wood from the stake as souvenirs.

European and Anglo-American legal codes generally distinguish between killings that are reckless acts of intense passion, often provoked (for instance the *crimes passionels* against a lover or spouse), and planned, 'intentional' killings. Harsher penalties tend to be given to the latter. Inevitably this principle raises difficult decisions, for example in the case of a murderer who intended to kill A but accidentally kills B. In many countries capital punishment has now been abandoned altogether. Others retain it as an option. In the USA, 150–200 people a year were executed in the decade before the Second World War, but the number fell after the war and there were no executions between 1968 and 1976. Although capital punishment was resumed in some states in 1977, in which year twenty-one executions took place, most states have seen none since the 1960s. Virginia has carried out forty-one executions by lethal injection since 1976, and Texas 131. In China, capital punishment is inflicted for a long list of offences, including bribery and the molesting of women. In 1996 4,367 people were executed in China, more than in the rest of the world put together.

One of the reasons behind the decline of capital punishment is that abandoning it does not lead to an increase in the number of murders. Another is that the law can make mistakes, and innocent people have been killed in this way.

Methods of execution have varied over time and place. *Hanging*, which has been practised since biblical times, was introduced into England by the Angles, Saxons and Jutes after AD 449, and lasted until the abolition of capital punishment for murder in 1965. Hangings were numerous in Tudor times: Henry VIII had 72,000 of his subjects executed in this way during the thirty-four years of his reign, and in the short reign of his young son Edward VI (1547–53) an average of 560 hangings a year took place at Tyburn in London. They took place in public, to convey a message to the people about the dire consequences of robbery, murder and rebellion; but they were also public spectacles, with people competing to get a good view, just as we today might go to see a royal procession or a football match.

The last public hanging in Scotland took place in 1865; the condemned man was a Glasgow physician, Dr Pritchard, who was convicted of poisoning and killing his wife and mother-in-law. Their sixteen-year-old housemaid was pregnant by him and had already had one miscarriage, but the motive was partly financial. Pritchard had used aconite, a drug extracted from monkshood or wolfbane root, and used to be applied externally to treat rheumatic complaints; he added it to other medicines that his victims took by mouth. He probably used tartar emetic as well, an antimony-containing drug. During the months before the murders he had bought more from the druggist than all the other physicians in Glasgow put together. Dr Pritchard was executed at Glasgow on 28 July, in front of a crowd of about 100,000 people. The last public execution in England was of Michael Barrett, who was hanged outside Newgate Prison, London, on 25 May 1868. The last hanging (behind closed doors) was of two murderers, Peter Anthony Allen and John Robson Walby, on 13 August 1964.

The apparatus used for hanging was the gallows. Before the eighteenth century the victim mounted a ladder to have the noose put in place, and then was allowed to dangle from it, death often being due to asphyxia. It was a gruesome business; if you were lucky people pulled on your feet to hasten your end. (The word 'gibbet' originally meant the gallows, but in the eighteenth century it referred to the upright post from which the bodies of executed criminals, in chains, were hung. The sight of corpses rotting in gibbets may well have deterred potential law-breakers.) In 1759 the ladder was replaced by a cart which was drawn away from under the feet – a possible improvement. The final and most humane method was the 'long drop'. The noose was placed carefully round the neck, the knot preferably under the chin, and the trap-door on which the victim stood was then opened. He fell freely for up to 8 feet before the sudden jerk and the end. The second and third cervical (neck) vertebrae were generally fractured and dislocated, with severe damage to the brain stem and spinal cord, and death was virtually instantaneous. The sudden compression of the main blood vessels to the brain would also cause immediate death. Stories of victims 'dancing on the end of the rope' arose because of twitching for a second or two after death rather than struggling while still alive. The hangman's art was a skilled one and often passed down from generation to generation. Henry Pierrepoint trained his brother, Tom, and Tom trained Albert Pierrepoint (who dispatched 450 people). Nevertheless there were occasional mishaps, such as when the head came off, or when hanging failed. There are many apocryphal stories. One man, John Lee, did indeed survive three attempts to hang him at Exeter gaol, England. On 23 February 1885 the trap-door failed to open three times over the course of about five minutes. Lee's sentence was therefore commuted to life imprisonment; he was later released, and in 1917 emigrated to the USA, where he married and lived until 1933.

Some of today's hangings have a medieval flavour about them. On 13 August 1997 an Iranian serial killer was put to death in front of 20,000 frenzied, chanting onlookers in Tehran. Many of the crowd clambered up trees and road signs, and some had camped out overnight to get a good view. The condemned man was a taxi driver, and his nine night-time killings had terrorized women in Iran's capital. First, blindfolded and with his hands tied, he suffered a ten-minute whipping by male relatives of the victims. His neck was then roped to a gigantic mobile crane and he was hoisted high into the sky for all to see, with legs flailing.

Lynching is a form of hanging carried out by a mob, without the backing of the law. It is named after Charles Lynch, a Virginian planter who headed an unofficial court formed to punish loyalists during the American Revolution. In the USA between 1882 and 1951 a total of 4,730 people were lynched, 1,293 of them white and 3,437 black. But it is not a uniquely American institution, nor a long defunct practice: a suspected murderer, a Pakistani, was lynched in London on 27 October 1958.

Beheading with sword or axe was a common method of execution in earlier centuries. King Charles I of England was decapitated before a large crowd on 30 January 1649. The efficiency of this method depended on the executioner, and it was often bungled, so that repeated blows were required to sever the head. In Tudor England the head was often impaled on spikes, for instance at London Bridge or Temple Bar, for the people to see. The last victim in the UK was Simon Fraser, beheaded on Tower Hill, London on 7 April 1747. It remains the standard method of capital punishment in Saudi Arabia.

Crucifixion was used by the Persians, Jews and Carthaginians. In 519

BC in Babylon, Darius (King of Persia) crucified 3,000 of his political opponents. The Romans practised it in the sixth to fourth centuries BC, until Constantine the Great (the first Christian Emperor) abolished it in AD 337. Crucifixion was a slow death, with maximum pain and suffering, and the Romans used it for political and religious agitators (such as Jesus), for pirates, foreigners and slaves – but not for Roman citizens, except deserters from the army. The word 'excruciating' means 'out of the cross'. The procedure, as carried out by the Romans, was as follows. The victim was first flogged ('scourged'), and then made to carry the cross or the crossbeam to the crucifixion site outside the city walls as part of a military procession. The cross would have weighed about 300 lb (Jesus was too weak to carry his, so Simon of Cyrene bore it for him). Usually the upright was waiting in the ground. The hands were either tied to the crossbar, or nailed with 5–7 inch iron spikes passing between the bones of the wrist. Then the feet were fixed with rope or nails, often to a wooden footrest, and to get a good nailing result the knees were flexed. The victim was offered a drink of wine with myrrh as an analgesic (Jesus refused it). The survival time on the cross varied from three or four hours to three or four days, depending on the stamina of the victim and the severity of the flogging. Death was due to interference with breathing, blood loss, dehydration, shock and exhaustion. To breathe out properly, you had to lift yourself up, pushing on the feet. The soldiers could hasten death by breaking the legs below the knees, which made breathing impossible. They stayed with the victim until he was dead, then pierced the body with a sword or spear before releasing it to relatives. Jesus was handed over for flogging and crucifixion by Pontius Pilate, and died after between three and six hours, on Friday 7 April in AD 33. He was already dead when they came to break his legs.

As a result of Jesus' death in this way, crucifixion has been an important subject in Western art since the early Middle Ages onwards, its depiction focusing attention on Christ's suffering and its spiritual significance. Details of its portrayal have changed over the centuries; depending on fashion and on the artist, the image may include the two thieves also crucified at the same time, or the Roman centurion with his spear, or the mourners, or the sun and the moon (both eclipsed on this occasion). Some artists have dwelt on the shocking physical detail; others have produced more stylized representations.

Garrotte is an old Spanish method of execution by strangling. Originally a cord was fixed round the neck of the victim and twisted with a stick (Spanish *garrote* = a stick), but in a later refinement the condemned man, usually seated, was fastened to an upright post by an iron collar round the neck: the collar was tightened by a screw and lever system and the spinal column eventually dislocated. It was much used in the seventeenth century, and was not a good way to die. General Lopez was garrotted in 1851 for trying to gain control of Cuba. An interesting variation of the method, as suggested by a certain uneducated man, was recorded by the French priest Jean Meslier (1664–1733). The ingenious idea was that all the great men and the nobility could be strangled with the guts of priests! The word was also used to describe a form of highway robbery (punishable by flogging) which became common in London in the 1860s, in which one thief strangled the victim while another rifled his pockets.

The *guillotine* is arguably the most reliable, painless method of enforced departure from this life. The heavy, oblique-edged knife fell swiftly down, severing the victim's head, which rolled into the waiting basket or sack. Doctors have noted that the eyes may move for up to a minute after decapitation, but tales of them following a moving object for ten or fifteen minutes seem improbable. The guillotine is so called because it was promoted in France as a humane method of execution by Dr Guillotine; a German named Schmidt constructed the machine. Experiments were carried out on dead bodies from hospitals; it was first used in earnest in 1792, to execute a highwayman. The guillotine saw plenty of action during and after the French Revolution, a period in which Charles-Henri Sanson executed 3,000 people, including King Louis XVI. It was finally abolished in 1982. The last guillotining in public was at Versailles on 17 June 1939, and the last time it was used was on a Tunisian murderer in Marseille in 1973. The guillotine was not exclusive to France: similar methods had been used in Germany and Italy, under different names, since the Middle Ages; the Society of Antiquaries in Edinburgh still displays the 'maiden', last used in 1710, and the 'Halifax gibbet', another decapitation machine, was in use in England until the seventeenth century.

Shooting by firing squad is a traditional method, especially in a military context. Life is reliably extinguished when many bullets tear into the heart. It is said that not all the guns had live ammunition, so that no one was sure who did the killing. A beautiful painting by Manet shows the firing squad at the execution of the Archduke Maximilian, Emperor of Mexico, in 1867.

The *electric chair* was supposed to be a tidier, yet efficient and painless method of execution. First introduced in Auburn State Prison, New York in 1890, it came to be used in twenty-four US states, and also in the Philippines and the former Republic of China. Unfortunately the results have been less than perfect. In 1940 Julian Rosenberg and his wife Ethel, members of the Communist Party, began attempting to transmit information about nuclear weapons to the USSR. They were spies: Rosenberg had acquired this information while working in the US Army Signals Corps. After being tried and sentenced to death, the couple were finally executed on 19 June 1953. The electric chair did not perform too well, and it took more than one 'jolt' to complete the process.

Chemical execution was intended to remedy the deficiences of the electric chair, with its unacceptable failure rate. The release of an immediately lethal gas (cyanide) into the death chamber, or the injection of a lethal cocktail of drugs into a vein, was at least more reliable. The gas chamber was introduced in Nevada, USA in 1924, and later taken up by eleven other states. A lethal injection is used in Texas, the death penalty capital of the free world, with a recent record of four executions in a single month. It is essentially the same method as that used by veterinary surgeons to put down old, sick or unwanted animals.

What is the best way to be executed – assuming the the object of execution is to remove a dangerous person from the world, rather than to be revenged on him and make him suffer for the evil that he has done? A good case can be made for the following procedure:

- first, not to be told that it was going to be done;
- then to have a strong hypnotic drug incorporated into a last, excellent meal;
- finally, once unconscious, to be injected intravenously with one of the many absolutely reliable lethal drugs.

Infanticide

Archaeological evidence shows that infanticide or child sacrifice goes back to the city state of Jericho in 7000 BC. The usual method was by exposure to the elements. The Vikings, Gauls, Phoenicians, ancient Romans and ancient Greeks allowed the killing of weak or deformed infants. Plato and Aristotle supported this practice, agreeing that such infants were a burden to society.

As recorded in the Bible, King Herod ordered the killing of all newborn infants (the 'slaughter of the innocents') at the time of Christ's birth, and the Pharaohs ordered midwives to destroy all male Hebrew infants, fearing them as a future military threat.

Infanticide, especially of female infants, was common in medieval

England and is still common in developing countries today. Poor families shared a bed, and stifling by 'overlaying' was an easy and non-violent method of disposing of an unwanted baby. Overlaying of infants is also mentioned in the Bible (1 Kings 3). Of course, it is only sometimes a method of infanticide: it can be accidental, and what looks like infanticide can often be a case of Sudden Infant Death Syndrome (SIDS or 'crib death'). Infanticide remained common in eighteenth- and nineteenth-century England, when disapproval was directed more against the associated fornication and illegitimacy than against the act of killing itself. 'Baby dropping' (abandonment in a public place) was popular, but this was not necessarily intended as infanticide because the mother often hoped that someone would take pity on the pathetic little bundle. Foundling hospitals (for found infants) were established in London in 1741. Other babies were killed simply by neglect.

The *Lancet* in 1861 recorded 1,130 murders of infants less than two years old between 1856 and 1860, in London alone. The true figure must have been much higher, because until 1926 it was not obligatory to register a stillbirth. Moreover, investigations of infant deaths either did not occur or were inadequate until 1860, when the County Coroners Act gave coroners proper pay and assistance. As for those cases which did come to court, the law was firmly on the side of moral propriety: the killing of an illegitimate baby was deemed to be murder, whereas married women were generally acquitted. Concealment of a birth was a crime because it often meant infanticide. Nevertheless convictions were rare, and the last execution of a woman for infanticide was in 1849: public opinion moved against the death sentence as people began to accept that a mother's mental state could be disturbed by birth and lactation. The Infanticide Acts of 1922 and 1938 said that as long as the balance of the mother's mind was judged to be disturbed by the birth or by lactation, killing her child before it reached one year old counted as manslaughter rather than murder (most infanticide takes place soon after birth). The maximum sentence under British law is still five years' imprisonment.

In considering infanticide in the past, we have to remember that in Victorian England the social and economic consequences of unmarried pregnancy were daunting. The unfortunate woman would lose her job, her self-respect and her place in society. Small wonder that in an era when abortions were dangerous, infants were often abandoned or killed. The number fell markedly once the stigma on unmarried pregnancies was removed and there were better facilities for single mothers and their children. Today, too, many infants that would have died in the old days are enabled to survive because of advances in medical science. In the USA 7 per cent of newborns have physical or mental handicaps. The brain especially is involved, because while serious defects in other organs may be fatal at an early stage of development, the embryo can survive in spite of major abnormalities in the nervous system. Examples include severe forms of spina bifida, anencephaly and hydrocephaly. Difficult decisions arise, in which parents and physicians must be involved. A ruthless, matter-of-fact view would be that such infants are not yet people, will be a burden to society, and can be disposed of. On the other hand, many would argue that the interests of the child, however 'abnormal', must always be foremost, and that it is better to be born handicapped than never to have been born at all.

As modern China came to terms with its population problem, strict rules were imposed concerning the number of children allowed in each family. Because female children were less valued than boys, infanticide of girl babies (and the abortion of female foetuses) began to be widely practised, although now this is said to be rare. There are, however, very large numbers of infant girls in China's orphanages.

In deciding whether or not a child's death is a case of infanticide, three key questions have to be addressed:

- *Was the child killed?* This is obviously the question for the law and the 'forensic' pathologist (see chapter 16). In England the prosecution has to prove that the infant was *not* stillborn. It can be very difficult to tell whether the child was killed or died from to other causes. For instance, head and limb injuries can occur during the birth itself, and strangulation can be caused by the umbilical cord round the neck. The commonest cause of death of the newborn is brain damage due to asphyxia before or during birth. Stabbing injuries, on the other hand, signify infanticide.

- *Was the child born dead (stillborn)?* Had it taken a breath? This used to be tested by whether the lungs of the dead infant would float on water. However, several breaths are needed to achieve this effect, because when blood begins to pass through the lungs they at first get heavier. The flotation test is now considered inconclusive. If the lung sinks in water it is taken to mean that the infant was stillborn; if it floats, however, it could be because a breath or two had been taken during passage through the vagina, or it could be because of post mortem decomposition.

- *Was the newborn infant capable of life?* Nowadays even a twenty-two-week-old foetus can be kept alive and saved after premature delivery. The eighteen-week-old foetus, however, is alive in the uterus but incapable of surviving in the outside world because the lungs are not yet developed.

Abortion

If the foetus is a person, then from a moral standpoint abortion is murder. If abortions are included in the list of causes of death they dwarf the other categories, the numbers being comparable with those of deaths due to war. The subject arouses very strong feelings on account of its scale as well as its moral aspects. In 1994 a gynaecologist in Vancouver who performs abortions was shot by a sniper. Yet is it not odd that people feel so strongly about abortion, as they do about infanticide and euthanasia, but often seem less concerned that hundreds of thousands continue to die due to car accidents, wars, starvation and infectious disease?

Abortion rates are worked out as numbers per thousand women per year. Table 2 shows figures for 1979–84. The highest rates, not unexpectedly, are for women under the age of twenty-five. The total number runs into many millions per year (160,000 a year in the UK).

Table 2   Abortions in selected countries, 1979–1984

	No./1,000 women/year
West Germany	6.8
England and Wales	12.6
France	14.1
Denmark	19.3
Italy	22.4
Japan	22.5*
USA	26.9
Hungary	35.5
Cuba	47.1

*This is an underestimate of the actual figure.

In ancient Greece and Rome, abortion was an accepted method of family limitation, and in most countries it did not become unlawful until the nineteenth century. China still uses abortion on a large scale as part of its population control policy. In Japan abortion is said to be readily obtainable and used as a method of family planning. The Muslim view on abortion is that it is permissible if the foetus is abnormal and will give rise to a disabled child, as long as it is done before 120 days, after which time the soul is present.

What is abortion?

Abortion means a foetus being removed from the womb too early to be capable of independent life, in other words before the twentieth week of gestation, when the foetus weighs about 400 g. Spontaneous abortion (often called miscarriage) is very common: about 10 per cent of all pregnancies end in this way. At least half of these abortions are due to faults in development, and sometimes occur so early that the mother, not knowing she is pregnant, experiences no more than a slightly delayed or heavy period. Other causes are hormonal imbalance or illness in the mother. Only one in a thousand is due to physical injury.

An induced abortion is one that is brought on deliberately, perhaps because the pregnancy threatens the mother's life and health or because the foetus is abnormal. On the other hand, it may be merely because the pregnancy was an unwanted accident.

The main methods of inducing abortion are as follows:

- aspiration, in which the neck of the womb is dilated, a thin tube inserted and the contents of the uterus sucked out using an electric pump;
- curettage, in which the contents of the womb are scraped out;
- injection of fluid into the uterus to stimulate uterine contractions;
- injection of substances such a prostaglandins into veins to stimulate uterine contractions;
- taking hormone-like substances by mouth to induce menstrual bleeding.

The more advanced the pregnancy at the time of the abortion, the greater the risk to the mother.

In Britain and the USA before the nineteenth century the law allowed abortion, as long as it was done before the mother felt the 'quickening' (movement) of the baby in the womb, which occurs at about fifteen weeks. Occasionally women prisoners claimed that they were pregnant to avoid the death penalty. Special juries of women were formed to feel the prisoner's stomach and test for quickening, because at that stage of pregnancy they could not be executed. An advertisement in *Bell's Court and Fashionable Magazine* in September 1807 (quoted in the *Lancet*, 21 May 1932) read:

lost happiness regained – any lady of respectability involved in distress from any expectation of inevitable dishonour, may obtain consolation and security and a real friend in the hour of anxiety and peril by addressing a line (post paid) to Mrs Grimstone, No 18, Broad Street, Golden Square, London, where a private interview with the advertiser will be appointed. Ladies thus situated may depend on the strictest secrecy and motherly attention.

Abortion after fifteen weeks became illegal in England in 1829, when an Act of Parliament stated that aborting a woman quick with child, unless it was to save her life, was punishable by death, and for doing it earlier in the pregnancy the penalty was transportation for seven to fourteen years, with or without a public whipping. Nevertheless, in Victorian England, when contraception was unavailable to most women, abortion was very common as it remained the only way to avoid the terrible social consequences of unmarried parenthood. A few doctors were prepared to stretch the law and carry out abortions on insufficient grounds, usually for wealthy clients. The poor had to rely on unofficial practitioners like Dickens's Mrs Gamp, or on Widow Welch's Female Pills, lead plaster, nutmeg, the gin and gunpowder remedy, and other quack 'remedies', many of them dangerous to the health of the mother. In 1853 handbills were being addressed to female domestics (so vulnerable to unmarried pregnancy) telling them how they could get abortions. Even in the nineteenth century a few people were beginning to argue that the woman, rather than the state, had the right to manage her own pregnancy, and that if abortion stayed illegal, then illegal backstreet abortionists would take over, with much more risk to life and health.

Abortions were so common because so few women used or knew about contraceptive devices. Annie Besant and Charles Bradlaugh (a Bristol bookseller) challenged the law in 1876 by republishing a banned booklet, *The Fruits of Philosophy*, which gave practical details about birth control. They were convicted of obscenity and sentenced to two years' hard labour, but made a successful appeal against the verdict, and in the following four months 125,000 copies of the booklet were sold. Yet in spite of such attempts at education, the poor remained largely ignorant of how to avoid conception. It was not until 1922 that Marie Stopes opened the first birth control clinic in Britain. She was a pioneer and her arguments were only very gradually accepted; as recently as 1937 the British Medical Association estimated that there were 44,000–60,000 *illegal* abortions carried out in the UK each year. The actual figure was probably about 100,000. After this the numbers stayed about the same, but more and more were legally performed, and after the Abortion Act in 1969 nearly all were legal.

In the UK at present about 160,000 legal abortions are performed every year (177,225 in 1996). It is at least arguable that most of these are due to carelessness or irresponsibility. Nevertheless it is a pity that for many women faced with an unwanted pregnancy, the choice is between having an abortion and giving birth to the child and having it adopted, when adoption is such an unsatisfactory business. The red tape of the procedures, doubtless designed for the benefit of the child, takes such a long time that it may be a year or two before the child finds a home. There is no shortage of prospective parents, but in the meantime the child is 'in care', in other words being looked after by social services.

A MORI survey carried out in Britain in March 1997 showed that over the past twenty years support for abortion has generally increased. Of 1,888 adults questioned, 64 per cent agreed that abortion should be legally available for those who want it, whereas in 1980 the figure was 54 per cent. However, only 67 per cent approved of abortion when the child was likely to be born mentally or physically handicapped, compared with 83 per cent in 1980.

In the USA, as in Britain, laws against abortion were passed in the early nineteenth century. It has been calculated that in 1840 there was one abortion for every five or six live births, most of them by unofficial practitioners and most of them resulting in death of the mother. Large families were still common: in 1800 the average woman bore seven children, but after this the number fell, reaching an average of 3.6 by 1900, mainly as a result of contraception and easier access to abortion. Women genuinely believed that the foetus was not truly alive until quickening. Nevertheless, many states passed anti-abortion laws between 1860 and 1880. Before then, pills for abortion had been quite commonly, if discreetly, advertised. For instance: 'Married ladies who have any reason to believe themselves pregnant are particularly cautioned against using these pills as they will cause a miscarriage.'

As in Britain, US laws on abortion began to be relaxed after the Second World War, but the debates on birth control continued right up to the 1930s, against an actual background of about 2 million abortions (many of them criminal) a year. It was still prohibited to bring books on the subject into the country, and the medical profession did not officially support birth control until 1937. In 1973 the US Supreme Court legalized abortion on demand; by this time the number of terminations performed, for various reasons, had begun to fall. However, in response to the relaxation of federal law, anti-abortion organizations sprang up and in many states abortions became more difficult.

Eugenics

The word 'eugenics' was coined in 1883 by Sir Francis Galton, a cousin of Charles Darwin who devoted his life to the study of heredity, and worked out the method of crime detection by fingerprints. Eugenics means the promoting of those hereditary qualities that aid the development of the human race. It implies improving the genetic stock of the species by selective breeding or other methods. Yet selective breeding, applied with such remarkable success to domestic animals, is almost a taboo subject for the human species. Who is to decide what qualities should be selected and bred for? And once this has been decided, what happens to those who possess features *not* so selected – the 'undesirables'? As this section makes clear, what seems a very sensible idea can have terrible implications for human life and death.

It is easier to identify the things that should be eliminated when we consider inherited diseases. An appalling mount of human misery is caused by genetic defects. Certain diseases have a well-known genetic basis, and can to a large extent be avoided. But the best of us have bad genes tucked away in the chromosomes – 'tucked away' because they are recessive and rarely get a chance to have any effect. The average human has so many bad genes that, if they were all added up and all came into action, there would be about four times the amount required to kill.

Many famous biologists of the twentieth century, including Arthur Keith and C. D. Darlington, have had no doubts about the need for eugenics. Unfortunately it sometimes led them to argue for the inborn supremacy of certain groups and consequently to advocate blatantly racist ideas. In 1953 Darlington maintained that 'some men are born to command, others to obey, and others are intermediate', and that the governing classes 'derive their dominant position from the fitness of their genetic character'. The opposite extreme, that all men are born genetically equal, is likewise a fallacy. No one doubts that men and women like Leonardo da Vinci, Einstein, the Brontë sisters, Darwin, Newton, Beethoven and Jane Austen have enriched our species, and that they carried a special constellation of genes. We might like to have a few more people like them; but not millions of them, and not at the expense of less outstandingly gifted individuals.

Over the centuries an occasional writer or philosopher has suggested that the poor are genetically inferior, and that because they have larger families this could lead to a deterioration in the human race. As the poor outbred the better off, bad genes would accumulate. Today we know that the poor are *not* genetically inferior, so views like this are rarely voiced. On the other hand, when someone carries a gene that is known to cause a serious handicapping disease, it is accepted that they should be counselled about not having children; and if a foetus is found to be carrying such a gene, it can be aborted. These are reasonable procedures for preventing inherited diseases. In China, however, a new law passed in 1995 seems to go further. It stipulates that couples with 'genetic' diseases of a serious nature are allowed to marry only after being sterilized or undertaking to practise long-term contraception. Children with such genetic diseases are going to be a burden to the state, and steps should be taken to prevent them being born. It is not clear whether and how this 'eugenics law' – which has been opposed by many geneticists and physicians – is to be enforced.

The USA was in the vanguard of the eugenics movement. A law in Indiana, dating from 1907, allowed for the sterilization of confirmed criminals, 'idiots' and rapists. Similar laws were passed in many other states, and tens of thousands of sterilizations were carried out between 1907 and the 1970s. In Japan more than 16,000 people (two-thirds of them female) with mental handicap or hereditary diseases were sterilized, with government approval, between 1945 and 1995. Their Eugenic Protection Law was abolished in 1996, but this attracted little attention. Until the 1970s, too, certain European countries, including Switzerland, Sweden, Finland and Norway, practised compulsory sterilization of mentally handicapped men and women (a total of 60,000 were sterilized in Switzerland over fifty years), and Austria is said still to be doing this. In some cases the practice extended to morally or racially 'unsuitable' or 'inferior' individuals, but just how far is not clear.

Eugenics takes on a more sinister tone when it is applied to traits like intelligence or physical appearance. This is when the alarm bells ring. We remember with horror the Nazi beliefs about the racial superiority of 'Nordic' types with blond hair and fair skin, and the need to restrict the breeding of other groups of people. Yet in Singapore recently the selective breeding idea resurfaced with officially sponsored schemes to encourage marriage and childbearing in college-educated individuals.

As far back as 1908 German scientists had been worried about people of mixed blood, and in the German colony of South-West Africa those entering into mixed marriages were deprived of their civil rights. In 1923 Hitler, while a prisoner in Landsberg, read a German textbook on race hygiene and incorporated the ideas into his own book, *Mein Kampf*

Rats and the welfare state

C. P. Richter was a biologist and psychologist who, working in the 1950s, compared rats (*Rattus norvegicus*) bred in the laboratory (for about 100 years) with wild rats. In the laboratory there is abundant food, water, shelter and mates. The struggle for survival has been eliminated. These rats are less resistant than wild rats to disease, stress, and fatigue. Their brains are smaller, their adrenal and thyroid glands are less developed, their sex glands mature earlier and they are more fertile. These features, in addition to the fact that they are tamer and less aggressive, make them well adapted to life in cages. In the wild they would not survive for long in competition with their ancestor, the wild rat. They are nicely fitted to cage life.

Impressed by these observations, Richter turned to humans. Tens of thousands of years ago our ancestors struggled for existence, and the strongest and cleverest survived, handing down their genes to their descendants. Modern humans are the result of that survival of the fittest, which is the evolutionary process. But in the past century the selection process has been halted. With lowered infant mortality, and fewer deaths due to environmental dangers and infections, the genetically unfit are able to survive and breed. Medical care, better housing and generally improved conditions have arguably reduced modern humans to the status of the laboratory rat.

Although Richter feared for the future of human evolution, we can argue that humans are better adapted to modern life than our distant ancestors would have been. Indeed, many of our sociological and medical problems arise from the fact that much of our genetic inheritance, while fit for a 'wild' (prehistoric) hunter and gatherer, is no longer suitable for present-day life. Features like aggression (leading to criminal activity, 'road rage', heart disease, high blood pressure) and the ability to store fat in the body, are unfortunate relics of an evolutionary process that produced such successful hunters and gatherers. In modern civilization they are handicaps.

Richter's worries will no doubt be fully addressed when humans begin to look after and plan their own evolution. Science will make this possible before too long,

---

(*My Struggle*). The National Socialists' (Nazis') beliefs on eugenics and racial supremacy were promulgated by psychiatrists, doctors and anthropologists, and there was no shortage of eminent men to support Hitler's programme. Hitler became Chancellor of the Reich in January 1933 and in the same year proclaimed a law allowing compulsory sterilization of schizophrenics, manic-depressives, alcoholics, and those with congenital blindness, deafness or mental defects. All German coloured children were sterilized in 1937, and between 1934 and 1939 350,000–400,000 other 'genetically unsuitable' people had been sterilized. Later experiments on women by a professor of gynaecology showed that with a newly developed X-ray machine a single doctor could sterilize hundreds of women a day.

The programme did not stop at preventing birth: soon it moved on to actual killing. Mental hospital patients were being starved to death in 1938, and the 'euthanasia' scheme began in 1939. By 1940 a total of 70,723 mental patients had already been killed with carbon monoxide gas, and professors at the Kaiser Wilhelm Institute of Brain Research had studied 500 of the brains. Victims were selected by psychiatrists and anthropologists, who were happy to support the programme although they did not want to be the ones who actually did the killing. The first gas chamber was built at Auschwitz in January 1942, using Zyklon B (hydrocyanic acid), and the new crematoria had a capacity of 4,756 persons a day. Initially the corpses from the gas chambers were buried, after removing all valuables such as rings and gold teeth, but later in the war, when the SS thought the peace negotiations would be affected if the buried bodies were dicovered, they were burned. When the Germans invaded Russia, special murder squads accompanied the troops, either shooting the Jews, gypsies and mental patients or killing them in custom-built gassing vans. Russian prisoners of war were an unacceptable burden for the Reich, and altogether 3 million of them were killed or allowed to die through lack of care in captivity.

Altogether, 6 million Jews were killed by the Germans during the Second World War. Gypsies, too, fell victim to the vicious Nazi regime. At Auschwitz 20,943 gypsies were eliminated. The word 'gypsy' is said to be a corruption of Egyptian: this people – whose own name for themselves is the Romanies – were supposed to have been driven out of Little Egypt. They are a wandering people, and spread through Europe in the sixteenth and seventeenth centuries, eventually becoming scattered all over the world. Gypsies were good fortune-tellers, good musicians and highly adaptable, but with a reputation for dishonest dealing. Like the Jews, they kept themselves to themselves in a separate community with separate beliefs, and were therefore sitting targets for racial discrimination and violence. They were often accused of kidnapping children, and in Hungary in 1782, after supposedly eating a dead body, forty-five gypsies were hung, drawn and quartered.

The story of the Nazis' euthanasia programme illustrates the immense potential for evil when eminent scientists and their ideas are wedded to unscrupulous politicians. (It is interesting to note that Hitler was an enthusiast for physical as well as racial health: at the same time as his evil killing machine was at work on the 'undesirables', he was carrying

Hitler's programme of genetic and racial hygiene

Hitler believed that people with genetically determined diseases, including schizo-phrenia and 'congenital feeblemindedness', should be prevented fron breeding, as should those suffering from many other conditions, such as congenital blindness and deafness, physical deformities, manic depression and chronic alcoholism. Similar beliefs were held by many individuals, not just Germans, at that time.

Before the Second World War about 400,000 sterilizations, by vasectomy or tying of Fallopian tubes, had been carried out. Most of those sterilized were Germans. By 1939 it had been decided that those suffering from chronic mental and physical illnesses should be eliminated rather than supported by the state. So, between 1939 and 1941, 70,000 people were subjected to what was called 'mercy death' in killing asylums. The first victims were 5,000 congenitally deformed children, who were given lethal injections or starved to death. Then an adult programme began, using carbon monoxide gas. The nation was being 'cleansed' and 'improved' by compulsory euthanasia.

Hitler had long maintained that certain human stocks were inferior and needed to be eliminated, and he was backed up by numerous anthropologists and psychiatrists. These were the people who provided the intellectual and scientific justification for racial superiority, and they helped set up the system for sterilization, mass murder and genocide.

As the concentration camps filled with political and racial prisoners, those who were too sick for labour were culled. Gypsies and Jews, deemed racially inferior, were to be disposed of in the interests of national hygiene. The procedures used in the euthanasia programme were readily converted for this purpose. The Nazis already knew which gases were cheap and effective, and continued to employ methods like the disguising of murder-chambers to look like shower-rooms. The victims, it was said, were not only racially inferior but also morally corrupt, with criminal leanings; this was a type of community medical treatment where the life of the individual was being sacrificed for the good of the whole. Dozens of medically qualified doctors were implicated in the supervision and administra-tion of the mass killing programme. Innumerable Russians, Poles, Czechs, Yugoslavs, and later French, Norwegians and Belgians were also victims.

Since those destined to be killed were no more than the walking dead, it was decided that some of them could usefully end their lives as objects of medical research. At the turn of the century the Prussian government had pioneered the idea of informed consent, but times had changed. Dr Josef Mengele and others were now prepared to use the concentration camp victims in cruel experiments.

Hitler had also planned to depopulate Poland and settle it with Germans, and by 1939 had put his SS 'death-head' formations in place. We need the living space, he said, and we must relentlessly and without compassion send into death those of Polish origin and language.

The Nazi philosophy on eugenics had led to a series of steps, starting with compulsory sterilization, passing through involuntary euthanasia, and ending with genocide. A series of steps like that soon becomes a slippery slope, and we should never tread it again.

out vigorous anti-smoking campaigns to foster bodily purity in the German people.) Yet we have to admit that the eugenic idea has a sound core, when applied to genetically determined diseases. The danger is that it can lead logically to the sterilization of genetically inferior individuals, and this can be the first step on a slippery slope which ends with killing of anyone who is, for whatever reason, 'unwanted' by those in power. The Nazis ruthlessly slaughtered gypsies, Jews, the mentally handicapped and other 'undesirable' groups of people on a massive, unprecedented scale. The 'ethnic cleansing' perpetrated in the Bosnian war of the 1990s, albeit not on the same scale, comes into the same category.

The ethical dilemmas of eugenics, like those of euthanasia, are going to loom ever larger more in the future. The Human Genome Project (see chapter 11) will eventually tell us not only which genes determine disease, but also something about the genetic determinants of longevity, intelligence, musical and mathematical ability, beauty, aggression, docility and so on. There are bound to be strong pressures to apply such knowledge to human reproduction. Will there come a time when 'designer babies' are common?

Sacrifice

Sacrifice is certainly a type of homicide. The Aztecs of Mexico in the fourteenth and fifteenth centuries practised it on a massive scale, as described in chapter 14. For them, human sacrifice was not an act of mere cruelty but a response to the need to offer up plenty of blood to their gods to keep their world in a stable condition. Those who were sacrificed were either slaves or prisoners of war. Indeed, the purpose of war and fighting was to take as many live prisoners as possible, and man-sized nets were as important as spears or arrows. It was not easy to get enough captives, and the Aztecs were obliged to wage almost constant wars with neighbouring peoples to feed the demand for sacrificial victims.

Murder

Most people, when they think of homicide, think of murder: the killing of individuals – one or more – by individuals. In 1993 there were 300,000 murders recorded worldwide. Homicide rates are highest in Latin America, the Caribbean, Africa and certain Middle Eastern countries. In these parts of the world personal disputes are more likely to be resolved by violence, and in some countries homicide is the second leading cause of death in

15–24-year-olds. Top of the list, for 1980, were Guatemala, Thailand, Puerto Rico and Brazil, with more than 11 homicides per 100,000 people. The USA came close behind with 10.5. Down at the bottom of the list were France, Japan, Greece, England and Wales, and Ireland. Most countries also have a list of missing persons; some have disappeared without telling anyone, others have commited suicide and others have been murderered. In Sweden, for instance, there are about 1,000 missing persons each year, and 200 of these stay missing.

In the USA homicide is a serious public health problem. In New York City between 1980 and 1983 homicide was the largest single cause of death in 16–45-year-olds, nearly all of the victims being males. If you count the number of years of potential life lost, its impact is greater than cancer, heart disease, drug dependency and accidents. In 1980 it reached its highest levels this century, when throughout the USA a total of 23,870 people were killed, most of them black or members of other ethnic minorities. The chance of being killed at some time are 1 in 240 for whites and 1 in 47 for other groups. In England the number of murders known to the police has stayed at 130–170 each year between 1920 and 1960.

Who are the killers? In all societies murder is mostly, but by no means only, committed by young males. The victim is often a member of the family, a friend or an acquaintance. Murderers are often spouses, and the wife more often a victim than the husband. Generally the courts treat with sympathy the battered wife who kills her husband after suffering assaults and injuries over a long period. It is estimated that in the USA 1.5 million women each year seek medical attention after being assaulted by their male partners. Robbery-related homicide, however, is largely committed by strangers.

A particular kind of homicide, prevalent at certain periods in the past, arose from the practice of duelling. A duel (from the Latin for 'two') was an arranged combat between two persons to settle a specific issue when there was no legal alternative. It can be regarded as a type of ritualized killing. Combat between two people to decide great questions goes back a long way in history – right back to the contest between Hector and Achilles in Homer's *Iliad* or between David and Goliath in the Bible. Duelling became common in north-western Europe in the first millennium AD, and in 516 the Burgundian king Gundobald actually legalized it as 'judicial duelling', to take the place of a legal trial: God decided who was right. The popes condemned it, but the practice was not formally abolished in the UK until 1818.

Duelling in its modern form, especially prevalent in France, was a method of resolving a private dispute or avenging an insult in such a way as to assert and maintain 'honour'. It grew to such an extent that Richelieu, in 1626, tried to put a stop to it by confiscating the property of duellists and banishing them from France. In England many historic duels were fought, including those between Pitt and Tierney, and between the Duke of Wellington and Lord Winchilsea. For the most part the practice was confined to the upper and the military classes. In Germany before the First World War duelling was common among military officers, but a lethal outcome was unnecessary. Honour, as defined by the conventions of the period, could be maintained by carrying out the ritual. It may seem incomprehensible now, but in those days duelling scars were emblems of manhood.

Duelling can be regarded as a hazarding of one's own life, and an incident in 1741 combined gambling, duelling and suicide. At a casino in London a man called Nouse had a violent disagreement with Lord Windsor and challenged him to a duel. The peer refused. Nouse knew this was because Windsor did not consider him a social equal, and his honour was so offended that he went home and cut his throat.

Another unusual form of homicide was that practised by the Thug- gees, a group of robbers and murderers who preyed on travellers in India from the sixteenth to the nineteenth centuries. After joining parties of travellers on their journey and giving all the signs of friendship, the thugs killed them all, ruthlessly and skilfully, by strangulation. They then buried them, to the accompaniment of semi-religious rites. About two million travellers were killed in this way. The British began to suppress Thuggee (thuggery) in the 1830s, but it persisted into the beginning of the twentieth-century. At the trial of one Thug (Buhran) it was established that he had strangled more than 900 victims between 1790 and 1830. The novel *The Deceivers* by John Masters contains exciting accounts of their activities.

In order to be convicted of murder, a person has to be shown to have intended to kill. However, there has been much interest in the question whether it is possible to kill without being conscious of doing so. In 1774 a Viennese physician named Mesmer began to practise what we now call hypnosis to treat his patients. He used pieces of apparatus including wires and magnets, suggesting that his successes were due to the transference of 'magnetic fluid', but these objects were later shown to be unnecessary. He gave us the word 'mesmerism'. In 1841 Dr Braid, a physician in Manchester, England, found that a person could be entranced by gazing at a bright object, and he coined the term 'hypnotism' (from the Greek *hypnos* = sleep). Monotonously moving objects, even a ticking clock, are also used, together with verbal suggestions from the hypnotizer, to induce the trance state. Those who have been trained to obey, such as soldiers and schoolchildren, appear to be especially susceptible. The physician Liebault claimed 1,700 successes in 1,756 subjects. In a suitable person hypnosis could be used to induce sleep, relieve pain, and to cure blindness, paralysis or loss of speech when these things were due to hysterical (mental) disturbances. Many other physicians practised it, public demonstrations were given, and people began to get exaggerated ideas of its power. Under what is called post-hypnotic suggestion, someone could be told, say, to sit on the floor when the clock struck three. On waking from the hypnotic trance the person would have no recollection of the suggested action but, when the clock struck three, would invent some implausible excuse for sitting on the floor. There was great interest in the possibility that in this way a person could be induced to commit a murder under the influence of a hypnotist. It was a great topic for fiction, and one or two fascinating legal trials took place in France in the late nineteenth century. But it became clear that a hypnotized individual could not be made to do things that were normally abhorrent or immoral. In other words, you could not hypnotize someone to commit a criminal act such as murder, unless they were already inclined in that direction.

Can you commit murder when you are sleep-walking – a somnambulistic crime? The subconscious takes over when we are asleep, but murder under these circumstances is exceedingly rare. On 14 August 1963, Jo Ann Kiger had a bad nightmare and thought that a madman was running loose in her suburban house in Kentucky, USA. She took two loaded revolvers, went into her parents' bedroom and fired the guns, killing her father and injuring her mother. She was arrested, but acquitted because of her history of sleep-walking.

What are the murder weapons most commonly used? In the USA, most killings (64 per cent in 1980) are by firearms. There seems little doubt that this is because guns are so readily available: the rate is 175 times that in England and Wales. Forty-eight per cent of American households have one or more guns, compared with 4.7 per cent of British households (even fewer since the 1997 handgun laws). (But the British are more likely to use guns for suicide than for murder; seven out of every ten gun deaths in Britain are suicidal.) Less commonly, knives, other cutting instruments, blunt instruments, hanging and strangulation are used. In recent years glass has frequently been used as a cutting weapon, especially the end of a broken bottle or beer glass. In the UK there are now about 5,000 such attacks a year, but they are not often lethal. In strangling, death is often due to a nervous reflex that stops the heartbeat, rather than to stopping the blood supply to the head or interfering with breathing (asphyxia). The original meaning of the word 'mugging' was throttling by squeezing the neck after trapping it in the crook of the elbow. An accomplice would then steal any cash or valuables. There was an outbreak of 'muggging' in London in 1862, which came to a head when an MP called Hugh Pilkington was attacked in Pall Mall. The word is now used to mean any form of robbery with violence.

Poisoning used to be commoner than it is today. There was something sinister, even magical, about poisoning. It was a sort of deadly kitchen craft, commonly used by women. During the reign of Henry VIII a certain Richard Roose tried to kill Bishop John Fisher and members of his household by adding arsenic to their porridge. As a result of this a special type of execution, being boiled alive, was devised for poisoners, which lasted until the end of Henry VIII's reign.

One of the most celebrated later cases of poisoning was that of Marie Lafarge, who in 1840 poisoned her husband with arsenic. Well brought up but with no fortune of her own, she had met Charles Lafarge through a matrimonial agency, and she had accepted his proposal of marriage because she thought he was rich. When she discovered that he lived in a decrepit house in an out-of-the-way village she decided to escape by poisoning him. She was seen adding a white powder to his food, and when he became ill one of the doctors said he was dying of poison. At her trial the experts disagreed, and the French public split into those who thought she was guilty and those who did not. But she was convicted by a new test for arsenic, developed by an English chemist, Marsh, and spent ten years in prison. Then there was Mary Ann Cotton, in northern England. Using arsenic, she poisoned three husbands and at least a dozen children, and the main motive sems to have been the insurance handouts. Her first husband developed diarrhoea and died in 1865, after he had insured himself for £35. At her trial in 1873 she said she had bought the arsenic to kill bugs, but was convicted and hanged in Durham gaol.

One of the most famous poisoners in history is Cesare Borgia (1476–1507). After starting his public life as the Bishop of Pamplona at the age of sixteen (his father became Pope), his violent and headstrong nature led him to an astonishing career as a skilled soldier and administrator. During his meteoric pursuit of power he ruthlessly killed those who stood in his way. He had many stabbed or strangled, but his favourite method was by poison, such as arsenic, which he added to the victim's glass of wine. Yet this cruel despot was also a patron of the arts, befriending Leonardo da Vinci.

There are plenty of poisonous substances, but not all are suitable for murder or easy to get hold of. Arsenic has long been favourite; in 1972 an Englishman, Graham Young, was convicted of two murders using the more obscure poison thallium. This case was unique because the poison was detected in the ashes of the victim after cremation. Morphia is a valuable drug, unfortunately addictive and potentially lethal. The first case of murder by morphia is described in the box.

The first case of murder by morphia

In Paris in 1823 Dr Edme Castaing, a twenty-seven-year-old physician, was accused of murder. The alleged victims were Hippolyte Ballet and, at a later stage, his brother Auguste Ballet. Dr Castaing was an extravagant young man and was always short of money. Hippolyte suffered from pulmonary tuberculosis and Castaing was his physician. Auguste had paid Castaing 100,000 francs to destroy Hippolyte's will and forge a new one which left a large sum to Auguste. But Castaing, still hungry for francs, managed to persuade Auguste to leave him all his money.

One morning at an inn at St Cloud, near Paris, Castaing ordered a glass of warm wine for his friend. He had laced it with a large dose of morphia, adding a lemon to disguise the bitter taste. Auguste only drank some of it because it was so bitter, but the next day he was unwell and stayed in bed. Castaing then gave him a glass of milk, which brought on vomiting and diarrhoea. A local physician noted that he had contracted pupils and difficulty with breathing, now known to be classic signs of morphia intoxication. He died the same day.

At the trial it was revealed that Castaing had bought large quantities of acetate of morphia from a druggist, saying that it was for experiments on animals. In 1823 this was a new drug and very few physicians had any experience with it. Unfortunately neither the vomit nor the stomach contents were available for testing, although the methods would have been crude. Nevertheless the jury found Castaing guilty of murdering Auguste (but not Hippolyte), and he was sentenced to death and executed.

Drowning is uncommon as a method of homicide; but in large cities a corpse can conveniently be thrown into a river or canal, and it may then be difficult to distinguish between homicide and suicide. One of the characters in Charles Dickens's novel *Our Mutual Friend* makes a living by recovering dead bodies found floating in the Thames in London. That was in the nineteenth century, but the Thames, like other great urban rivers, still carries its quota of corpses.

Alcohol, although not the instrument of murder, is often associated with murder, as it is with all types of violent behaviour. The cold-blooded calculating murderer may not need it, but it can trigger off murderous action in others, and has always done so. In Paris in 1885, 27 per cent of a total of fifty-seven seven cases of murder and attempted murder were alcohol-related, and the figure rose to 35 per cent in 1895.

Much public attention is attracted by sensational homicides: multiple and serial murderers, or murder together with mutilation, torture, cannibalism or occult practices. In all cases the murderer is mentally deranged. All are uncommon and some extremely rare, but they attract a disproportionate amount of news coverage. It seems that people would rather hear about a murderer who kills on a large scale, or who eats, chops up or sexually abuses his victims, than about one who merely shoots an unfaithful spouse.

Jack the Ripper was an unknown person who ushered in our age of violent crime. Between August and November, 1888 he murdered at least seven women, all prostitutes, in the Whitechapel district of London. After cutting his victim's throat he mutilated her body in a manner that showed he had some knowledge of human anatomy and was handy with the knife. The abdomen was ripped open (hence 'the Ripper') and the womb removed. Mrs Martha Turner was the second victim. She lived in Commercial Road and her husband had left her thirteen years earlier. Initially she received twelve shillings a week, but when this was reduced to two shillings and sixpence, she took to prostitution. When her body was found it had thirty-two stab wounds. The fourth victim had her entrails tied round her neck and, according to an account in the *British Medical Journal* (29 September 1888) the parts removed by the Ripper were the womb, two-thirds of the bladder, and the front of the abdomen including the navel. The killer wrote taunting notes to the police, calling himself Jack the Ripper, and once sent them half a human kidney, presumably from one of his victims. The killings horrified and fascinated the nation; the newspapers made the most of it, and a dark, foggy street in the East End of London became a classically spooky setting. But the Ripper was never identified, and the failure to find and arrest the murderer caused such a public outcry that the London Police Commissioner resigned. The Metropolitan Police had been refurbished in 1884, and 20,000 special whistles issued, but it was still not very effective. The episode stimulated several horror novels and movies, and about a hundred books have been written on the subject, often with suggestions as to the identity of the killer.

In 1953 England was gripped by the death-house murder cases. At 10 Rillington Place, London, the bodies of three young women were found in a cupboard that had been covered with wallpaper. Another older

The universal human drug – alcohol

All human societies discovered at an early stage how to ferment sugars from grains, rice, potatoes or other substances to produce alcohol. Beer was in use more than 8,000 years ago and the vine was cultivated 5,000–6,000 years ago. The drug has undoubted beneficial effects. It softens the stresses of life, and we drink to be merry, to forget our tribulations. When taken in the form of moderate doses of red wine it seems to reduce the incidence of heart disease.

But not everyone can drink it. About a fifth of Chinese and Japanese people have an inherited lack of a certain enzyme that helps deal with alcohol. Normally the body turns alcohol into acetaldehyde, and without that special enzyme the acetaldehyde builds up in the blood, causing severe headache, nausea and collapse. Those Chinese and Japanese people therefore cannot handle alcohol and do not drink it.

For the rest of us, unfortunately, alcohol has a downside, and today it does great damage to our physical, social and psychological health. In the UK, about 30,000 premature deaths each year are related to alcohol consumption (at least 300 of them due to liver cirrhosis), and alcohol is associated with:

> 80% of suicides;
> 50% of murders;
> 30% of fatal road traffic accidents;
> 15% of drownings.

Alcohol also makes a great contribution to divorce, family violence and child abuse.

And yet between 1950 and 1980 in the UK the relative price of alcohol halved, and consumption doubled. There are now about 200,000 drinkers in the country who are periodically or chronically intoxicated, have an uncontrollable craving for it, need large doses, and are damaging themselves and society.

Perhaps alcohol is like the motor car. It inevitably causes damage and death, but because it is an essential part of our life it is tolerated. We are prepared to see it regulated, but we are not prepared to give it up.

woman was discovered under the dining-room floor. The corpses were quite well preserved and partly mummified. Vaginal swabs showed that the young women had had intercourse not long before death. The husband of the older woman, John Christie, had left a few days before the bodies were discovered. The police then found a human femur in the garden, and dug up two more female skeletons. The Home Office pathologist, Sir Bernard Spilsbury, found that all three of them had been strangled, with signs of carbon monoxide (coal gas) poisoning. Christie was apprehended a week later and confessed to all the murders.

It turned out that he had killed eight other people between 1943 and 1953. Unfortunately, another man – the innocent, mentally subnormal Timothy Evans – had confessed to two of the murders and been hanged for them. Christie used to entice women to the house when his wife was away and, being incapable of sex with a fully conscious partner (he had been unable to have intercourse with his wife for two years after their marriage), his usual strategy was to stupefy them by getting them to inhale gas before raping them. Afterwards the victim had to be strangled, and the corpse hidden. Christie was hanged on 15 July 1953. The house was later demolished and the name of the street, which had become so well known, was changed.

For sheer numbers of victims, H. W. Mudgett must hold the record. His 'castle' in Chicago, USA was equipped with chutes, a furnace, acid bath, dissecting table and surgical instruments. Altogether he killed about 150 young women before he was convicted and hanged in 1896.

In recent years there have been a few dramatic cases of mass murder. In so far as there is a motive it is often one of publicity, or revenge against people or institutions. The killer is likely to be obsessed with guns or other weapons and to have a good collection of them.

Mr Banks was a prison officer in Pennsylvania, USA. He had once said he had thought of shooting the inmates, and was sent to the mental health centre, but without being treated. Fascinated by weapons, he had devised plans for protecting his family in case of civil disaster or warfare. One day he took his semi-automatic rifle and ammunition, put on a Civil War cap, and shot and killed a total of thirteen people, including women and children. He then locked himself in an abandoned house and held the police at bay for seven hours before surrendering. He was convicted and sentenced to death.

Mass killings in Dunblane, Scotland and in Port Arthur, Tasmania captured the headlines and horrified whole nations. It is difficult to kill on this scale by strangling or stabbing, and in each case the murder weapon was a firearm. Public revulsion in Australia forced through anti-gun laws that would otherwise have been bitterly opposed, but so far there have been less drastic legal responses in the UK.

Mass killings by terrorists are now, sadly, a regular feature of life. Explosive devices of one sort or another are the usual methods. When terrorists smuggled explosives on board a Pan-Am flight one day in 1988 the result was unforgettable. Nearly 300 bodies fell out of the sky on to a quiet Scottish village.

Mass killing could be achieved on a far greater scale by releasing a powerful poison into the air, food or water supply. This was attempted by the Aum Shinrikyo cult in Japan in 1995, when nerve gas was released in the Tokyo underground system, and the threat of poisoned food was used recently in one or two unsuccessful attempts to extort money from big food retailers.

The prospect of a new method of serial killing was raised in June 1997 when a thirty-five-year-old Italian woman was alleged to have deliberately infected eight men with HIV. She faced an attempted murder charge.

Manslaughter

There is more than one type of killing. The victim is dead whatever the killer's intentions; but a premeditated, deliberate killing is nevertheless different from an accidental one at the hands of a the person who had not really meant to kill, and where there had been no 'malice aforethought'. Both legally and morally the distinction is worth preserving. In English law, unpremeditated killings are classed as manslaughter ('culpable homicide' in Scottish law) instead of murder, and the penalties for the respective crimes reflect this difference.

The dividing line between manslaughter and murder can be a fine one when the injury causing death happened in the heat of the moment. A fight suddenly flares up between two men and one receives a blow on the head which subsequently proves to be fatal. Neither had been carrying a weapon or intending to inflict such damage. Most cases of manslaughter, however, are due to culpable negligence: a labourer high on a building site throws down some debris and kills a passer-by on the pavement below; a careless driver knocks down and fatally injures a pedestrian. A verdict of manslaughter would also be brought if the killer was mentally defective and deemed incapable of malice aforethought.

Sometimes it is very difficult for the jury to reach a decision on a charge of manslaughter. Supposing a parent refuses to allow doctors to give life-saving treatment to his seriously ill child because it is forbidden by the religious sect to which he belongs; or a pregnant woman takes a drug or a poison that results in the death of her newborn child. Is this manslaughter? Moral feelings may have to be balanced against the letter of the law.

In the USA the terminology is different. Murder by design is called 'first-degree' murder and originally received the death penalty, whereas other kinds of murder are 'in the second degree' and punished by imprisonment.

4

Ageing and Death

When thou wast young, thou girdedst thyself, and walkest whither thou
wouldest: but when thou shalt be old, thou shalt stretch forth thy hands,
and another shall gird thee, and carry thee whither thou wouldest not.

John 21:17

Ageing is tied up with death, and the only escape from ageing is to die
early. Before embarking on a discussion of age and death, it is important
to define some terms: in particular, to distinguish between *life expectancy* and *life-span.*

Life Expectancy and Life-span

Life expectancy just tells you how long you are likely to live at any given
moment, generally at the time of birth. When someone talks about life
expectancy, you should really ask at what age it is being considered. If
you add it on to your age it tells you at what age you are expected to die.
It is an average figure. At birth you are not expected to live to the age
that would be expected later in life. This is because people are dying off
at all ages, and there is something about the survivors that makes them
tend to live longer. Thus in England and Wales in 1993 the life
expectancy was 73.7 years for males at the time of birth, 45 at the
age of 30 (meaning death at 75 yrs), 26.3 at 50 (death at 76.3) and 6.5 at
80 (death at 86.5). In 1825 Benjamin Gompertz, an English actuary,
discovered while studying death records and mortality that after the age
of thirty the likelihood of dying is doubled every seven years. It reaches
its highest point between sixty-five and eighty, after which it increases
less rapidly. The oldest are likely (on average) to live longest because the
'early diers' have by then passed away, leaving the long-lived types
(those whose genes, lifestyle or other factors give them longevity).

The threescore years and ten allotted to us in the Bible have long been exceeded in developing countries. In the USA, life expectancy at birth was 49.2 years in 1900–2, rising to 66.7 in 1946 and 75.7 in 1991. Improvements in medicine, hygiene and nutrition had resulted in fewer deaths in infancy and childhood, so the figure increased. Life expectancy at forty years has increased to a much smaller extent: once you had passed through the especially dangerous early period in life your chances (life expectancy) were not much greater than they had been earlier in the century.

Today, life expectancy at birth is highest in Japan (79 years) and lowest in Uganda (41 years). Females score higher than males, reaching about 82 in developed countries (in Japan 82.5 for women and 76.2 for men). High life expectancy goes with good medical services, good nutrition, clean water supplies, high levels of literacy and national prosperity – in other words, with wealth. If you list eighteen different countries in order according to their GNP per person, their life expectancy at birth decreases in exactly the same order. Thus per capita GNP in Japan (1993) is nearly $30,000, and in Uganda less than $200.

'Life-span' is something different. This tells you how long you would live if the all the environmental hazards were removed: if all the big killers – cancer, heart disease, stroke, accident – indeed, all the things itemized as causes of death in chapter 2 were miraculously eliminated, and you were left to die of other things – in other words, of 'natural causes'. It has been worked out that if cancer, heart disease and stroke were eliminated, life expectancy would increase by about seventeen years! But you would die in the end, not because of any particular disease or abnormality, but because you had reached the normal life-span for our species.

Each species of animal has its own normal life-span, which is determined by the genes and has developed in evolution as an adaptation to life. For the human species it is probably about 115 years – longer than that of any other mammal – and it is unlikely to have altered significantly in the past 100,000 years. Probably it is not changing, although the virtual absence of natural selection in human populations nowadays calls this into question. Our ability to keep the sick and the malformed alive means that nature no longer eliminates the 'unfit', and as a result genetic defects can accumulate in the human genome. Whether this is going to have any impact on life-span, or indeed whether it has had any impact at all, remains to be seen.

It is surprising that the subject of life-span and what determines it has been so neglected by scientists. Although we can say what the life-span *is,* we don't know *why* it is what it is. As we get older, various changes take place in our cells and tissues that are neither actual disease processes nor due to environmental or preventable causes. They are not by themselves obvious causes of death, and up to a third of those aged over eighty have no known cause of death when examined at post mortem. These are the changes of ageing.

Table 3   Life-span: an assortment of maximum ages

Species	Natural life-span (years)
Galapagos tortoise	175
Sturgeon	82
Carp	50
Human	115
Donkey	50
Elephant	69
Chimpanzee	56
Gorilla	47
Hippopotamus	54
Giraffe	28
Lion	25
Domestic dog:	
large (e.g. mastiff)	10
small (e.g. spaniel)	14
Domestic cat	27
Hump-backed whale	47
Killer whale	12
Vampire bat	13
Ostrich	50
House mouse	3
Guinea-pig	7
Texas rattlesnake	16
Beef tapeworm	35

Note: Animals of the same species will usually live longer in captivity than in the wild.

Ageing in Animals and Humans

Death through old age can have a simple mechanism. In animals such as the mongoose, the goat, the shrew and the African elephant, the teeth get worn down with age, and, left to themselves the animals starve. Fitting dentures, which is done for valuable stud animals, prolongs life. A few animals keep growing for as long as they live and do not age. This includes many fish, amphibians, tortoises and lobsters. Cold-blooded animals kept at warmer temperatures do not live so long, suggesting that ageing has something to do with metabolic activity.

In animals, ageing is something that happens only in captivity, when the usual hazards of life have been greatly reduced. In the wild, nearly all of them are killed by predators, starvation, infection or accident long before they have begun to age. They never have a chance to reveal how long they could have lived if they had been protected from the hazards of life. This also holds for our close relatives, the Neanderthals, few of whom passed the age of forty. So far we have discovered the remains of about 300 specimens and only one of them was old. The presence of this old individual suggest that the elderly were looked after by the others and that they therefore had some familiar human characteristics, a conclusion confirmed by the finding of a Neanderthal burial where flowers had been placed beside the corpse.

It is easier to understand ageing if you consider how animals adapt to their environment and evolve. Evolution favours the individuals who survive and reproduce. The fertile, not the meek, inherit the earth. When reproduction is over, then, from the ruthless point of view of selection and survival of the fittest, the fate of the individual is irrelevant. He no longer contributes to the species. There is nothing to be gained by keeping him young and fit, and indeed a few animals die immediately after reproducing. So the body degenerates, begins to malfunction and finally dies, upon which the body's components can then enter into the elemental recycling system on which life depends.

The queens in termite, ant and bee colonies, however, are an apparent exception to this principle. They live about a hundred times as long (five to twelve years) as other insects. This has something to do with the fact that the queen keeps on producing eggs, while living in a heavily defended colony, protected from predators. Because she is safe and still serves the species by producing eggs, evolution has endowed her with a long life-span, although we do not know how this is done. In animals that die off early, however, there is no call for evolutionary adaptations that preserve the body into old age, and such animals tend to have a shorter life-span.

Is ageing, then, just a result of evolution's indifference to old age? Each species of animal has a built-in life-span that is genetically determined, although, as mentioned above, this natural life-span is never normally reached. It looks very much as if ageing is a luxury we have brought about for our species by providing food, shelter and care for the elderly. It is an artefact of human civilization.

On the other hand, humans do have the longest life-span, and the longest post-reproductive life, of all mammals. This is something that has been acquired during evolution, and there are two reasons for it. First, we have developed big brains (in relation to body size), giving us the ability to learn from experience; and second, we learn from previous generations. The first calls for very slow development of the brain, and the second for people to survive into old age so that they can hand on their knowledge. Thus we can argue that even by nature's harsh rules, elderly humans are useful to the species.

Young humans undergo a very long period of development, up to about fifteen years, during which time the parents play a useful role in helping with food, shelter and protection. It makes sense for them to survive for at least another ten to fifteen years after the last children are born. Our success as thinking, tool-wielding primates has depended on knowledge ('culture') handed down from generation to generation, and it is old people who are the repositories and handers-on of this knowledge. Old people are worth keeping, at least for a while. Some have suggested that the word 'sageing' be used to describe that useful period between the end of reproductive life and senescence. But you could still argue that the best arrangement, from nature's point of view, would be for people to autodestruct at the age of, say, seventy. That way, after carrying out their 'sageing' function, they would be less of a burden to the young; and it would prevent overcrowding.

'Sageing' suggests a positive view of age and experience; but most writers and poets have not looked kindly on old age. Youth and vigour are worshipped and old age viewed as a regrettable ending to life. Shakespeare, in his incomparable verse and in depressing detail, describes the seven ages of man (*As You Like It*, 1599, Act II, Scene 7) and refers to the downward path from the full vigour of life, via the 'lean and slippered pantaloon' of the sixth age to the final dismal stage

> That ends this strange eventful history,
> Is second childishness, and mere oblivion,
> Sans eyes, sans teeth, sans taste, sans everything.

The American writer Mark Twain (1835–1910) had a novel way of looking at things and suggested that 'life would be infinitely happier if we could only be born at the age of eighty and gradually approach eighteen'. On the other hand, youth has its problems. Surely at no time in life can you suffer so acutely, so inconsolably. Jonathan Swift (1667–1745) said that no wise man ever wished to be younger. Perhaps really to understand the old you have to be old yourself, and it is true that most old people have learned to take a more stoical attitude. In their maturity they can relax and accept what has been dealt to them by fate. And it can still be a time of great achievements. Picasso was still drawing when he was ninety, Pablo Casals giving cello concerts at eighty-eight, Bertrand Russell still active on international peace committees at ninety-four, and in Britain the Queen Mother was still walking and meeting adoring crowds on her ninety-seventh birthday.

The world's oldest person, until she died on 4 August 1997, was Jeanne Calment. She lived in Arles, France and was 122 years old, with a birth certificate dated 21 February 1876 to prove it. In the end she was frail, blind and almost deaf, but she had outlived her daughter by sixty-three years and her grandson by thirty-four years. Humour and plenty of good olive oil, she said, was the secret, and she smoked and drank until she was about 120! A local lawyer wanted her house and had agreed to pay her a sum each year until she died if he could then have it. But she outlived him and he ended up paying her three times the value of the house. After Mme Calment's death the world's oldest person appeared to be Christian Mortensen, who is 115 and lives in a retirement home in San Francisco. He enjoys a cigar each week. Britain's oldest person in August 1997 was Lucy Askew, who lives in Essex and was then 113.

These ages pale into insignificance beside some claims of longevity that are made. One of the problems with the study of ageing, or gerontology as it is called, is the plethora of tall stories about the old. A degree of scepticism is necessary, about the age of animals and especially about the age of people. The subject is a minefield of legends and myths, or an Aladdin's cave, depending on your point of view. The Bible, for instance, refers to people living to great ages: Methuselah (969 years), Noah (950), Abraham (175) – but these figures probably signify months rather than years. Early in the twentieth century many stories were told about long-lived groups of people. They were usually in remote, often mountinous areas such as the Andes, the Himalayas, or the Caucasus Mountains. People began to accept these stories, and to believe in the magical measures said to be responsible, such as eating yoghurt, drinking fermented mare's milk, not using tobacco or alcohol. Unfortunately, careful investigation of these stories has shown that all of them were false.

Nevertheless, we are certainly living longer. In the UK in 1950 the average life expectancy at eighty was six years: this had increased to nine years by 1997. All this means more centenarians, and in developed countries the number are increasing at 8 per cent per year. By contrast, for 99 per cent of the time humans have been on this planet, their life expectancy at birth has been about 18–20 years. This was so in the Early Iron and Bronze Ages in Greece. In Europe by the seventeenth century it had increased to 33.5 years, and to 66.7 years in the USA in 1946. In developed countries it is now even higher, the highest being in Japan (82.5 years for women and 76.2 for men) while in the USA it is 78.6 for women and 71.6 for men. This is, as we have seen, largely a result of fewer deaths during the hazardous period of infancy and early child-hood, but the life expectancy of middle-aged and elderly people has also increased. Consequently there are more and more old people. The figures from the USA speak for themselves. In 1900 about three-quarters of people died *before* the age of sixty-five, whereas in 1996 three-quarters died *after* the age of sixty-five (and three-quarters of these deaths were due to heart disease, cancer or stroke).

But if medical science saves you from one disease, there are generally others lying in wait for you. Removing any one of them would have a surprisingly small impact on life expectancy. For instance, it has been calculated that eliminating all forms of cancer would add a mere 1.2 years to the life expectancy of sixty-five-year-olds, and only 2.3 years to life expectancy at birth. Eliminating all infectious diseases would add a mere 0.3 years at sixty-five, and 0.8 years at birth. A greater effect would result from removing all the main diseases of blood vessels, heart and kidney, which would add 15–17 years at sixty-five. Already we are beginning to move along the path of healthy ageing. But more diseases must be taken away if we are all to survive up to the full life-span of our species.

Ageing in the Test Tube

It is a common belief that cells in the test tube live for ever. The great pioneer and Nobel Prize winner Alexis Carrel (1873–1944) grew frag-ments of chicken heart in test tubes. When the number of cells had doubled, about once a week, he separated off half of the cells and kept them going. This he did for a total of thirty-four years, and it looked as if, under these conditions, cells were immortal. If he had managed to keep all the cells growing instead of discarding half of them each week, the total amount after thirty-four years would theoretically have weighed more than 20 million tons. But the conclusion about immortality was a false one. Each time he split the cultures he added a fresh supply of the nutrient fluid that bathed the tiny clumps of cells. The nutrients included an extract of embryonic tissue that, unknown to him, contained a few live cells, and these were the cells that multiplied and took over the cultures as the old cells died out. The regular addition of new cells had made the cultures seem immortal.

The truth about ageing in the test tube was revealed by the work of the American scientist Leonard Hayflick. He showed that cells from normal human embryos are able to divide about fifty times in the test tube, but then they die. There is a natural limit. It was clear that this was significant, because cells from middle-aged people divided a smaller number of times and there was even less division of cells from old people. The number of divisions made sense, moreover, when different animals with different life-spans were compared. Cells from newborn mice (life-span three years) could divide fourteen times, whereas cells from the Galapagos tortoise (life-span 175 years) divided about 110 times. The human figure of fifty divisions for a life-span of, say, eighty years fits into this sort of scheme. We do not understand how this limit to cell division is set, except that it must be controlled by genes.

The only cells that are exempt from this rule are cancer cells, whose multiplication, by definition, is not controlled in the normal way. At Johns Hopkins Hospital, Baltimore, USA, in 1952, Henrietta Lacks was being operated on for cancer of the cervix. A piece of the cancer tissue was given to an enthusiastic scientist, who managed to get the cells to grow in test tubes. They divided every day or two, have been maintained in culture ever since then, and were called Hela cells. Colonies of Hela cells were sent out to research laboratories throughout the world, and from them a great deal has been learned about cell biology. The combined weight of all the cells descended from that original piece of cervical tissue must be many times the weight of the patient herself, so she has achieved a type of immortality. (It is worth noting that, while people think of cancer cells as multiplying an at incredible rate, in terms of the increase in number of cells no cancer grows as fast as the developing human embryo.)

In a sense the germ cells, too, enjoy a type of immortality. When we die our germ cells are not all discarded and left to die like the rest of the body's cells. The lucky egg and the occasional even luckier sperm will have joined together and developed into embryos to form the next generation. And so it continues, from generation to generation. The line of germ cells is immortal, whereas the rest of the body dies at each generation (see chapter 1).

What about deep-freezing and immortality? Probably single cells such as sperm, eggs and fertilized eggs, and small pieces of tissue such as very early embryos, can be kept indefinitely at the temperature of liquid nitrogen (–178°C). So far, the longest time for which human cells have actually been kept frozen and capable of being brought back to life on thawing is 32 years.

What Happens to us during Normal Ageing?

To look at 'normal' ageing we have to miss out all those things that are in principle preventable, such as heart disease, cancer, stroke and accident. This raises problems, because doctors and pathologists are interested in diseases. They want to know whether the deceased had a cancer, a heart attack, a stroke or wounds incompatible with life. In other words, what did the person die of? To say nothing more than 'old age' is not good enough, and old age is not even included in the top fifteen causes of death listed each year by the WHO or the US government. 'Myocardial degeneration' (the heart gradually giving up) sounds better, and if it was pneumonia or some other infection that finally laid the elderly person to rest, this too would be acceptable. But here we are concerned with old age itself – and you certainly can die of it, although it has not been so intricately studied as the main diseases. Yet as no one escapes it, everyone should be interested. What are the features of the normal ageing process? It turns out that we can describe them in some detail, but we do not know why or how they take place.

Most of us have seen what happens to grandparents, uncles or aunts as they get old. An infrequent visitor can often follow the process better than can the old people themselves, because the changes are gradual. There is a sadness about it. In the words of an anonymous writes, 'Old age is when more things happen for the last time, and fewer and fewer things happen for the first time.' None of these changes individually is lethal, but underlying them are changes in organs and tissues that ultimately lead to death.

The Changes Brought by Age

Height decreases by about one-sixteenth of an inch a year from the age of thirty, due to thinning of the space between the vertebrae.

Body fat tends to settle more in the thighs and hips ('middle-aged spread').

Skin becomes less elastic and gets thinner. When you pick up a fold of skin on the back of the hand and then let go it takes a second or two to settle back into position. As the skin slackens, wrinkles appear at the corners of the eyes, round the lips ('purse-strings'), and elsewhere. The worst skin changes are on parts of the body exposed to sunlight, and are due to this, not to the ageing process.

Hair on the scalp decreases in quantity. A youth has about 100,000 scalp hairs. The general decrease with age is a different thing from the characteristic pattern of baldness that overtakes males, often quite early in life, which is attributed to genes and hormones. In males extra hairs may sprout from the nostrils, ears and eyebrows, and in females after the menopause from the upper lip and face. Hairs also get thinner, break more readily and lose their pigment.

Fingernails grow more slowly after about the age of about thirty, and tend to show ridges and other abnormalities. A distinguished American physician, Dr W. B. Bean, charted with meticulous care the growth of his left thumbnail over the course of thirty-six years. Over this period, if not cut, the nail would have grown to be several feet long. Among other things he noticed that his nails grew faster in a warm environment, stopped growing while he had mumps, and grew faster when he was younger.

Healing of wounds takes longer: a cut will not heal, or a bruise disappear, so soon as in the young.

Lungs become more rigid, less able to cope with exercise or chest infections. There is less in reserve, and this is why the elderly easily go down with pneumonia ('the old man's friend'), and why the influenza vaccine is recommended for them. An infection they would have shrugged off at the age of fifty may prove fatal at age seventy.

Muscle activity often declines: older men have reduced grip strength and reduced atheletic performance, but this is parly due to reduced activity. Seventy-year-olds who exercise regularly are in better shape than sedentary sixty-year-olds. However, peak stamina and muscle coordination, and peak performance in 100 metre and marathon races, is at twenty-five to thirty-five years.

Mental state may change. There is no change in personality, but memory for recent events and for proper nouns gets worse (senile forgetfulness). This is due to gradual loss of nerve cells and less intricate connexions between nerve cells in certain parts of the brain such as the hippocampus. The idea that from an early age we all lose thousands of nerve cells each day seems to be untrue. Alzheimer's disease is a quite different condition, with a much more severe and progressive type of change. The tragedy with severe Alzheimer's disease in a friend or a relative is that although the face and the body are the familiar ones, there is no one at home, mentally.

Reaction times become slower – including reactions to temperature change, so that old people are more susceptible to the cold (hypothermia) and to overheating.

Aches and pains in joints, and *rheumatism*, are common features, associated with *osteoarthritis*, which is partly due to damage to joints earlier in life.

Sleep patterns change: old people have less of the type of sleep associated with rapid eye movements and dreaming. They are more likely to snore because the palate falls back: two-thirds of males over sixty are habitual snorers. Old people spend longer periods awake, and are easily woken up.

Sexual life is affected by the menopause in females, and reduced sexual activity in males. In males there is no change in testicle size: initially, at least, the testicle continues with its usual prodigious output of about ten million sperm a day, enough in six months to populate the entire planet. But men do show a decline in testicular function with ageing, analogous to the decline in ovarian function in women. The testes of older men produce smaller amounts of the male hormone testosterone, and the big question is whether this is what causes the changes. It looks as if it is not so, because giving testosterone to old men doesn't make any difference. Also, the prostate gland enlarges and begins to cause problems with urination. The latest disheartening news (November 1997) is that old men suffer from a reduced volume of semen, reduced motility of sperm, reduced ejaculatory force and reduced sense of impending orgasm. They also tend to be less easily aroused by sexual stimuli, and need direct physical stimulation to produce an erection. The erection itself becomes the problem. Two-thirds of those aged seventy suffer from erectile dysfunction, even when they are interested in intercourse. Treatment used to be unsatisfactory, but the new drug Viagra, taken by mouth, acts on penile blood vessels and often solves the problem.

In older women the physical changes are based on the reduced output of sex hormones after the menopause. In the genital area, the vagina becomes less elastic and its walls thinner, the hood of the clitoris gets smaller and the pubic pad of fat atrophies. But there is no good evidence that sexual desire necessarily wanes, and reduced activity is often due to loss of a partner or reduced interest of the partner. Princess Metternich, when asked at what age a woman ceased to be capable of sexual love, said: 'I do not know, I am only sixty-five.'

Hearing becomes less acute in that we are less able to hear high notes after about seventy years. As these notes give the consonants their character, this can cause problems in following speech in the presence of background noise. Older people are also less able to hear loud sounds. Some of these changes could be due to exposure to noise earlier in life.

Smell also often becomes less acute, with reduced awareness of strong smells.

Sight usually deteriorates, with less ability to focus on near objects (presbyopia or shortsightedness). This is due to the lens getting thicker and less easily altered in shape during focusing. It is an almost universal change in old people, present in three-quarters of 65–75-year-olds and in more than 90 per cent of those older than seventy-five. *Cataracts* are also common; these are caused by changes in lens proteins and sunlight, including the indirect sunlight that comes from the side. Cataracts and shortsightedness can be dealt with, but the occasional unfortunate person with macular degeneration, which means death of the layers of cells that respond to light, cannot be helped. Blindness is then inevitable. (In India today there are 12 million blind people, mostly because of cataract, and 80 per cent of them are over sixty. The over sixties in India are expected to double in number by the year 2016, when there will be 112 million of them.)

Eating habits change. Usually old people take in less food, partly because they are less active physically, with less muscle on them, and partly because they have a lower metabolic rate. Old people easily become malnourished due to a combination of factors including poverty, social isolation, loss of taste and mell, loss of interest in cooking, and perhaps less absorption of food from the intestine. Eating may also become more difficult as teeth tend to drop out, often due to tooth decay (caries) or periodontal disease, but also because of the ageing process. We are probably the first species on earth that regularly lives longer than its teeth.

Artery walls thicken, making them narrower and more rigid ('hardening of the arteries'). This has an obvious effect on the supply of blood to vital organs. On top of this, and making matters worse, fatty patches may be deposited on the inner lining of arteries: this is called arteriosclerosis and is to a large extent due to diet or other environmental factors. But even people without arteriosclerosis nevertheless age, and arteriosclerosis does not seem to occur in very old animals.

Kidneys become less efficient at removing waste products from the blood, although this is partly due to actual disease in the kidneys rather than to a natural ageing process.

Bones get lighter after about the age of fifty; the loss of bone mass is more marked in women. The cause is osteoporosis, which is partly attributable to diet, partly to hormones, partly to a sedentary lifestyle and reduced exercise.

The brain weighs less in old people, and loses about 100 g between the ages of twenty-five and seventy. As discussed above, this is mainly the result of the gradual loss of nerve cells.

Having set out this catalogue of deterioration, of course, it must be remembered that people differ tremendously: some already look ancient at the age of fifty-five, while others are still hale and hearty at seventy and eighty.

It is encouraging to read the following: 'Is not old wine wholesomest, old pippins toothsomest, old wood burns brightest, old linen washes whitest? Old soldiers, sweethearts, are surest, and old lovers are soundest' (John Webster, 1590–1625).

What Causes Normal Ageing?

The traditional way of trying to understand aging has been to look at very old people or at animals that live a long time. We wonder about the Galapagos tortoise, which lives to 175 years, or about human centenarians. How do they manage it?

The oldest person reliably recorded was 122 (see above). It is simplest to concentrate on people who have reached 100 years (centenarians), but even so it can be hard sorting out fact from fiction. The sensible response to someone who claims to be 100 is scepticism. People want to be able to say they are 100, and unless there is a birth certificate it is best to be suspicious, because unfortunately there is no other method that gives a reliable indication of age in humans – nothing like the yearly rings on a tree trunk, the layers added each year to a fish's scales, or the growth zones of a whale's ear wax. In England, birth certificates were not compulsory until 1837, and this means that you should not accept anyone's claim to be 115 years old unless it was made after the year 1952. European birth records are the best: Belgium records five and Sweden seven centenarians in every 100,000 people, and at least three out of every four of them are females. Assuming that about fifty in every million people are more than 100 years old, we can go further and say that about one in forty of them are more than 105 and one in forty of these are more than 110 (which works out at about one in every 40 million Americans). Everyone knows that centenarians are becoming more and more common. In the USA in 1986 there were about 25,000, and if present trends continue there will be more than 100,000 by 2000 and more than a million by the year 2080. The same thing is happening in the UK, with 271 centenarians in 1951, 1,185 in 1971, 4,400 in 1991 and an expected 30,000 by 2030. The Queen, who sends a congratulatory message to those who reach their 100th birthday, is going to need a lot of envelopes.

Attempts to try to find out the secret of this longevity are usually disappointed. The centenarians tend to be independent, forthright people, but they share no common dietary or other habits. Some have never smoked or drunk alcohol, but others have and many still enjoy doing so. They feel old, and of course have experienced many of the changes of normal ageing. The American physician and writer Oliver Wendell Holmes (1809–94) commented on this:

> Little of all we value here,
> Wakes on the morn of its hundredth year,
> Without both feeling and looking queer,
> In fact, there's nothing that keeps its youth,
> So far as I know but a tree and truth.
> (This is a moral that runs at large,
> Take it—You're welcome—No extra charge.)

One thing is clear, and that is that longevity runs in families. Many centenarians have close relatives who have reached very old age.

At the other end of the spectrum are unusually short-lived people. Several conditions may lead to an abnormally early death:

- *Progeria.* This is a very rare condition, of genetic origin. The sufferer may look normal in infancy and mental development seems normal. But physical growth is greatly retarded, and the face is birdlike. The skin soon wrinkles, the hair falls out and the arteries harden. The average age at death, which is usually due to a heart attack or stroke, is twelve to thirteen years.
- *Werners syndrome.* This also is rare, although about twice as common as progeria. It, too, has a genetic basis, with an abnormality in one of the genes that ensures the wellbeing of DNA. The familiar ageing changes such as loss of hair or white hair, cataracts and skin wasting occur early in life. Females have an early menopause.
- *Downs syndrome.* Those with Downs syndrome (another genetic condition) show many features of premature ageing, and few survive beyond fifty years.

Do these conditions tell us anything about normal ageing? Probably not, although the faulty genes that accelerate ageing have not been identified. At least we can draw one conclusion. If ageing were due to nothing more than wear and tear, it is difficult to explain why in these diseases it occurs so early in life. The difficulty in researching the causes of ageing is that if you focus on individuals with a short life-span this often turns out to be due to special harmful mutations, which are not involved in normal ageing. This is why we have not learned much about normal ageing from studying diseases like progeria and Werners syndrome.

Scientists have nonetheless come up with a variety of theories as to what causes normal ageing. The very fact that there is a longish list indicates that no one is sure. But we are the longest-lived mammals, so perhaps we should be asking why this is so. One factor is brain size. Animals with larger brains live longer, and while we do not have the largest brain, we do have the largest brain in relation to body size. This has something to do with life-span because large, adaptable brains take a longer time to mature, and the also, as noted above, the old are still useful to the species.

Rate of Living

According to this theory, an individual starts with a certain limited amount of potential energy. If you live fast and use it up fast, you die young. Consider the elephant and the mouse. The former lives for up to sixty-nine years, and the latter for three years. Yet during their lives each of them burns about the same total number of calories per pound of body weight, and the heart of each beats approximately the same number of times. There are exceptions, however, and there is no evidence that for a given animal an increased rate of living (increased energy expenditure) means faster ageing.

Can you alter life-span by altering the rate of living, either by diet or by temperature control? The answer from animal experiments seems to be that you can increase life expectancy by dietary means. It has been known since the 1930s that rats kept on low-calorie diets, but with adequate intake of essentials like vitamins and minerals, live longer. They are undernourished but not malnourished, and the more protein in relation to carbohydrate the better. Why do they live longer? It has been suggested that rats in the laboratory are fed as much as they can eat, whereas in the wild state the food supply is irregular and rarely plentiful. Perhaps too much food kills them off earlier, and the rats on a restricted diet are living no longer than they would have done in the wild. But there must be more to it than this, because the rats on a restricted diet are infertile. Mice fed on a diet containing about half the calories they would normally consume show a 50 per cent increase in longevity, and there is a smaller effect even when the low-calorie diet begins in middle age. The elderly mice are in good shape, still doing about a kilometre a day on their running wheel. The same lengthening of life is seen when other rodents, fish and certain insects are maintained on restricted diets. However, it would be interesting to know whether the brain and intelligence develop normally with these restricted diets.

What does this mean for humans? In developed countries a lamentable number are overweight. Because overweight people have reduced life expectancy you would expect to increase their longevity by dietary restriction. But what about mild dietary restriction for those who are not overweight? Would they live longer? Unfortunately there is no good evidence as to what would happen in humans. Monkeys on a low-calorie diet show delayed physical and sexual maturity, but it is too early in these experiments to say what the effect is on longevity. Tests on primates are more likely to give results meaningful for humans than are rats or mice, but they are expensive, and keeping them for long periods to study ageing poses too many problems for research workers. In any case, it would be difficult to reach conclusions about life-span.

Temperature is another possible factor. If you look at animals that hibernate, such as hamsters, they live longer if they spend more of their time hibernating. Going cold-blooded for the winter adds extra years. Humming-birds live for up to eight years, in spite of their small size and intense metabolic activity, and this is perhaps related to the fact that their body temperature falls considerably at night. Bats do the opposite and cool down in the daytime, and the Indian fruit bat lives for thirty-one years. Cold-blooded animals do tend to live longer. It is a big jump from insects to humans, but water fleas live longer when you lower the temperature and die earlier when you raise it. It's an equally big jump from spiders to humans, but a certain species of spiders lives seventy years in captivity. The sturgeon (nearer to us but still pretty remote) reaches eighty-two years, although reports of carp living for hundreds of years have been discredited. Cold-blooded animals also live longer at lower temperatures as long as they are kept cold all the time, which happens when they live near the North or South Pole. But animals often show complex metabolic responses when you alter their environmental temperature, and most mammals increase their metabolism when they are kept cold because they try to keep warm. The meaning of some of these observations on animals is therefore uncertain.

Accumulation of Waste Products

This theory about ageing started with the old and popular belief that many illnesses were due to toxins. If people constantly absorbed into the body the toxins formed by the billions of microbes in the intestines, no wonder they sometimes got ill. The mere sight and smell of bowel contents made it obvious that here was a formidable source of disease. It followed that constipation was a dreaded condition, needing relief by a veritable army of laxatives. The toxin theory of disease reached such a pitch in the 1920s that a famous London surgeon used to remove large portions of the intestine as a treatment for vaguely defined illnesses manifested in headaches, backaches, tiredness, insomnia, depression and so on. Even today ritual washing out (irrigation) of the large bowel has its devotees.

Could toxins and other harmful substances build up in the body and cause the dysfunction of ageing? The answer seems to be no. We know that with ageing certain pigments (e.g. lipofuchsin) accumulate in nerve, heart and muscle cells, but they don't seem to do any harm.

Changes in Proteins

As rats get old the collagen (a protein) in their tail becomes less elastic, due to gradual 'cross-linking' (binding together of separate lengths) of the protein. When this happens, proteins begin to function badly. Also, proteins that are constantly turning over, constantly being broken down and built up again, aren't turned over so often in older people, so that abnormalities accumulate. When this happens to the special protein in the lens of the eye, for instance, it causes cataract. This is not a bad theory, but examination of cells from old people does not reveal abnormalities in proteins that could account for ageing.

Free Radicals

These are fragments of molecules formed in cells by the action of oxygen. Burning glucose by means of oxygen is the basic energy source and life-support system of cells, but there are undesirable side-products called free radicals or oxidants. They are unstable and readily join up to other molecules and damage them. A good deal is known about these free radicals. One powerful one is called superoxide. It is possible to prevent them being formed from oxygen by means of antioxidants. Antioxidants are added to foods as preservatives because oxygenation is the path to decay, and certain substances naturally occurring in foods, such as vitamin E, vitamin C and beta-carotene, are antioxidants. A connection with ageing is suggested by the fact that if you feed mice or fruit flies with large amounts of antioxidants they live longer. In the body, these free radicals are destroyed by special enzymes, the tongue-twisting superoxide dismutase and catalase. Scientists tell us that longer-lived animals tend to have higher levels of these protective enzymes.

So what is the message about free radicals and ageing? The greater longevity of the animals fed large amount of antioxidants seems to be largely due to later onset or prevention of cancers, heart disease and degenerative diseases of the nervous system. In other words, because these diseases occur less or are delayed there is increased life expectancy, rather than an increase in true life-span. Banning cars would increase human life expectancy without affecting life-span. But most of us would be happy with increased life expectancy even if we did not live to be over 100. Second, there is a tendency for animals that live longer to have more of those enzymes that destroy free radicals.

Changes in the Immune System

As they age, humans show some waning in strength of the immune system. Antibodies are formed in slightly smaller amounts and they have a tendency to react with the individual's own tissues. Certain immune cells are less active. But this is not the case in all mammals, and it looks as if it could be a secondary result of changes in the nerves and hormones that control immunity, rather than an actual cause of ageing.

Misbehaviour by Genes

Ageing is undoubtedly tied up with the genes, and there are several ways in which this could work. In normal people, all the time, the genetic machinery (the DNA of genes) is undergoing damage (mutation) due to natural radiation or to bad copying of genes when cells divide. The DNA of a cell suffers thousands of small injuries each day, but the body normally repairs the damage or the error in copying. In older people, however, the repair system is less efficient, and the mutational damage accumulates (according to the theory), so that the beautifully balanced control of cells by the genes is interfered with. Most mutations are harmful and those occurring in key organs in the body ('somatic mutations') could make the cells malfunction. The second way in which genes misbehave is when pre-existing harmful genes come into action with increasing age. These late-acting genes are either controlled or are not active earlier in life. This sounds like a good basis for an explanation of natural ageing: the bad gene makes defective proteins, which cause malfunction in cells. But once again there is no evidence that this actually occurs.

Nevertheless, the one striking influence on life-span is undoubtedly the genetic one. Centenarians tend to have very old relatives, as mentioned earlier, and each species of animal has a life-span determined by its genes. Fruit flies (*Drosophila*), which breed rapidly in the laboratory, are favourite organisms for studying genetics and have been used in research on ageing since 1913. In a particularly interesting experiment, scientists over ten years used for breeding only the flies that survived until ten weeks old, and compared this colony with one maintained in the usual way by mating flies at the age of two weeks. They ended up with flies that had a mean life span of seventy-two days, in contrast to the 'normal' flies with a mean life-span of thirty-nine days. They had selected out the genes for an increased life-span.

If such an experiment on any animal gave the same sort of result, as seems quite possible, then the maximum human life-span could theoretically be increased from 110 to at least 150. If ever the experiment were contemplated it would certainly be a long-term one, lasting hundreds of years. We know more about the genetics of *Drosophila* than of any other animal, and about fifteen longevity genes have already been identified. Yet the actual mechanism of ageing of these tiny flies is still not understood, and in any case there may be a different set of rules for humans.

The basic message is that it is our genes that control our life-span. The likelihood of surviving to reach this normal life-span, however, depends on the diseases that kill us prematurely, and these diseases are mostly caused by lifestyle and the environment, as discussed in chapter 2.

Our Cells Cease to Divide

Human cells, as we have seen, are capable of a maximum of about fifty doublings. Do we age because our cells have run out of doubling capacity, can no longer divide? We know that the doubling capacity is reduced as we get older, as might be expected if it has biological significance. The figure of about fifty times is for embryo cells; this falls to twenty to forty by middle age and ten to twenty in the case of cells from old people. Cells from prematurely aged young adults suffering from Werners syndrome can divide only ten to twenty times. Theoretically, senescent cells unable to divide could accumulate in old people, but there is no evidence that this has anything to do with natural ageing.

George Bernard Shaw (1856–1950) once said that every cell in the body was renewed every seven years, which meant that at the molecular, cellular level a person changes completely every seven years. Should one therefore be responsible for things that the 'other person' had done more than seven years ago? We now know better. Some cells (nerve cells, heart muscle cells) never divide again after infancy, and the DNA of the cell does not 'turn over'. When you are sixty you have exactly the same nerve cells you had when you were six, though there are fewer of them. Cells in other parts of the body however (skin, intestines, bone marrow) divide continuously until the day we die. Our susceptibility to Alzheimer's disease, for instance, is due to the fact that nerve cells do not divide and therefore cannot replenish the loss that takes place in this disease. But we do not *die* because our cells can no longer divide. The maximum number of cell divisions bears a relation to our life-span but does not determine it. Nerve cells, for instance, don't divide once the brain has been formed in the foetus. We age and die because of changes in individual cells rather than because the cells stop dividing. Indeed, many of the things in the above list amount to the failure of normal maintenance.

The subtle difference between dying and not being able to divide is well illustrated by what happens in microbes. If a microbe cannot multiply it is regarded as dead, and the word sterility is used to describe a surgical instrument or a dressing that is free of germs. Sterile means barren, unable to have offspring. The microbe may be alive in the sense that it exists, but if it cannot reproduce itself it is for all practical purposes dead. Judged by the odd rules of microbiology, many of us are dead.

We do not, then, age because more and more of our cells die off. But in a book about death something needs to be said about the death of the individual cell. What kills cells? First, there are a host of cell-killing poisons and toxins coming from the outside world, or formed by invading microbes. Some microbes, moreover, can actually grow inside cells and kill them. Thus, many types of diarrhoea are caused by viruses and bacteria that kill off the cells lining the intestines, and polio viruses can cause paralysis by growing in and destroying the nerve cells that make muscles contract. Second, the body can kill off its own cells, and this is worth doing when the cell has a virus growing in it. A special type of immune cell (the T-cell) recognizes the infected cell and kills it, before the virus has a chance to finish its growth in the cell and spread further in the body. (This last sentence covers a large part of the modern science of immunology!). Third, the cell can kill itself (see box).

What does a dying cell look like? As might be expected, the cell that commits suicide does it in a way that is convenient for the waste disposal and recycling systems. Its DNA becomes disorganized and it adopts a spherical shape, but it does not disintegrate. Dying cells have common features, whatever the cause of death, just as in the case of the whole body (see chapter 5). The outer membrane of the cell begins to let through molecules that are normally kept outside or inside. The cell stops

---

Cells can commit suicide

When a tadpole loses its tail, the cells of the tail do not disintegrate and make a mess but round up, die in an orderly fashion, and are removed by phagocytes. A phagocyte is a large cell that engulfs the dead cell, digests it and disposes of it, in much the same way as that an amoeba takes up and digests its prey. The tail cells die in an orderly fashion because of an inbuilt suicide program that is switched on. It is called apoptosis (Greek = a falling away), and it plays a vital role during the development of the embryo. Certain structures that are formed in the early human embryo, such as gill slits and a tail, are to a large extent dismantled and discarded as the embryo develops. In addition, a great deal of remodelling and reshaping of organs takes place. Destruction goes hand in hand with construction, and unwanted cells have to be killed off and removed. Apoptosis is sometimes called programmed cell death, and it is the natural process by which the body controls cell numbers and rids itself of superfluous or redundant cells during development.

The cell also switches on its suicide program when it is infected with certain viruses – better to die than to support the growth of the virus with the formation of hundreds of extra invaders. The story of infection by viruses and other microbes is a story of a constant battle between the invader and the defences of the invaded individual (the host). The defences (the immune system and the phagocytes) are extraordinary, and we owe our lives to them, but the invader is always developing strategies for getting round or avoiding them. The host then responds with a countermeasure, and so on. It is an eternal conflict that has been going on for millions of years, with illness or death as the penalty for the unsuccessful host, and death (extinction) for the unsuccessful microbe. It has recently been discovered that many viruses, once they are inside the cell, have a sophisticated system for preventing the cell turning on its apoptosis program. The cell then is forced to stay alive and act as a breeding site for the ingenious virus. Fortunately the host has another trick up its sleeve, because there is still the possibility that a T-cell will come to the rescue and kill the infected cell, as described in the text.

making its enzymes, can no longer keep the glucose fires going, and finally rounds up into a ball and falls to pieces. A series of disastrous chemical events has unfolded.

Order Gives Place to Disorder

The mature individual is a model of order. But order is an unstable state of affairs, and since nature tends towards a state of equilibrium, things proceed from order to disorder. This is what scientists call the second law of thermodynamics, and ageing is a reflection of this fundamental law. As a theory of ageing, however, this is little more than a philosophical one, and there seems no way we could prove or disprove it. Nor does it help us understand what is actually happening during ageing.

The conclusion must be that we know a lot about what happens in ageing but we do not know why or how it happens. Perhaps in the coming years some of the mysteries may be solved. There could be a single mechanism, or there might be many different ones. The science of gerontology is a very young one, and until recently it had been neglected. It is a curious fact that although the medical speciality of pediatrics is well established in almost every country in the world, only Britain has a well-developed field of geriatrics. The world is unprepared, medically as well as socially and economically, for a time when old people outnumber children. In October 1997 the subject of ageing was highlighted in 100 medical journals from over thirty different countries in an attempt to alert people and governments to the problem of ageing, and draw attention to our lack of understanding.

How to Cheat Death

So far, no one has ever proved that we can do anything to increase the human life-span. Life expectancy, in contrast, can indeed be increased, and that is the *raison d'être* of most biomedical research. Vaccines, antibiotics, the prevention of heart disease, stroke and cancer, have already made a massive impact on our life expectancy. Over the ages, various treatments and rules of life have been proposed to prevent ageing and make people live longer, if not for ever.

Myths and legends abound. There are many tales of magic drinks or potions that confer immortality: fruit from the tree of life in the Bible; 'ambrosia' (the food of the gods) in ancient Greece and Rome; also magic charms, ceremonies, precious stones, fire. The elixir of life was the goal of the alchemists of the Middle Ages. They were convinced that base metals like lead could be transformed into noble material (gold), and, knowing that caterpillars turned into butterflies and tadpoles into frogs, they strove to discover the elixir of life or the philosopher's stone that would transform the old man into a fresh-faced youth.

Another formula for rejuvenating an old man was to have a young virgin lie in his arms, as tried out for King David in the First Book of Kings in the Bible. Alas for King David; although he may have enjoyed the treatment, he derived no lasting benefit and soon died. From this remedy it was a logical step to try injecting man with extracts of first animal and later human testicles. The great physiologist Charles Brown-Séquard (1817–94) injected himself with an extract of monkey testicle and said that he felt much better for it. Many old men were treated in this way, and Serge Veronoff, the Russian biologist, grafted monkey testicles into people. Grafting foreign tissue in this way caused complications and it was soon abandoned. But the quest continued, and doctors in San Quentin prison, California, tried transplanting the testicles from executed felons into other prisoners. The testicles would presumably have been rejected, and this too was abandoned. More recent pursuit of the elusive 'fountain of youth' has included the injection of blood serum, with no effect, and of the male testicular hormone testosterone, also with no discernible effect.

Other attempts involved bacterial brews, based on the stories of centenarians in remote places partaking of special diets. Yoghurt from cows and fermented mare's milk (koumiss) were considered, and the great Russian biologist and Nobel Prize winner Metchnikov (1845–1916) reckoned that the special bacteria (lactobacilli) in fermented milk would do the trick by replacing the harmful bacteria present in the normal bowel. Once again, though, no effect could be proved.

Many have recommended dietary restriction and exercise. Here we have to remember that difference between life expectancy and life-span. In a large-scale study of London Transport employees carried out many years ago, the bus drivers, sitting long hours at the wheel, were found to die earlier than the conductors, who were always walking upstairs and downstairs to collect fares. The conductors had greater life expectancy. Doctors agree that regular exercise, partly by increasing the blood supply to the heart, makes you less likely to have a serious heart attack and also protects against high blood pressure. You have a greater life expectancy. Yet there is no good evidence that athletes live longer. Perhaps in their cases other life-shortening factors come into the picture. It may be irrelevant, but fruit flies die earlier when their activity level is raised.

The effects of dietary restriction have been referred to earlier. You live longer if you are not overweight, but otherwise there is no evidence about the effect of reduced food intake. For the time being the animal experiments (living longer on restricted diets or lower levels of activity) stand alone. We have to conclude that while exercise and diet can have an effect on our life expectancy, there is no evidence about life-span.

Do We Want to Cheat Death?

People have always wanted to live longer, and as far as life expectancy goes it is already happening. But we want to stay well and in possession of our mental faculties. An alert mind in a suffering, senile body in as awful a prospect as a demented, destroyed mind in a healthy body.

When the French painter Eugène Delacroix (1798–1863) was elderly he wrote in his diary: 'This strange discrepancy between the power of the mind which comes with age and frailty of the body which is also its consequence has always impressed me and seems to me to be a contradiction in the decrees of nature.' In Indian philosophy, the old are said to take an evening view of life. It is a time for repose and serenity. Yet many old people want to have new challenges, enjoy new experiences and tread new paths. Dr Samuel Johnson would have been pleased to hear about our Open University and colleges of further education (although he would have been a headache for the teachers). When he was seventy-two he wrote in his diary: 'I have resolved to plan a life of greater diligence . . . My purpose is to pass eight hours every day in some serious employment.'

There seems no doubt that deaths from cancer, stroke and heart attack will continue to fall as we take up healthier lifestyles and attend to the environment. Serious infections will be kept at bay, although there is always the possibility that a major pestilence will one day emerge and decimate our crowded species. It would be the sort that *spreads fast by droplets* and would makes AIDS seem slow and old-fashioned. But this is only a theoretical possibility, and for the moment we are all living longer. Senile dementia (Alzheimer's disease) will no doubt be preventable (although not curable) before long.

But this increase in life expectancy is bringing about stupendous changes and will cause tremendous problems in developed countries, In the USA today about one in eight people are more than sixty-five years old. By 2020 there will be one in six, and by the year 2040 it will be an unbelievable one in four or five. Science, however, is unstoppable; and in any case, who could argue against permitting people to live out the full life-span of our species? It is the morally correct thing to do. Premature deaths are by definition preventable and as long we can keep the elderly fairly fit in mind and body they should not be allowed to die prematurely.

The Burden of Longevity

What worries governments, even now, is the *financial burden of old people* who no longer work. They have to be supported, either with their own or with state pension schemes, and their illnesses have to be tended. In industrialized economies state-provided pensions account for about half of all social security expenditure, and old people consume more than half of publicly financed health care. There is the dilemma that when you reduce deaths due to cancer, stroke and heart attack you actually increase the financial burden. This is because people now survive to suffer the disabling diseases of old age (dementia, osteoarthritis, loss of vision and hearing) which are more costly. Yet the aim of health care is not to save money but to prevent suffering and premature death. The financial problem will not go away, and will get worse over the next few decades, unless we can do something to prevent the disabling diseases of old age. If people contribute to personal pensions and personal health insurance it may help the cash flow problem but the underlying imbalance will remain. Does this mean that the young will have to work harder and harder to support a massive elderly population? Probably not, because immense increases in working efficiency are taking place. People's lifetime working hours necessary to provide basic needs (food, shelter) will soon be less than half what they were a century ago. Present trends indicate that the working week will be reduced to between twenty-eight and thirty-two hours, and the average working life to between thirty and thirty-five years. This is partly because less work is actually needed by society and partly because people want more leisure. An optimistic view, therefore, is that in spite of the fact that people work fewer hours, in spite of the fact that the average period spent in retirement (a period with greater demands on healthcare) will increase to about thirty years, the system will be economically sound. Whether it is socially sound will depend on better education for leisure, and on allowing old people to continue to work if they wish to do so.

In the past, people tended to live together in large families, and the wisdom and status of the elders were accepted. Then in the twentieth century, families began to split up more frequently, so that today in the UK a large proportion of all households have only one resident. Old people are not wanted at work and where possible are shunted off to retirement homes. There is an unchallenged dominance of younger persons and their values. If the old are reintegrated into the life of society, and old age is transformed into a third age of self-fulfilment, with a happy balance of leisure and useful work in the community, then the pendulum will have swung back. Attitudes to the elderly do seem to be changing, and a unique request to the US government in 1997 illustrates this point. About 1,500 chimpanzees are now living in research laboratories in the USA, most of them specially bred for the study of AIDS. But only a single chimp out of about 200 infected with HIV has succumbed to AIDS, which means that they are poor models for AIDS and are not needed. Scientists, through the National Research Council, have expressed concern about their welfare. Chimps live for thirty to fifty years and the scientists want the government to take over responsibility for them and guarantee them a secure old age. Laboratory animals like mice or rabbits are normally disposed of after use by euthanasia, but chimpanzees are not like other laboratory animals. They are like us. Ninety-eight per cent of their hereditary material (DNA) is identical to our own, and their use in research is regulated with very special care. In the laboratory they have been providing vital answers to questions about AIDS vaccines, cancer, Creutzfeld-Jakob disease, etc. They deserve better than euthanasia.

If by some miracle we do learn how to increase the human life-span, it will cause grave problems for life on the planet. Even now the population explosion threatens world stability (see chapter 2). If our life-span were increased to, say, 125 years, the extra burden of people and the colossal increase in the elderly would dwarf all other issues.

Apart from the practical problems that emerge when people live longer, do we really *want* to increase our maximum life-span? I have argued for the necessity of death in chapter 1, and when it comes to it surely no one wants to be immortal. Living for ever would be an ordeal beyond our comprehension. Perhaps most of us would welcome the increase in life expectancy that would give us a healthy old age up to 110, as long as society could afford to support us, make provision for our needs and perhaps even employ us, as discussed above. Old people often say it is the happiest time of their life. Yet for many of us it might be too much of an ordeal to live for 110 years, even if we retained health and vitality. Surviving to that age would mean that we had learned how to complete our life-span without dying of infections, cancer, heart disease, etc. We would enjoy uninterrupted bodily health. But what about boredom? In a world with about 10^{10} (ten thousand million) people it may prove to be impossible for us all to enjoy a full life. Our species has evolved and been successful because it was so well adapted to a life of danger, with episodes of cold, hunger, infection and injury. We have been genetically engineered by nature to deal with such challenges and stresses. Perhaps we need the ups and downs, the constant striving, the stimulation provided by new problems. Without them, with no physical threats and few worries, might we succumb to boredom? A man without any stresses would surely be like a violin without strings. We might end up with epidemics of deviant behaviour and most of us in need of a psychiatrist.

The Anglo-Irish writer Jonathan Swift (1667–1745) told a story about longevity in his famous satire *Gulliver's Travels*. During his adventures Gulliver encounters the Struldbruggs, a special race blessed with immortality. But he finds that the Struldbruggs grow old and frail, and degenerate into bickering, avaricious senility. The moral is that they are cursed with immortality, not blessed with it. An ingenious short story by Bertil Martensson (1986) entitled 'Myxomatosis Forte' describes a future society where people live for about 200 years, but in which they begin to commit suicide because there are no challenges, no dangers, no sicknesses. After the psychiatrists report that 'mankind is too healthy' it becomes possible to buy mild (e.g. common cold) disease-producing packages from pharmacists. They are much sought after. A doctor's prescription is needed if you want a more serious disease, and there are heavy fines for having an illicit disease.

PART II

What Happens to Corpses?

5

The Body After Death

It matters not how a man dies, but how he lives. The act of dying is not of importance, it lasts so short a time.

Dr Samuel Johnson (1709–84)

The present life of man on earth is like the flight of a single sparrow through the hall where, in winter, you sit with your captains and ministers. Entering at one door and leaving by another, while it is inside it is untouched by the wintry storm; but this brief interval of calm is over in a moment, and it returns to the winter whence it came, vanishing from your sight. Man's life is similar; and of what follows it, or what went before, we are utterly ignorant.

The Venerable Bede (673–735)

The human body is mostly water, which accounts for about 60 per cent of body weight. The exact figure depends on the amount of fat; most tissues are 80 per cent water but fat is only 10 per cent. About 18 per cent of body weight is protein, nucleic acid and carbohydrates. Protein provides structural material and metabolic machinery in the form of enzymes, nucleic acids carry genetic information and direct the activities of the cell, and carbohydrates are energy sources and contribute to cell structure. Another 15 per cent is fat, and 7 per cent is mineral (assorted salts). That is all we are, chemically speaking.

When the corpse is fed on by microbes, worms or carnivores, these substances are broken down and digested into simpler ones. Large molecules like proteins are converted into smaller ones (amino-acids) and directly re-utilized. If, however, the body is burned, water is driven off and there is more extensive breakdown of molecules. Carbon is the basic material of life, forming the backbone of proteins, carbohydrates and fats, and when the body is burned it is converted into the gas carbon dioxide. All that is then left behind is a mixture of simple salts containing sodium, potassium, nitrogen, calcium and phosphorus. This is the ash. It weighs 5–7 lb and amounts to only about 4 per cent of the original body weight.

What Time of Day Do We Die?

A regular twenty-four-hour cycle is built into most of our bodily activities. We call it the biological clock. It is an ancient adaptation to life on this planet, and is controlled by a special part of the brain. The pattern of sleeping is part of this twenty-four-hour cycle, and it takes a while to adjust to a new cycle, as those starting night-shift work or suffering from jet-lag well know. During sleep the brain and the body have an opportunity to rest and recover, and the pituitary gland at the base of the brain discharges pulses of growth hormone into the blood. Before we wake up each morning the pituitary gland shows fresh bursts of activity, this time increasing the amount of steroid hormones in the blood, getting us ready for the day's activities. Then, after waking, our body temperature rises about half a degree Celsius and a host of other body processes, such as secretion of urine by the kidney, are increased.

In the small hours of the night our metabolic activity is at its lowest, and our response to stress is poor. This is when people tend to die, when life is at a low ebb. It is true statistically, and it makes sense intuitively. Nevertheless, we must remember that when an event such as death takes place it is not necessarily due to, or in any way connected with, what happened at around the same time. We can, after all, say with complete truth that everyone dies after their last meal without implying that they were poisoned.

Death as a Physical Event

Departure from life can be silent and peaceful, or so it seems to the observer, but sometimes the final moments are more turbulent. As the dying process gets under way muscles may go into spasm, and it may look as if a real death struggle is in progress. The face twitches, breathing becomes difficult and a 'death rattle' is occasionally heard. There may be a brief convulsion before, with a series of heaving gasps and a final expiration, the body relaxes and all is still. The person 'has expired'. Hearing is the last sense to go, so it is kind to hold a dying person's hand and talk to him.

That last death struggle is sometimes called a 'death agony', from the Greek word *agon*, a struggle, and we refer to an event as agonal when it occurs during dying. If the heart has not yet stopped beating, it soon does so. The unseeing eyes take on a dull appearance, and the pupils dilate. Now that the machinery of life has ceased to provide warmth, the body starts to cool. The most unnerving, unforgettable things about the dead body of a loved one are its coldness, its perfect stillness and unresponsiveness. The human being has come to resemble a sack of potatoes, a roll of carpet.

Certainly there is no doubt about death having occurred when the body is stiff and cold, but in the early stages it can be more difficult to recognize. This is because death is a process rather than a single definable event. Cut off from their life-giving supply of oxygen and glucose as soon as the blood stops flowing, the tissues and cells begin to die – but at different rates. Skin can be used for grafting at twelve hours after death, and skin cells taken into culture after up to twenty-four hours are found to grow normally. Contrary to common folklore, nails and hair do not continue to grow after death, although the beard may seem more prominent because of drying of the skin or contraction of hair muscles ('goose pimple' muscles). For a few hours after death, muscles will still contract on electrical stimulation. Intrepid observers in the nineteenth century noted that in those who were guillotined the knee-jerk reflex was still present for up to twenty minutes after decapitation. White blood cells move about for at least six hours. Different cells in the body survive for different periods after their blood supply ceases. Transplant surgeons are acutely aware of this, as mentioned in chapter 12.

On the other hand, nerve cells in the cortex of the brain die a mere three to seven minutes after the heart stops beating and their oxygen supply is cut off. Even a temporary fall-off in the blood circulation through this part of the brain causes loss of consciousness. The brain is only about 2 per cent of body weight yet needs 20 per cent of the oxygen, and hungriest of all are the nerve cells in the cerebral cortex. They need a lot of energy to keep them going, and this comes from using oxygen to burn glucose. There are no local reserves and so they are exquisitely vulnerable to lack of oxygen and glucose. As the guardsman stands rigidly on parade, blood accumulates in the veins of the legs and after a while the amount of blood reaching the heart falls. If he is susceptible, the blood pressure now falls, and his cerebral cortex, momentarily deprived of oxygen, begins to malfunction. He feels faint and soon falls to the ground unconscious. In the horizontal position the circulation to the head is restored and the embarrassed guardsman comes round. For exactly the same reasons you can feel faint when you get up out of a hot bath. If the nerve cells of the cortex are deprived of blood for a longer period they begin to die. As the great physiologist J. S. Haldane pointed out in 1930, depriving these cells of oxygen 'not only stops the machine but wrecks the machinery'.

Nerve cells in the stem of the brain survive a little longer; they need only half as much oxygen as the cells in the cortex. Brain-stem cells control breathing and heartbeat and it is possible to have a 'cortical death', in which the patient is in a deep coma because the cortical cells are dead, but with the brainstem still functioning and heart and lungs still working. We call this the 'persistent vegetative state'.

Does it work the other way round? Can you get damage to the brain stem with the cerebral cortex more or less unaffected? In the old days of poliomyelitis, before a vaccine was available, the virus sometimes damaged the brain-stem centres responsible for breathing, causing 'bulbar polio'. Under these circumstances the cortex was intact and the conscious patient was put into an 'iron lung' to provide artificial ventilation until the brain-stem centres had recovered.

Death and the Law

Death is taken seriously by the law and by the state. It is unlawful to dispose of a corpse until the cause of death has been satisfactorily ascertained. The medical practitioner attending the terminal illness, if there was one, must do this 'to the best of his knowledge and belief'. He signs the death certificate to confirm that the person is dead, to record that death was due to something reasonable rather than to homicide or any other unnatural cause, and to provide statistical information. Once the death has been registered, the registrar can issue a certificate for the disposal of the body. Special arrangements are made for registering deaths at sea or overseas, for deaths in the armed forces or in prisons, and for stillbirths. It is not always a simple matter.

If the doctor has not seen the person within two weeks of death or if the death is not satisfactorily explained, the coroner is called in. He may decide to go no further, but could ask for a post mortem, or find it necessary to hold a public inquiry (inquest). The inquest generally ends with a verdict of accident, unlawful killing or suicide, but it may be that there is no definite conclusion (an open verdict). Sometimes both the identity of the dead person and the cause of death are unknown. The position of coroner dates from Saxon times, his full title being Coronae

Curia Regis (Keeper of the Royal Pleas), and his job was to maintain the private property of the Crown. He was responsible for investigating accidents such as shipwrecks to see what money could be obtained for the royal purse. Later he was responsible for all sudden, unexpected deaths. Coroners are qualified lawyers or doctors, responsible only to the Crown, and in the UK they are needed in about a quarter of all deaths. In other countries the task is undertaken by medical examiners, magistrates, judges or the police.

When Has Death Occurred?

For hundreds of years people have worried about this because of the fear of being buried alive (see chapter 6). One of the old, rough and ready signs of death was the cessation of breathing, as indicated by failure to move a feather or cloud a mirror held in front of the mouth and nose. But people can be revived an hour or two after they appear to have stopped breathing. Absence of a discernible heartbeat is a more reliable sign, and for hundreds of years the heart had been accepted as the pre-eminent organ, source of the vital principles that define life. But this may require a stethoscope and careful clinical examination. The stethoscope was not invented until 1819, and in the eighteenth century doctors used Balfour's test, in which needles with flags attached were inserted through the skin into the heart, a heartbeat then being visible as a waving of the flag. Still, though, people can be revived after their heart has stopped beating, although brain damage is inevitable after about six minutes. 'Irreversible cardiac arrest' is perhaps a more acceptable definition of death.

The question of when a dying person can be said to be actually dead has come to the fore again because of two recent advances in medicine: intensive care and organ transplantation.

Modern intensive care can keep the rest of the body functioning during a temporary failure of heart, lungs or brain. It can also do this when the brain has died and there is no hope of recovery. After a person has been maintained on life-support systems for many months, with no hope of independent existence or awareness, the question of switching off the machine has to be faced. Can this person be regarded as dead? The idea of brain death was originally developed as a guide to management under such circumstances. This was in the interests of the patient and family, although in the era of organ transplantation the interests of the recipient were also considered.

Transplantation requires fresh organs taken as soon as possible after death, and requires a precise definition of death. At what stage can the dying person be considered dead, so that organs can be removed from the corpse and used for the benefit of sick people? The question is ethical and legal as well as medical.

The dilemmas created by modern intensive care are exemplified by the case of a brain-damaged, comatose fifteen-month-old boy, kept alive by a life-support machine in a hospital in California in 1989. The boy's father wanted the machine to be switched off, but the doctors refused to do so because he was not brain-dead. The distressed father then took things into his own hands, held hospital staff at gunpoint and disconnected his son, who soon died. He was arrested, but the County Grand Jury refused to indict him for murder, and he was instead convicted of unlawful use of a firearm and placed on probation for one year.

A difficult decision about brain death and pregnancy

An eighteen-year-old woman, fourteen weeks pregnant, was involved in a bad traffic accident. In intensive care she showed no spontaneous breathing, and on the third day it was decided that she was brain-dead. But if she was disconnected from the life support machine the unborn child would be killed. The machine was kept going. The doctor, after discussing the matter with the woman's parents, assembled an advisory council consisting of an anaesthetist, a neurosurgeon, a pediatrician, an obstetrician, a professor of legal medicine and a lawyer. But they could not agree on a decision, and so the mother remained on the life support machine.

The ethical problem was whether the 'cadaver' or corpse of the officially dead mother should be used as a sort of incubator, to breed an unborn child. Does the unborn child have a 'right' to be preserved in this abnormal, expensive manner?

Brain death criteria have not been worked out for the unborn. As it happened, the mother developed an infection and became feverish, and the child died before it was capable of independent life. But children have been born and survived under similar circumstances. One brain-dead pregnant woman was kept on a life-support machine for sixty-four days to increase the chance of delivering a live baby.

The Idea of Brain Death

If dying is a process, it might be expected that it is not going to be easy to construct an exact definition of death. Is it like trying to define night as opposed to day, having to pick an arbitrary cut-off point at some stage during that transition period of dusk and dawn? Or like arriving in

London by air: when is the actual moment of arrival? Is it when you enter London airspace, or when the plane touches down, or when you set foot on London soil? Hearts can stop beating and start again. Lungs can stop breathing and start again. But neither can carry on without assistance after the brain has died. The term 'brain death' was introduced in 1965, following a report of a kidney removed for transplantation from a seemingly brain-dead donor whose heart was still beating. Definitions were modified over the years, but in the USA in 1981 the Uniform Determination of Death was adopted as statute by all states. This said that an individual who has sustained either (1) irreversible cessation of circulation and respiratory functions, or (2) irreversible cessation of all functions of the entire brain, including the brain stem, is dead. The essential thing is that the changes must be irreversible. When you look at the warm, breathing body without a brain, you realize that what has been developed by modern technology is really a second type of corpse, the brain-dead corpse.

Brain death itself is a complex process. Consider first of all death of the forebrain (the cortex of the brain), thought to be the seat of consciousness. Does this mean death? How does one check for it? The child born without a forebrain (anencephaly) or the patient in a persistent vegetative state (see above) may not be conscious in the way that the rest of us are, but the latter may live for a long time and recover consciousness up to three years later. On the other hand there is death of the brain stem, which means that the nerve centres at the base (stem) of the brain controlling heartbeat and breathing are out of action. Although one could argue that the person with brain-stem death may possibly have feelings, dreamings, this could not continue for long. Once the heartbeat and breathing have ceased the process of dying begins and is soon completed. Death of nerve cells and swelling of the brain takes place within a few minutes. The person has now vanished, leaving organs such as heart and kidney still capable of life if connected up to the blood supply of a living individual.

Few would disagree that a person who is brain-dead is dead, when brain-dead signifies the irreversible absence of forebrain and brain-stem function. This is the 'flat line' on a recording device such as an electroencephalogram. But organ transplantation, with the need for fresh organs taken as soon as possible after death, has raised the need for a yet more precise definition of death. For most purposes brain-dead means brain-stem dead, and in most countries brain death as spelt out below is now accepted as legal death, even when the heart is still beating. Unfortunately the criteria are not universally agreed upon, and legal definitions vary. It is possible to be brain-dead in one country but not in another. To get round the problem of such decisions about death, a software program called 'Ryadh' has been introduced. It is used in intensive care units in certain hospitals in the UK and Germany, and computes the outlook for the patient in relation to the cost of maintenance on the life-support machine. But not everyone agrees about its reliability.

The usual criteria for brain death are as follows:

1   The person must be in deep coma (unwakeable) which is not due to a condition such as drugs, poisoning, low body temperature (hypothermia), shock or diabetes. Even the electrical brain waves (electroencephalogram or EEG) can be reversibly suppressed under these circumstances.
2   The person must be on mechanical ventilation, spontaneous breathing being inadequate or absent.
3   The basic cause of the condition must be clear. In other words, the doctors must know that it is due, for instance, to a head injury or to brain haemorrhage.
4   The special diagnostic tests for brain-stem death must be positive, as determined by two doctors and another independent physician. To avoid bias, none of them may be part of a transplant team if an organ donation is a possibility.

The tests for brain-stem death include:

1   The absence of about ten basic reflexes, such as the reaction of the pupil to light.
2   The absence of respiratory movements when the ventilator is disconnected and carbon dioxide given as a stimulus to breathing.
3   The absence of circulation of blood in the brain. This is the ultimate, 'gold standard' test. Ideally this means injecting a fluid that will show up on X-ray into the aorta or into the four blood vessels supplying the brain. This is not easy, needs special equipment, and is not always possible.

The Death of the Person

We have been considering death as a biological event; what about its philosophical and spiritual aspects? Death involves the end of consciousness: the end of the ability to communicate and reason; the end of spontaneity, control, integration. It means that the biological foundation for that complicated living creature, the person, has gone. The body is empty and is beginning to disintegrate. The soul or spirit is now either dead or existing separately from the body.

What, exactly, is this essence of personhood that has departed from the dead body? Suggestions by great philosophers have included:

- the ability to think (Descartes);
- the awareness of progress and persistence over time (Locke);
- psychological characteristics (Hume);
- rationality, basing actions on moral principles (Kant);
- self-consciousness and directed behaviour (Sartre).

Unfortunately, none of these definitions is clear enough to be of practical use, or measurable. At what stage does the spirit or soul of a person depart? This question is a religious and philosophical one, and answers must be sought in the teachings of the great religions. These are discussed in more detail in chapter 16. The family of Miguel Martinez, however, were clear about this. Miguel was a Spanish athlete; injured in a soccer game in 1964, he survived in a comatose state for eight years. His family then announced that 'Miguel died at the age of thirty-four, having lived twenty-six years.' One of the heart-rending features of death is that the just-dead human being on the bed, or the patient on a life-support machine, can look like a living one; asleep perhaps, but alive.

Near Death Experience

This chapter deals mainly with observable physiological events. Yet a description of dying would be incomplete without mention of what it may feel like to die, and this is hinted at in what are called 'near death experiences'. There have been studies of hundreds of people who very nearly died, for instance during a heart attack or pneumonia, but then recovered and were able to say what it was like. It seems that there is a central core experience which is much the same in people of different races, cultures and religions, and which is perhaps characteristic of human beings. NDEs are described in more detail in chapter 16.

From Rigor Mortis to Putrefaction

It is fortunate that corpses are degraded and decomposed so efficiently by natural processes. If it were not so, the surface of the earth would be covered with dead bodies. What are the bodily changes that take place after death, and what are the naturally occurring processes of decomposition?

At the time of death or soon afterwards it is quite common for the stomach contents to be regurgitated into the mouth or air passages, for urine to be passed and semen emitted. The skin becomes pinkish or purple on the underside of the body where blood accumulates, although it is paler in places where the weight is resting and the blood has been squeezed out. Muscles undergo contraction (rigor mortis), beginning after one to four hours in small muscles (jaw, fingers, eyes, mouth) and after four to six hours in larger muscles of the limbs and trunk. At this time the body is still warm, but stiff. After thirty-six to forty-eight hours, muscles relax and the body feels softer but is now cold. The exact times vary, and are not a reliable guide as to how long the corpse has been dead. Rigor mortis, for instance, appears earlier after physical exertion or in warm weather, but later in the cold and in old or feeble people. As the body dries the cornea becomes cloudy and the eyeballs more sunk in their sockets. Rigor mortis takes place in the smallest of muscles. The pupils are affected, and also, as noted above, the tiny muscles (*erector pili*) that make the hair stand on end, and whose contraction gives the appearance of gooseflesh. The fact that this makes the hairs more prominent, together with the drying and shrinking of skin, has given rise to the false idea that hair continues to grow after death.

The rate of cooling of the body has always been of interest to forensic pathologists, who have to give opinions about the exact time of death. It is a complicated matter, as described in chapter 14. Only a rough estimate is possible. For instance, events will be different under water or underground. Decomposition in air is about twice as fast as in water and about four times as fast as underground. Corpses are preserved longer when buried deeper, as long as the ground is not waterlogged.

The countless microbes in the intestines are still alive and some of them (the clostridia, the coliforms) take the opportunity to spread through the body, invading the normally prohibited parts. Because of their activity the first parts to putrefy are the intestines. Also the body undergoes its own intrinsic breakdown (autolysis) which is caused by the enzymes and other chemicals released from dead tissues. The pancreas, source of so many of the gut's digestive enzymes, soon digests itself. Green substances and gas are produced in the tissues, as a result of which the skin, beginning with the abdomen, takes on a green or bluish colour and develops blisters. The blisters may expand into large sacs of fluid, with the skin slipping away underneath them. The front of the body swells up, the tongue can protrude from between the teeth, fluid from the lung may trickle out of the mouth or nostrils, and the corpse becomes an unpleasant sight. At this stage, reached in temperate countries four to six days after death, there will be a disagreeable odour due to release of the gases hydrogen sulphide and methane, and traces of mercaptans. The smell of a decomposing corpse is remarkably persistent and penetrating. Although Charles IX of France remarked that 'the body of a dead enemy always smells good' it was surely, even for the most bloodthirsty, the sight rather than the smell that gave satisfaction. Putrefaction is inhibited at about the freezing point of water or when it is very dry, but in the tropics can take place within hours.

Consumption of the Corpse

Left to itself, a corpse above ground is soon used as food by insects and animal predators. A host of different insects feed on dead bodies, including beetles, ants and wasps, depending on the season and the part of the world. Ants are early arrivals, feeding round eyelids, lips and knuckles, and forming little ulcers. The most common insects are the larvae (maggots) of *Diptera* flies. The carrion flies (flesh eating or 'sarcophagus' flies) are specialists and soon arrive, attracted by the smell of the corpse. Eggs or larvae are laid in strategic positions such as the orifices of the body or skin wounds. The first and most numerous are bluebottles or 'blowflies' (*Calliphora erythrocephala*), the sort that make meat 'flyblown'. A single bluebottle lays up to 2,000 eggs, in clusters of 50–100. Next come the specialized flesh-flies such as *Lucilia*, the greenbottle. There is great competition, and those that are viviparous, able to lay already-hatched larvae that can start feeding straight away, are at an obvious advantage. The larvae secrete a powerful enzyme that dissolves skin. Most people have seen the sequence of events when a dead mouse or rabbit lies undisturbed for a few days.

In the tropics, particularly if there are wounds to give easy entry, a corpse becomes a moving mass of maggots within twenty-four hours. When I worked in Africa, my first visit to a post-mortem room in 1953 was to see the body of a young man who had been murdered the day before. The flies had done their work and the skin of the dead man was moving gently as the maggots fed on the tissues beneath it. The proverbs and the poets talk about worms, but neither the common earthworm nor the white nematode worms in the soil show much interest in decomposing flesh. The 'worms' referred to since the Bible and Shakespeare are in fact the insect maggots. They are worm-like, and as long as there is access for the adult fly they are a regular feature of the rotting corpse. If it were not for these fast eaters we would see more corpses, undergoing a slower decay of the flesh. The maggots are the unseen undertakers of the world.

An easy way to prepare an animal skeleton is to enclose the body in wire netting to exclude animal scavengers, leaving the insects and ants to clean it up. In cooler countries the common house fly may lay its eggs if the flesh is already decomposed. This is an alternative to its preferred site of garbage or manure. An interesting corpse lover, the 'coffin fly' (*Conicera*) may appear in the coffin or emerge in large numbers from the soil above a buried corpse. How they get there is not clear. Perhaps the eggs are laid in the soil and the larvae travel down to the corpse, or the female fly herself moves through the soil to deposit the eggs, or (more likely) the eggs are laid on the corpse before burial.

A definite sequence of insect invasion takes place in parts of the world like northern Europe, and seven to eight different stages can be distinguished. Forensic pathologists use this progression to help work out how long the body has been dead. After the flies and their maggots come an assortment of beetles including skin beetles (*Dermestes*), then mites, and also the type of beetles and moths found on fur and fabrics in museums, which consume the hair. In a year or two the body has been converted into a skeleton.

Animal predators include rats, mice, dogs, foxes, flesh-eating birds (vultures, crows, etc.), and sometimes larger mammals. Parts of the body may then be spread over a wide area. During the Battle of the Somme in the First World War, the dead had to be left as they fell, in no-man's-land, for as long as three to four months, before the Allies advanced and were able to attend to them. By this time decomposition was at an advanced stage and well-fed rats had often made their homes in the chest cavities of the corpses.

How the Corpse Decomposes

When there is no animal predation the hair, nails and teeth become detachable within a few weeks, and after a month or so tissues become liquefied, as a result of autolysis and microbial action, and the main body cavities burst open. Putrefaction is generally slower in a coffin, and if it is dry the body remains identifiable for many months. Cracks appear in the skin due to shrinkage and the skin surface begins to look like old paint.

Moisture helps decay. In places where the body rests against the coffin the skin decomposes, exposing the back of the skull, the back of the shoulderblade, and the tips of the vertebrae. Whitish fungal growths may appear on the skin, often arising from fungal spores that were present in the cosmetic materials used by embalmers. Tendons and ligaments resist decomposition, and also, for unknown reasons, the uterus and the prostate gland, which may survive for a few months. Bones and teeth are even more resistant, and within a year or so all that is left is a skeleton. The bones are still a bit greasy and contain organic material, so that there is a smell of burning if they are sawn. It takes forty to fifty years for them to become dry and brittle in a coffin or in fairly dry soil.

What happens after the body has been finally reduced to a skeleton depends on the circumstances. Studies of twenty-three victims of the Vietnam War showed that decomposition of bones is hastened by acid, moisture, warmth and shallow burial. If the soil is acid, peaty, the bones are gradually dissolved, as in the case of the 'bog-men' described in Chapter 10. This can take 25–100 years if it is warm and wet, or 200–500 years in dry cool soil. In neutral soil bones stay in good condition for much longer. Bones of young adults last longer than those of children and the elderly. This was clear from examination of skeletal remains of Native Americans buried in Purisima Mission cemetery in Lompoc, California between 1813 and 1849, and of bones in a prehistoric cemetery in California. This is probably because children's bones and old people's bones have less calcium. A corpse can mummify when it is very dry, for instance in the sands of Egypt or in a cool, dry loft or barn. Tissues become leathery and brown, and the skin stretched tightly over the face. The body has dried up before the microbes have had time to produce their moist putrefaction.

When the corpse is in water, the head and limbs sink lowest, but gas formation in the skin and in the abdomen may later bring the body to the surface, generally floating face upwards. After about two weeks, the skin peels off. The body sinks again as small aquatic creatures like shrimps and fish strip it to the bones. An interesting diatom test can sometimes help determine whether the person died by drowning or entered the water after death. Diatoms are microscopic algae with a silica skeleton that is almost indestructible, and they are present in sea and in natural fresh water. During drowning they are inhaled into the lungs, enter the blood and are carried to distant parts of the body such as the brain, kidney, liver and bone marrow. They can be seen under the microscope after digesting organs with strong acid. Presence of diatoms in these organs suggests drowning. Diatoms entering the lungs after death, in contrast, do not spread to other organs. But the diatom test, which has been in use since 1904, is still controversial. There are about 10,000 species of diatom, and some of them are present in foods like shellfish, and could perhaps spread through the body after being eaten.

Occasionally, especially in wet conditions, the body fats turn into a waxy, greasy material, often greenish-white in colour, called adipo- cere. It takes months to form and has a rather rancid, earthy or cheesy smell. It eventually turns into a chalky brittle material, and in this way subcutaneous and abdominal fat may be preserved for decades, even centuries.

Fossils and Petrification

The vast majority of plants and animals undergo decay and destruction, but very occasionally bones are 'mineralized'. What this means is that minerals from the soil gradually impregnate and replace the original bone without obscuring its structure. The bone, embedded in sedimen- tary rocks, is now a fossil (from the Latin, meaning something that is dug up). It takes at least 10,000 years for this to happen, and it is a rarity. Fossil primates are especially rare, which is why our own evolutionary history is so incomplete, so often being revised.

The actual tissues of a dead plant are sometimes replaced, particle by particle, with minerals like silica, forming a stony structure. This is called petrification (Greek *petros*=a rock or stone), as seen in the petrified forests of the south-western USA.

Fossils are generally formed when the corpse lies on the floor of an ocean, river or lake where particles are being deposited. As the particles build up, the body is buried. Being buried under a dust storm can have the same effect. Occasionally animals are trapped in a pool of sticky tar or asphalt. Predators are ensnared as they try to catch them, and birds may alight and become stuck. This happened during the Pleistocene epoch a million years ago at Rancho La Brea in southern California, where there is a rich collection of well-preserved fossils.

Teeth are the toughest parts of the body, highly resistant to decomposition. Such a large proportion of fossils are teeth that paleontologists have jokingly suggested that mammalian evolution can be looked upon as the evolution of teeth: the teeth mate with each other, and over the ages give rise to dental descendants with slightly altered teeth! Teeth tell a great deal about an animal and its lifestyle.

Some paleontologists are happy to draw a picture of the complete animal just from examining the teeth.

Are Corpses Dangerous?

The putrifying, decomposing corpse looks gruesome but generally presents no threat to human health, as long as death was not due to a lethal infection like plague, tetanus, anthrax, hepatitis or AIDS. In the past, the unpleasant odour and appearance of a rotting corpse led naturally to the belief that it could be a source of physical illness. A fear of contagion was added to the basic apprehensions about the person's dead body and spirit. This was reflected in the 1855 UK Burial Act, prohibiting burials within 100 yards of existing homes without consent.

When, in the 1760s, a new cemetery was proposed near the Petit Luxembourg, a densely populated area of Paris, the Procurator General opposed it, pointing out that 'the fetid odors emitted by cadavers is a sign from Nature who is warning us that they should be moved to a distance . . . the impure exhalations . . . cling to the walls which they impregnate with a noisome essence and . . . may carry unknown causes of death and contagion' (quoted by Philippe Aries, *The Hour of our Death*, 1981). People living next to cemeteries complained that their meat and wine soon spoiled in the pantry and that metals lost their shine. In Paris in 1779 air from a large common grave 50 feet deep in the cemetery of Les Innocents was said to be leaking into the cellars of neighbouring homes. People felt ill, lamps would not light, and as a result of a public outcry the grave was eventually opened and treated with lime and quicklime. Les Innocents was the first cemetery to be closed by the police, in 1780.

Nature's Need to Recycle the Corpse

It is not only convenient from a physical point of view for corpses to be broken down and decomposed. A recycling of elements such as nitrogen, carbon and phosphorus is an essential part of the earth's economy. All living things, when they die, must be decomposed and the component elements released so that they become available for future plants, animals and people. Elementary textbooks of biology set out these cycles, and Shakespeare refers to them in *Hamlet*: 'To what uses may we return, Horatio! Why may not imagination trace the noble dust of Alexander til he find it stopping at a bung-hole? Imperial Caesar dead and turned to clay, might stop a hole to keep the wind away.' Bodies eventually re-enter these elemental cycles whether they are left on the ground, buried, drowned or burned. Embalming and freezing, however, interfere with this process, as described in chapters 9 and 10.

Preliminary Treatment of the Corpse

Whatever the nature of the death, steps must be taken to dispose of a human corpse. Most animals are content to leave their dead lying on the ground, where they will be dealt with by natural means. This is not good enough for humans. For one thing, we live in permanent settlements and it would be inconvenient to allow the dead to accumulate in our daily living space. A corpse smells bad as it rots. The natural solution has been to bury, because it is then out of sight and nature will do the rest.

But there are more important considerations. We differ from animals, first in knowing that we must die and second in our unique capacity to think about the future and worry about it. We can therefore think about death and about the fate of the dead. We can contemplate an afterlife for ourselves. It follows from this that a corpse must be acknowledged and responded to. As Ruth Richardson has said, 'A corpse has a presence of its own. It resembles the dead person yet it is not that person.' It must be disposed of in some special way. At all times humans have moved their dead to a place of burial, of burning, or of exposure to scavenging animals. It is one of the hallmarks of humanity. The Neanderthals were almost human because they appear not only to have looked after their sick and injured but also to have buried their dead. For instance at Le Moustier, France, a teenage boy was buried in a pit with a pile of flints as a pillow, and at Teshik Tash, Central Asia, the skeleton of a Neanderthal child lies with six pairs of ibis horns forming a ring round his head.

Remembering the Dead

A complex and amazing variety of rituals accompany these different methods of corpse disposal. The rituals express fundamental religious and social beliefs. Some of the variations can be regarded as mere shifts in fashion, but throughout the world and in all ages funeral rituals contain a characteristic core, revealing our concern about death. The universal, unavoidable question is 'What happens to me when I die?' To the ancients it seemed natural that they should continue to live on in some form or other after the body had died. Arrangements were sometimes made to ensure that the spirit could leave the body; food and possessions were placed with it for future use. Most people in the world today still believe in an afterlife (see chapter 16) and the rest of us, however irreligious, do not always find it easy to accept that one day we will be snuffed out. Monuments and tombs owe their existence not only to the wish of survivors to honour and remember their dead, but also to the concern of humans about their own death and anonymity. The tomb of the Black Prince (d. 1376) in Canterbury Cathedral bears the simple reminder (in French) 'Such as you are, I was once: such as I am, you will be.' People like to feel they have left behind some lasting, tangible object to help others remember them, to give reassurance in the face of that ultimate eclipse of the self. This is so whether the monument is the Great Pyramid of Cheops, an initial carved on a tree trunk, or an ephemeral graffiti message on a wall. Many other left-behind tokens of life and work come into the same category. Some of the satisfaction of leaving descendants and posessions, of creating books, music, works of art, or having a portrait painted, stems from the same worry about anonymity and extinction.

But people *are* forgotten, and the even the greatest of monuments eventually crumble, as set out so unforgettably in Shelley's poem 'Ozymandias':

> I met a traveller from an antique land
> Who said; 'Two vast and trunkless legs of stone
> Stand in the desert . . . Near them on the sand,
> Half sunk, a shattered visage lies, whose frown,
> And wrinkled lip, and sneer of cold command,
> Tell that its sculptor well those passions read
> Which yet survive, stamped on these lifeless things,
> The hand that mocked them, and the heart that fed:
> And on the pedestal these words appear:
> 'My name is Ozymandias, king of kings:
> Look on my works ye mighty and despair!'
> Nothing beside remains. Round the decay
> Of that colossal wreck, boundless and bare
> The lone and level sands stretch far away.'

Less illustrious individuals are more rapidly forgotten. The buried remains of ordinary citizens are either moved or built over within a few generations. They have sunk into oblivion. It can be explained in biological terms. We are all intensely interested in our children and grandchildren, concerned for their well-being and the state of the world they will have to live in. But after this there is a distinct cut-off. Most of us will not see our great-grandchildren, and the thread of family love and intimacy is broken at this point. Not many of us visit our great-great-grandfather's grave. It is not that we are indifferent about our more distant descendants but that our concern becomes less intense, less personal. We are interested but not emotionally involved. Perhaps this is why so few people have very strong feelings about global warming or the exhaustion of fossil fuels. These events seem too far ahead. It is as if the loyal concern that we feel for our own kith and kin extends to cover only the world we live in at the moment. In a poem written nearly 400 years ago, George Wither (1588–1667) laments the damage being done to his beloved England. It might have been written yesterday:

> When I behold the havoc and the spoil
> Which, even within the compass of my days,
> Is made through every quarter of this isle,
> In woods and groves, which were this kingdom's praise;
> And when I mind with how much greediness
> We seek the present gain in everything,
> Not caring (so our lust we may possess)
> What damage to posterity we bring . . .
> What our forefathers planted, we destroy:
> Nay, all men's labours, living heretofore,
> And all our own, we lavishly employ
> To serve our present lusts, and for no more.

A consequence for corpse disposal (among people who bury or 'memorialize' their dead) would be that, with the exception of the famous, corpses should be finally rendered 'anonymous' after two or three generations. All recognizable monuments should be removed as a routine, and any remains or ashes scattered. This would at least help deal with the problem of crowded cemeteries.

Laying Out and Tidying Up the Corpse

The immediate practical tasks are to close the eyes, sometimes with a coin laid on the eyelids; to close the mouth by placing a pillow under the chin or by tying a piece of cloth round the chin and head; and to straighten the legs and position the arms crosswise over the chest or stomach. The dead person is thus made to look peaceful and natural. It was a common practice in ancient Rome and in other Mediterranean countries to place a coin in the mouth. This was the fee for Charon, the one who ferried the souls of the dead across the rivers of the lower world.

These things are best done before the onset of rigor mortis. Also, for hygenic and aesthetic reasons, the orifices (rectum, throat, nasal passages) are plugged with cotton wool and the head raised on a pillow to prevent regurgitation of fluids. The corpse is washed and dressed.

Traditionally, before the rise of the undertaker in the seventeenth and eighteenth centuries, the 'laying out' of the dead body was a family responsibility, carried out by women. Among Sikhs and Muslims the washing remains the duty of family members. Laying out has a practical purpose but also a ritual significance, because in most societies there is a period during which the body is available for viewing.

Viewing

In the past, when death took place in the home rather than in hospitals or institutions, everyone, including children, was acquainted with the dead body. People expected to die in bed at home, often with their family and friends present. Death was accepted rather than fought against, and at times dying became an almost public ceremony. The rituals of death and paying the last respects included seeing, perhaps touching or kissing the dead person. But death has now become institutionalized. In the UK in 1992 71 per cent of deaths took place in institutions (54 per cent in hospitals, 13 per cent in nursing homes and 4 per cent in hospices). Viewing the corpse has become less common in the UK, but it is still important in the USA. The dead person, washed, made up, dressed, often shampooed and manicured, is displayed in an open casket during the funeral service.

Under certain circumstances the corpse used to be exhibited in a more public way. In nineteenth-century Paris the city morgue received about 1,000 unidentified cadavers a year. After initial cooling to $-15°C$ they were stored at $-2°C$ and could be inspected by those who might have known who they were.

Big public displays of corpses are reserved for the famous and the great. If the display is to be for more than a day or two the corpse has to be preserved from putrefaction, usually by embalming (see chapter 9). The ancient Greeks went in for lengthy lamentations, funeral rites and viewing of corpses, and this led to legal time-limits being laid down by Solon in 590 BC. There was also the problem of putrefaction. One royal solution was to construct an effigy of the dead person, made of wood or wax, and built up from a death mask. It made less of an impact than the real corpse, but was something concrete that could be used during the lying in state before the official funeral. Wax effigies of William III (d.1702) and of Queen Mary 11 (d.1694) were displayed in Westminster Abbey. The corpses of various European kings and queens have been exhibited in public for a period after their death. Lenin, embalmed and wearing his khaki jacket, was exhibited in a glass sarcophagus in Red Square, Moscow, for more than sixty years. Grace Kelly (Princess Grace of Monaco) lies in state, embalmed and looking good, in the Palatine Chapel.

Subsequent Treatment of the Corpse and Embalming

One of the most upsetting, unpleasant things about a corpse is the smell emitted during putrefaction. Many of the rituals and customs for disposal of the dead can be looked upon as practical measures that have been taken at various times to reduce this offensive reminder of the event. These measures are described in more detail in the following chapters. They can be summarized in sequence as follows:

1　Keep the body cool.
2　Adorn it with sprigs of rosemary or other pleasant-smelling herbs or spices.
3　Bury or otherwise dispose of it as soon as possible.
4　Remove and bury the viscera, to reduce the rate of decomposition (see chapter 14). If the flesh is then boiled off, the skeleton can be kept for a long period.
5　Embalm it. The process of embalming is fully described in chapter 9. It involves injecting an embalming fluid containing the preservative formaldehyde into arteries, plus or minus cosmetic and other treatments of the corpse. The idea is to make the corpse look presentable and peaceful. Funeral directors have cold storage rooms, and some have deep freezes so that they can if necessary keep the untreated body for a few weeks. In the USA nearly all bodies are embalmed, usually with cosmetic treatment of the face. In the UK some funeral directors always embalm, but others only as required.
6　Put it in a double (or triple) coffin.

7 Seal the coffin thoroughly. For instance, line it with waxed paper or pitch, or use putty as a counterseal over screw-holes. If the money is available use a lead or steel coffin.

8 Alternatively, encourage rapid underground putrefaction by using a readily biodegradable coffin. Wicker coffins enjoyed a limited popularity in the late eighteenth century, and nowadays cardboard, even newspaper coffins have their enthusiasts.

6

Burial

What must the king do now? Must he submit? . . . I'll·give . . .
. . . my large kingdom for a little grave,
A little little grave, an obscure grave;
Or I'll be buried in the king's highway,
Some way of common trade, where subject's feet
May hourly trample on their sovereign's head;
For on my heart they tread now whilst I live;
And buried once, why not upon my head?
                    William Shakespeare, *Richard II*, Act 3, Scene 3

Religion Makes a Difference

From as far back in history as we can see, in different countries and with different religions, burial has been the preferred method for disposing of corpses. The first human known to have been buried died about 400,000 years ago, and the body was discovered in a cave north of Beijing, China. The physical preservation of the body was a central feature of belief in ancient Egypt and among early Christians and Muslims. When buried, the dead person is seen to be intact, initially at least, and is deposited in a defined place of rest. Things needed in the afterlife can be conveniently buried with the body. These are known as grave goods and include food and drink, along with the necessary containers, and weapons. Kings and nobles needed servants, horses, pet dogs, possessions, even wives. These could not be expected to die a natural death at the required time, so they had to be killed. Queen Shubad of Ur in ancient Sumeria was buried about 4,500 years ago together with sixty to eighty attendants including harpists, ladies-in-waiting and grooms. Her husband Abargi was also well provided for; he was buried with ten female attendants, laid in two rows with head-dresses of gold and lapis lazuli, who had the honour of serving the king in his afterlife.

The body is nearly always placed in a horizontal position, usually supine (on the back). Muslims are placed with the right side facing Mecca, Buddhists with the head to the north. In some Native American tribes and in pre-dynastic (before 3100 BC) Egypt the 'foetal position' (on one side with the knees under the chin) was adopted. In England today there is a Voyager Trust (originally the Pagan Hospice and Funeral Trust) whose logo depicts a corpse in the foetal position. The Trust promotes environmentally friendly funerals, and many of its supporters would like to be buried in the foetal position, with plenty of flowers, after being smeared with red ochre (representing the life force). Unfortunately they have had difficulty getting recognized as a charitable trust because paganism is not officially a religion and is not acceptable under Britain's archaic charity laws.

Vertical burial is uncommon. For one thing it means a lot more work for the gravedigger, and for another the bones, when the body is reduced to a skeleton, tend to sink into an amorphous mass instead of roughly retaining their human form. But in some parts of the world warriors, who should be ready for instant action, have been buried standing up.

Ancient Civilizations

The ancient Romans at first favoured cremation rather than burial, but during the second century AD burial became more common, perhaps because of changing ideas about the place of the individual in the afterlife. In Babylon and Sumer an interesting distinction was made between the king or other exalted person, buried in a sleeping position, and the servants, who after being killed were often buried in a crouching position, ready to respond to any royal command.

Judaism

Classical Judaism held that life alone matters, and death closes the book. 'A living dog is better than a dead lion,' says the Bible (Eccles. 9:4). But as the belief took hold in ancient Greece that the dead would one day arise and enjoy everlasting life, it came to be important that some part of the body at least was preserved. Official Judaism developed the idea that there was a bone called Luz, just below the eighteenth vertebra, that was indestructible and never died. God used this bone to achieve resurrection of the body. After being breathed on by the divine spirit the Luz fused with other bones to form a new body that could then arise from the dead. But in 1543 the great anatomist Vesalius noted that the bone did not exist. Nowadays, Orthodox Jews in Israel bury their dead, on the day of death if possible, and in a shroud but no coffin. Liberal Jews in the UK, however, practise cremation, generally with no embalming or viewing of the body.

Christianity

Death has always been at the core of Christianity. The early Christians believed in a mass resurrection of the dead at the second coming of Christ, and therefore burial was indispensable for the corpse. For Christians, burial was the seventh work of mercy, the seventh of the acts of compassion towards fellow creatures (Matt. 25: 35–40). In seventeenth-century England graves were aligned in an east–west direction with the head on the west side so that when the corpse sat up on the day of resurrection it would be facing east.

Roman Catholics have been especially particular about who should be allowed a proper religious burial. People who were unbaptized, apostate (had renounced religion) or excommunicated, or who had died in a duel, were all denied burial in a Catholic churchyard, and there could be no burial rites or masses.

In England suicide was for centuries a criminal offence and until 1823 suicides (unless they were mad) could not be buried in consecrated ground. Their property could be confiscated by law and they were to be buried at a crossroads with a stake through the heart. The last of these suicide burials took place in 1823 at the junction of Grosvenor Place and King's Road, Chelsea. Certain primitive tribes have special disposal methods for the bodies of suicides, including decapitation or dismemberment; suicides are treated like other social deviants (sorcerers, murderers), war captives, or those struck by lightning.

In the seventeenth century stillborn babies were not always buried because of the belief that life did not begin until the moment of baptism. Hence during difficult labours baptismal syringes were occasionally used in desperate attempts to give the unborn but living infant this birthright.

Islam

For Muslims, too, the proper disposal of the dead body is by burial. The soul leaves the body via the throat, as long as a month after death, and undergoes a judgment conferring rewards (heaven) or penalties (hell). The dead person is ritually bathed, covered in a sheet, and buried as soon as possible, within twenty-four hours of death, a promptness that is advisable in hot countries. The head is turned towards Mecca, and so that the earth does not fall directly on the body it is overlain with a few boards or placed in a 'chamber' hollowed out on one side at the bottom of the grave. Only Muslims should handle the body. There is no coffin and the corpse must not be violated by cremation, post-mortem examination or dissection. Hence Saudi medical students study anatomy on corpses imported from other countries. Organ donation, however, has been permissible since 1982.

Buddhism

Buddhists believe that the conscious principle remains in the body for about three days after death (longer in the case of holy men). The Buddha himself, a prince with a wife and 500 concubines, was transformed after seeing three things: first a sick man, then an old man, and finally a corpse. He then turned away from mundane things and for the next forty-five years was a teacher. Buddhists bury their dead in coffins, letting off fireworks to scare away evil spirits. Things that may be needed in the afterlife, like money and cars, are provided, but they are made of paper and are burnt during the ceremony. The initial burial is for up to ten years, after which the coffin is dug up and the bones cleaned, placed in a pot and reburied.

Hindus and Sikhs

Hindus are always cremated, as described in chapter 7. In modern crematoria it is the eldest son of the deceased who presses the button or sees the coffin into the cremator.

The Sikhs, who became separated from the Hindus in the fifteenth century, are also cremated. The hair must be left uncut, kept in place by the ritual comb, and the dead person should wear a steel bracelet on the wrist and a small symbolic dagger.

Is Burial Natural?

For those who bury the dead there has always been apprehension about any disturbances or violations inflicted on the cadaver. Perhaps the living imagine themselves in the position of the dead. Many poems express the hope that the earth will lie lightly on the corpse (or the ashes). The face in particular needs protection. To some extent this apprehension applied also to embalming and burning, as well as in the past to what were seen as outrageous abuses of the corpse such as dissection or post-mortem examination (see chapter 11). Queen Elizabeth I of England did not like the idea of her body being opened and worked on after death and said that she did not wish to be embalmed (although she was, in the end). To be buried, therefore, seems natural, in spite of the fact that the body must ultimately decompose and become unrecognizable. Burial was thought of as a slow and gentle as well as a natural process, although today we could look upon it as relatively violent when considered at the microscopic level.

However surprising it may seem to Westerners now, a large fraction of humanity has taken a different attitude to the dead body and found burial abhorrent. More than 4,000 years ago the peoples of Mesopotamia (Sumer, Babylon, Assyria) held out no prospect of an afterlife, and although they buried their dead they made no attempt to preserve the body. Hindus believed that the individual soul (*atmen*) is a mere fragment of an all-pervasive cosmic principle. It is temporarily trapped in the human body and can be released by fire. The dead body is offered to Agri, the god of fire, in a ritual incineration (see chapter 7). Children less than two years old, however, are buried, and ascetic holy men are buried upright with salt round the body. Tibetan Buddhism, which was absorbed from India in the seventh century, retains much of its original character and lays particular emphasis on death and dying. The *Tibetan Book of the Dead* (in the Evans–Wentz translation) came to the attention of the Western world in the 1930s and was adapted for Westerners by Soygal Rinpoche in 1992. Understanding death and accepting the ephemeral nature of life is believed to be central to spiritual well-being. Death is a doorway, not a blank wall, and the state of mind at death is important. The corpse is offered to birds (see chapter 7) as a final act of charity, and is not buried unless diseased or unsuitable for other reasons. Holy people, such as the Dalai Lamas, are mummified.

Simple Pit Burial

Throughout most of human history common people have been buried as they died, lightly clothed or covered, in a simple pit in the ground. Anything more complicated was reserved for important individuals (nobles, kings), who enjoyed more elaborate interment, sometimes complete with chariots, horses and full regalia. In most parts of the world burial is still simply a matter of digging a hole in the ground. Natural caves are made use of in western India and Sri Lanka, and at Abu Simbel in Egypt the caves have been extended 55 metres into the rock. The impressive burial chambers in Paraca, Peru, are hewn from solid rock, extend 5 metres below the surface, and can accommodate 400 corpses and their belongings. Tombs cut out of rock to accommodate the corpses are essentially troughs, but in some of the Etruscan chamber-tombs they are sculpted to look like beds, with pillows and legs. Rock-shelters and caves played a large part in burials throughout the last Ice Age in Britain, but as a site for a shallow earth grave in the cave where people lived, rather than as a rocky coffin. Radiocarbon dating of a skeleton buried in this way in the Paviland Caves, near Swansea, Wales, showed a date of 16,500 BC.

Before the eighteenth century in England, simple burial without a coffin was the lot of the poor. In the late eighteenth and nineteenth centuries a particularly ignominious resting place was in a pauper's grave. A pauper's funeral was greatly feared by common working people. Quite probably there would be no shroud and although there would be a coffin, it was the cheapest available. The grave was likely to be a communal one, with many coffins one on top of the other, up to within a few feet of the surface. Graves for these 'pit burials' were up to 25 feet deep, and the top coffins were easy targets for grave robbers. The shortage of land in churchyards meant that after ten years or so the bones could be dug up and transferred to a charnel house (from Latin, *carnale* = flesh) or ossuary (bone-house). But coffins, although expensive, are convenient. The carriers of the corpse avoid physical contact with the dead body, and not even its outline is visible. This is an example of the distancing of the living from the dead, referred to in chapter 15. Partly as a matter of charity and partly, perhaps, for convenience, parishes often had 'common coffins'. They can still be seen in some churches, as for example in Easingwold, Yorkshire. The poor enjoyed the dignity of a coffin until the moment when the body was tipped into the grave, the container being kept for future use. The modern equivalent is the so-called 'rent-a-coffin' obtainable from funeral directors in the USA and Canada.

Burial places can turn out to be unsatisfactory. In 1996 it was discovered that the noted Urdu poet Sheikh Mohammad Ibrahim was buried under one of Delhi's largest public lavatories. There are calls for the toilets, built in 1961, to be replaced by a less inappropriate monument.

Mass Graves

In times of war, after natural disasters, during plagues and after massacres (see also chapter 2), there are too many bodies for regular methods of disposal. There is no time (in war), or not enough healthy people (in plagues) to dig separate graves, and the dead are thrown into large pits that function as mass graves. After massacres the perpetrators are also in a hurry to cover up evidence of their evil act. After natural disasters the dead lie where they fell, and in unusual circumstances the event is permanently on the record. This was the case in the eruption of Mount Vesuvius in 79 AD (see chapter 7).

Plague

Outbreaks of the bubonic plague (Black Death) in Europe from the fourteenth to the seventeenth centuries placed a great strain on the provisions for corpse disposal (see also chapter 2). Bubonic plague was spread by people being bitten by fleas from infected rats. Everyone, rich and poor, was exposed. In London, with an average of one rat family living in each household and three fleas per rat, the number of dead overwhelmed public health services. In the year 1349, 20,000–30,000 died out of a total population of 60,000–70,000. Some of the outbreaks in later centuries killed people on a similar scale. The plague pit in Aldgate, London, was 40 feet long, 15–16 feet broad, and up to 20 feet deep. In 1665, between 6 and 20 September, it received no fewer than 1,114 bodies of plague victims.

War

During trench warfare in the First World War mass graves were an everyday occurrence, ranging from a handful of bodies in a shell hole to the interment of hundreds of dead soldiers after major battles. Officers were often buried separately from the ranks. More than a million died at Verdun and the Somme in 1916 (see chapter 2). The scale of the slaughter can still be seen in the war cemeteries, whole landscapes filled with endless rows of identical crosses. The Commonwealth War Graves Commission looks after about 1,750,000 graves on thousands of sites in more than 140 different countries.

A less bleak reminder of the lethality of war is provided by the Battle of Bronkurst Spruit in the First Boer War (1880). Before the battle the men of the 94th Regiment had raided a nearby orchard and stuffed their pockets and knapsacks with peaches. Afterwards, the dead were buried on the spot in their uniforms and ten years later it was noted that there were 'two uncommonly fine orchards' at that site (see J. H. Lehmann, *The First Boer War*, pp. 118–19).

Large numbers of decaying bodies, of course, give off evil smells. After the Battle of Sedan (1870), when the French emperor Napoleon III was defeated by the Germans, the hastily dug graves were filled to over-flowing with hundreds of corpses. As they rotted they began to emit offensive, pestilential fumes. Sedan was near the border with Belgium and the Belgian government called in a chemist, M. Creteur, to deal with the problem. His practical, down-to-earth remedy was just what was needed: he opened the graves, poured in tar and kerosene and set fire to the contents.

In recent wars it has been civilians as often as soldiers who have died in their thousands and whose remains have had to be disposed of by survivors. Examples include the 25,000 victims of the Allied fire-bombing of Dresden (1945), the 1.3 million victims of the German army seige of Leningrad (1941–2), and the 80,000 victims of the nuclear bombing of Hiroshima in 1945. One of the largest cemeteries in the world, in Leningrad, contains the bodies of 500,000 Russian war victims, and a mausoleum in Hiroshima Peace Park contains the unidentified and unclaimed ashes of tens of thousands of the Japanese dead. Another 180,000 Japanese war dead are collected in a tomb in Okinawa, Japan.

Before the days of air transport those killed in wars in foreign lands could only be brought home for burial after their bodies had been embalmed or otherwise preserved. Nelson's body was preserved in rum on board the *Victory*. More commonly the corpse was eviscerated, then boiled to free the bones, which could be taken home for burial. After the Battle of Agincourt (1415) the bodies of the Duke of York and the Earl of Oxford were boiled until the bones separated, and the latter were then shipped back from France to England and buried (see chapter 14). The Americans, since their Civil War, have always brought back the dead in 'body bags' for burial at home.

Massacres

Massacres are a recurring and terrible feature of history. Humans have a unique capacity for large-scale killing of other members of their own species. In the past the main pretexts have been war, usually fanned by ethnic, economic or religious flames, the expansion of populations into new territories, and religion. Religion certainly has a lot to answer for. In 1572, Catholics killed tens of thousands of Huguenots in Paris, and also at least 100,000 Albigensian heretics in thirteenth-century France under the authority of Pope Innocent III. There are well-recorded details of the cold-blooded, calculated genocide (systematic massacre over a period of time) of Jews and gypsies carried out by the Nazis during the Second World War (see also chapter 3). For example, about 1.2 million Jews were killed in 1942 in the extermination camp established by the Nazis in Maidanek on the outskirts of Lublin, Poland. All clothes and belongings were removed from the victims and later sorted, classified and exported to Germany. The naked men, women and children were then driven into dark concrete boxes about five yards square, 200–250 people in each box. Hot air containing crystals of Zyklon B (hydrocyanic acid) was showered down from above. The lethal crystals rapidly evaporated and everyone was dead in five to ten minutes. Zyklon B was used in the first gas chambers built at Auschwitz in January 1942. The Nazis had six of these gas chambers built so that 1,500 could be killed at a time. Corpses were loaded on to lorries and disposed of in the crematorium. The white ashes and small bits of bone, when added to manure, yielded luxuriant crops of cabbages. Other methods for extermination were considered but rejected. At a conference in the winter of 1941–2 there was discussion of a plan to drown 30,000 German gypsies by sending them out into the Mediterranean sea on ships and then bombing the ships. The sea would do the cleaning up.

Royal Burials and Palaces

A final type of mass grave is the sort that long ago accompanied the building of a new palace or the burial of a great king. It was usually single victims who were killed and buried to celebrate these occasions, but the practice was taken to extremes during the Shang period in China (1500–1028 BC). To mark the construction of the new palace, Hsaio T'un, a total of 852 humans, 18 sheep, 35 dogs and 15 horses were sacrificed and buried in the foundations.

Shrouds and Coffins
Shrouds

Thinking what it must be like to be dead, imagining oneself in the position of the corpse, it seems improper that the face and body should lie in direct contact with the earth. Some sort of covering is needed. In the prehistoric era (before written history, which began about 5,000 years ago) corpses were often buried without coverings, but this may have been because fabrics were not plentiful. In pre-dynastic Egypt skins or matting were used to protect the body from surrounding sand, and it was not until the early dynastic period about 5,000 years ago that coffins came into general use.

A rectangular pit burial at Ur (about 3000 BC) has the body wrapped in matting secured by a long copper pin. At a later period in England, Egypt and Peru cloth or reed mats were used. In more recent times corpses have been buried unwrapped only in unusual circumstances, for instance after battles, massacres or plagues, as described above.

Muslims bury their dead enclosed only in a shroud (*kafan*). The traditional covering in Europe has been a shroud or winding sheet. In recent centuries white was the preferred colour, and until the eighteenth century it was tied or sewn above the head and below the feet, completely enclosing the body. Sprigs of rosemary were often placed on the body before enclosure. Rosemary was not only a symbol of love and remembrance but had the added virtue that its smell counteracted any putrefactive odours from the corpse. Corpse-dressing goes back at least to Elizabethan days; other plants used included yew, rue, and box. The head was sometimes left uncovered until burial so that the dead person could be seen by mourners.

Shrouds were ideally made of linen, the material said to have been used to enclose the body of Christ, but in most cases an ordinary household sheet was used. In 1600 it became a legal requirement in England that winding sheets be made of wool (flannel), in an attempt to encourage the declining English wool industry. Another Act imposed a £5 fine for defaulters, but these Acts were repealed in 1814.

In the late eighteenth century the winding sheets were reduced in extent, and were attached to the sides of the coffin. They could then be pinned or sewn into place after the body was in the coffin. The corpse was less trussed up but still lay in an orderly fashion because the legs were tied at the ankles and the arms to the side of the body.

Coffins

The use of coffins (the word comes from a Greek word meaning a basket) goes back to prehistory. About 1,800 years ago a man of the 'beaker' folk (so called because of the characteristic pot they buried with their dead) was buried in Cartington, Northumberland, England. He was in a four-foot-long coffin hollowed out of the trunk of an oak. Making a stone coffin must have been even more demanding, but at about the same period another man was buried inside a sepulchral chamber of stone (cist) under a mound of earth (barrow). One type of coffin made of limestone was originally given the name sarcophagus (Greek 'eater of flesh') because it was thought to dissolve the flesh.

Reed coffins were used for humble people in early dynastic Egypt, and wood and reed coffins are known from the first dynasty. The ancient Romans and Greeks, however, buried their dead in shrouds, without coffins, and there is no exact word for coffin in these languages. Giant pots (*amphorae*), normally used as containers for wine or oil, served as a sort of coffin in ancient Rome and in parts of the Middle East 2,000 years ago. In the ancient city of Zarathan (Jordan) many people were buried with their heads only in pots; perhaps the poor man's version of the amphora, perhaps reflecting the wish to give special protection to the head.

Making a coffin takes time, trouble and resources. They have been disapproved of by those who maintain that dead bodies should be allowed to decay directly in the earth so that the products of decomposition can more readily be absorbed into the surrounding soil. In the Sussex town of Rye in England in 1580 the town council decreed that the poor should under no circumstances be buried in coffins or chests, and anyone making such a coffin was to be fined ten shillings. Coffins did not come into general use in England until the eighteenth century. They were usually made of wood, especially elm, and some of the the finest ones were made in England between 1725 and 1775. During the nineteenth century, when everyone wanted to be buried in a coffin, there were many to choose from. An 1838 catalogue from J. Turner of London offered no fewer than thirty-three different types, including fourteen for children only.

In the constant attempt to protect the corpse from destruction, people have made coffins consisting of several layers, generally of wood. Perhaps the record is held by the duke of Wellington, who in 1852 was buried in four coffins. For the corpses of kings and wealthy people there have been lead coffins, so heavy that six men were needed to carry them, and iron coffins. Zinc and lead-lined coffins are mentioned in undertakers' catalogues in the early twentieth century. Churches and cemeteries charged a higher fee than for wood, because metals did not rot and the ground could not be re-used for a long time. In more recent years an occasional steel coffin has been used, something that presents a problem in the crematorium, and eccentric individuals in their wills have arranged for coffins of bronze, glass or celluloid – even rubber. The cult of the coffin or casket reached a peak in recent years in the USA, where they are constructed, finished and polished with as much care as a piece of furniture intended for the sitting-room.

In the eighteenth century the body was generally placed in the coffin on a layer of sawdust, wood shavings or bran, sometimes with a mattress. This had the practical merit of soaking up the liquids of putrefaction. Perhaps it also seemed kinder, because living people could imagine themselves as corpses and felt that it was wrong for the body to rest directly on a cold, hard wooden surface. Not everyone wants something so expensive, and the 'rent-a-coffin' has an inner cardboard container in which the body is cremated, so that the outer wooden coffin can be re-used. Chipboard containers serving as inner coffins are available in the UK for £30–50. As a further, greener step towards simplicity, some use recycled newspaper.

As for shape, rectangular coffins modified to take the horizontal human form are commonest. Occasionally they have been moulded into a more human shape, the 'anthropoid' coffin. These were used by the Philistines in the twelfth-century BC, and Egyptian mummies were often placed in such containers. A cubic coffin is appropriate when the corpse is buried curled up, sitting or lying in the foetal position. The Buddhist or Shinto coffins of Tibet and Japan are of cubic shape, and the corpse sits or crouches in them.

Among the Aboriginal people of Arnhem Land, North Australia, the bones of special people, after decomposition, were cleaned, painted with ochre and placed in log-coffins. The log was from a tree that had been hollowed out by termites, and after ritual singing and dancing the coffin was left to the elements.

In Ghana, West Africa, certain master carpenters have recently been making coffins in every conceivable shape, to suit the customer: cows, chickens, lions, fishing boats. All are painted and lifelike, although some of them are not acceptable in local Christian cemeteries.

An irreverent comment on coffin shape comes from the English playwright Joe Orton (1933–67), one of whose characters remarks to someone of small virtue: 'You were born with your legs apart. They'll send you to the grave in a Y-shaped coffin!' *What the Butler Saw*, Act 1).

Coffins can be given a very personal flavour. George II, who died in 1760 (in the lavatory), left instructions in his will to be buried with his wife Caroline of Anspach. The sides of the two coffins were to be struck away so that their ashes might mingle. A more visible token of married love is provided by the tomb of the Earl of Arundel in Chichester Cathedral. The knight and his lady appear to be lying conventionally, side by side, with the traditional dogs at their feet, until you notice that their hands have crept across and are entwined. This image prompted Philip Larkin's poem 'An Arundel Tomb'. Another more obviously hand-in-hand tomb is that of Ralphe Greene (d. 1417) and his wife at Lowick, Northamptonshire, England.

The Fear of Burial Alive

Once you decide it is a possibility, it is easy to become obsessed with the fear of premature burial. In Edgar Allen Poe's short story 'The Premature Burial' this preoccupation is spelt out in terrifying detail. The subject, who suffers from catalepsy (seizures and trances) and has often worried about being buried alive, wakes up on one occasion and is at first convinced that he has undergone this fate. He is in a closed container, he can smell soil, and he can feel the undertaker's bandage round his jaws. Then he remembers that he is in fact on board ship in a very narrow berth, the smell is from the garden mould the ship is carrying, and he had tied a silk handkerchief round his head in place of the usual nightcap.

Fears of premature burial, fuelled by horrifying eye-witness accounts, reached a peak in nineteenth-century England. One or two of these accounts were probably true. Fear of premature burial was even given a name (taphephobia) and 120 books in five different languages were written about it and the methods of distinguishing life from death. One of these, written by Franz Hartmann in 1895, consisted of a collection of more than 700 accounts of premature burials. The evidence included noises from coffins or vaults (sounding, for instance, like the hissing of geese), the finding of bodies displaced, shrouds torn, splinters found under fingers in presumed desparate attempts to escape, and even tales of people giving birth after burial. Possible explanations include rigor mortis leading to changes in the position of the body, or sounds due to corpses bursting open or escape of gases (hissing of geese) from the putrefying abdomen. The book was dismisssed with a scathing review in the *British Medical Journal* of 29 February 1896, but people were easily persuaded that there was a real risk of premature burial.

There is probably a core of genuine cases. Rapid burial was common during major pestilences such as cholera, plague and smallpox, and occasional interment of a still-living sufferer was likely. There are, moreover, a few medical conditions that might look like death, such as catalepsy, coma and hypothermia. The heart would still be beating, and an occasional breath taken, but these signs of life could be difficult to detect.

The strong and widespread fear of premature burial was one of the reasons for waiting a few days before burial, just in case the corpse showed signs of life. In hot climates, however, rapid putrefacion made this difficult to carry out. Municipal arrangements were made in nineteenth-century Munich, which had ten 'waiting mortuaries'. Here, bodies were kept for seventy-two hours before burial, in close proximity to antiseptic fluid and camouflaged with sweet-smelling flowers. The fingers, tied to an elaborate system of cords and pulleys, caused a bell to ring in the porter's lodge if the corpse moved. False alarms were common.

To deal with the problem *after burial* more complicated pieces of apparatus were available, for the wealthy. In 1896 Count Karnice-Karnicki invented a device consisting of a tube passing vertically into the coffin and attached to the chest of the corpse. The slightest movement at the end of the tube activated the ringing of a bell and the display of a flag above the ground. Francis Douce, an antiquarian and collector who died in 1834, had a more drastic strategy for avoiding premature burial. His will contained these words: 'I give to Sir Anthony Carlisle two hundred pounds, requesting him to sever my head or extract my heart from my body so as to prevent any possibility of the return of vitality'. Carlisle was a well-known surgeon (Litten, 1992). Many others arranged for arteries to be severed before burial, just in case.

One notes that a guaranteed method of avoiding premature burial is to be cremated, although it has to be admitted that the viable (but hopefully unconscious) corpse would then be burnt alive.

Delayed Burial and Reburial

Most burials take place shortly after death, but sometimes after a preliminary period of decomposition or artificial treatment. The Dyaks of Borneo kept the body of a dead chief in the longhouse with the people, treating it as if it were still alive, providing it with food, drink and company before burial. Corpses exposed in trees ('tree burial') or in Parsi Towers of Silence (see chapter 7) can be 'reburied' once they have been reduced to bones by maggots and vultures. Tree burial was carried out by the Naga people in India, the Sioux of North America, the Melpa in the Highlands of New Guinea, and certain Aboriginal groups in Central Australia. Burial after mummification is another possibility. Greek Orthodox Church burials are for an initial three to five years. After this period the body is dug up at a family ceremony with the priest in attendance. The bones are taken and washed (sometimes in the Greek wine retsina), left to dry in the sun, and finally put in a casket and placed in a mausoleum of the columbarium type (see below). This seems a practical method for letting nature do the preliminary cleaning up of the corpse. It takes a few years for the flesh to rot away from the bones.

One of the oldest reasons for delayed burial is to make quite sure that the person is truly dead, as described above. Another practical reason is when the death is in a foreign country and the corpse is to be shipped home for burial. In the old days before the refrigerator and the aeroplane, something had to be done to avoid decomposition en route. The Bishop of Hereford died in Italy in 1282. His body was cut into pieces and boiled in vinegar until the fat and flesh had separated from the bones. The skeleton was then shipped home for burial and the rest deposited (a 'flesh burial') at the place of death. Similar treatment was needed for Henry V after his death in Normandy in 1422. The flesh, the fat and the bones were sealed, together with assorted spices, in a leaden case, and carried back to England. The final burial took place in Westminster two months later. In both of these examples we assume that the viscera were disposed of immediately. Further instances where different parts of the body ended up in different locations are described in chapter 14.

Death in foreign lands has become commoner with the growth of *tourism*. Between 1973 and 1988 a total of 952 Scots died abroad, the young ones mostly in accidents and the elderly mostly of heart attacks; their bodies, embalmed or refrigerated, were brought back to Scotland for cremation. Delayed burial is very occasionally due to confusion or uncertainty about the place of burial, as in the case of Charles Darwin (see box).

Sometimes the corpse is truly reburied, that is to say dug up and then replaced in another grave. According to Buddhist beliefs, the conscious principle remains in the body for about three days after death and only then can the soul enter into the land of the dead. A second burial, with accompanying ceremonies, marks this occasion. The Balinese reckoned that the soul had left the body after forty-two days, and the buried (or mummified) corpse could then be cremated. The second burial often becomes the definitive burial, and is accompanied by all the funeral ceremonies. Periods of as long as six or seven years between primary and secondary burials were recorded in a Chinese village in the 1970s.

Secondary burials seem to go back a long way in time. In a study of seventeen neolithic burials in Franchthi Cave, Greece, the disposition of the bones, in simple pits, showed that while children were buried soon after death, adults had been buried after a period of temporary burial or exposure. We do not know why this was done. Perhaps there were ceremonies to be carried out on the cleaned-up corpse. Beliefs and customs of early humans are wrapped in mystery. Nothing but a time machine would give us a full understanding of these things. In neolithic burials elsewhere it is common for the mandible to have been removed fron the

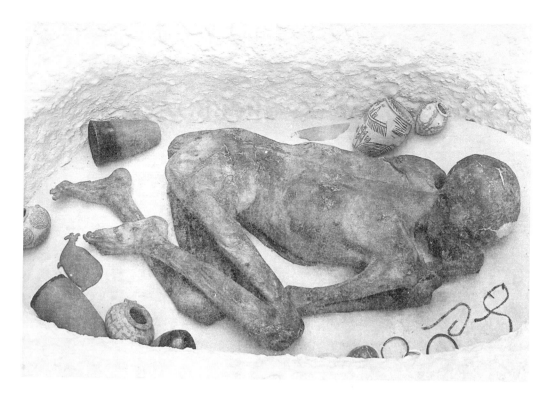

Above: Sand dried body surrounded by artefacts. From Gelebein, about 3300 BC.
Below: Reconstruction of a chariot burial excavated at Garton in the Woods, North Humberside, in 1985.

Left: Scenes of the plague in London; nocturnal burial of plague victims. Chalk drawing by Edward Matthew Ward (1816–1879).

Below: Dissection of a woman, directed by J Ch G Lucae (1814–1885). Chalk drawing by Hasselhurst, 1864. Part of a study for students of Anatomy and of Art; the body is that of an 18 year old suicide victim.

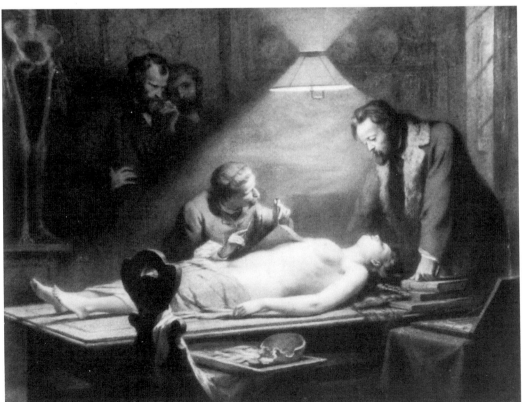

Above: Ossuary in crypt of St Leonards Church, Hythe, Kent, England. The crypt contains about 2000 skulls and 8000 thigh bones, churchyard remains dating from before 1500. *Below left*: Surgeon struggling to rescue a naked woman from the embraces of death. Etching by Ivo Saliger (1894–1986). *Below right*: A woman stealing the teeth of a hanged man. Etching with aquatint by Goya (1797). In the C18th–early C19th false teeth were often obtained illegally in this way.

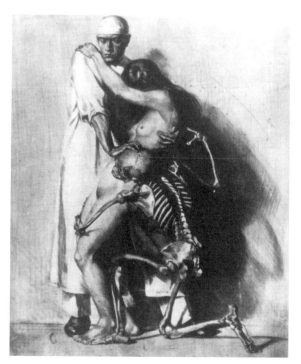

Face half dead, half alive. A device used in C18th–C19th European art; one half of the image alive and well dressed and the other half dead and decaying. C18th oil painting on copper.

skull. Sometimes the skull only is retained. The Melpa people in New Guinea kept ancestral skulls in 'skull houses' until 1964 when, following the teachings of the Christian missionaries, they began to use coffins.

A last-minute place in Westminster Abbey for a freethinker

Charles Darwin (1809–92) was one of the greatest scientists who ever lived, the propounder of the theory of evolution. He was a naturalist and an indefatigable collector, and at the age of twenty-two jumped at the opportunity of sailing round the world on a three-masted ship *The Beagle*. It was a tiny ship, 90 feet long and with only two cabins, and he shared his with an officer. It was a mere 10 × 11 feet in size, and he, at six feet, had to stoop to enter it. They sailed on 10 December 1831 and the voyage took five years. *The Beagle* visited South America and the Pacific, and the enthusiastic Darwin came back armed with:

> a 770-page-long diary
> 1,751 pages of notes on zoology and geology
> 1,529 biological specimens in spirit
> 3,907 labelled skins, bones and other dried specimens.

He now settled down in the village of Downe, Kent, married his cousin and began to develop his theories. The epoch-making book *On the Origin of Species* came out twenty-three years later, in 1859. He also did experiments in his garden, some of which formed the basis for a well-known book about earthworms.

Darwin died at home on 19 April 1882, and he had expected to be buried with his infant children in the local churchyard at Downe. But his friends started a campaign to get him buried in Westminster Abbey. His work had brought honour to England and he was a landmark in English history, so it was only fitting that he be buried with other great Englishmen, not in an obscure grave in the countryside. Letters were written, with appeals to the government and the Dean of Westminster Abbey. The campaign gathered strength, and it was finally successful.

Now there had to be a sudden switch of undertakers, and a smart, expensive coffin instead of the rough oak box in which the body still lay. A big funeral was hurriedly organized, with full ecclesiastical pomp. Friends and relatives, the great and the good (but not the Queen or the Prime Minister), and 'the greatest gallery of intellect that has ever been brought together in our country' were in attendance. And so, a week after he had died, Darwin the freethinker, the scourge of the bishops, was buried in Westminster Abbey close to other great men such as Sir Isaac Newton the mathematician, and Sir John Herschel the astronomer.

Second burials can be for purely practical reasons. In South American and European countries the reason has usually been the shortage of burial space. The grave is leased for a number of years, after which the body can be dug up (exhumed) and the bones either reburied in a common grave or placed in a niche or ossuary. The bodies of the poor, as would be expected, are more likely to end up in this way. The New Poor Law in England (1834), which delivered the ultimate blow to working people, made strict rules to reduce the cost of funerals that were supported by public funds. Common graves, which were often on land attached to workhouses, were crammed with coffins up to twenty deep. Only numbers, not individual names, marked the burial site and quicklime was applied liberally. The bones are said at times to have been sold for fertilizer and the coffins chopped up for firewood. Details of the indignity, anonymity and shabbiness of the pauper's funeral are given in Ruth Richardson's book *Death, Dissection and the Destitute*. As late as 1938 nearly one in ten dying in Greater London were buried in paupers' graves by Public Assistance Committees.

In eastern Europe in the eighteenth century those suspected of being vampires were sometimes dug up after burial. If the telltale signs, many of which sound to us today like normal post-mortem changes, were present, the vampire was killed by driving a stake through the heart. The digging up of corpses and selling them for anatomical dissection is described in chapter 11.

For the Victorians in particular, burial signified a final laying to rest, a desirable and peaceful end for the body. People wanted to be buried deep enough, in a quiet place, and certainly did not want to be disturbed. Alfred Lord Tennyson expresses the anxiety about deep burial in a peaceful place in the poem *Maud*:

> Dead, long dead,
> Long dead!
> And my heart is a handful of dust,
> And the wheels go over my head,
> And my bones are shaken with pain,
> For in a shallow grave they are thrust,
> Only a yard beneath the street,
> And the hoofs of the horses beat,
> Beat into my scalp and my brain,
> With never an end to the stream of passing feet,
> Driving, hurrying, marrying, burying,
> Clamour and rumble, and ringing and clatter;

And here beneath it is all as bad,
For I thought the dead had peace, but it is not so
To have no peace in the grave, is that not sad? . . .
O me, why have they not buried me deep enough?'

The case of Dante Gabriel Rossetti's wife, Elizabeth Siddal, is a unique one. She had died in 1862 (of a laudanum overdose) and he had placed a book of manuscript poems with her in her coffin in Highgate Cemetery. Years later he wanted to publish the poems but had copies of only some of them, so in 1869 he got permission from the Home Secretary to open the grave. His lawyer was needed to answer the objections of the cemetery authorities. When the coffin was opened and the book lifted out, some of his wife's golden-red hair came away with it. The book, which had an evil smell, was immediately saturated with disinfectant and later dried page by page. Rossetti himself was not present, having 'stayed away in a state of agitation and torturing suspense'. The poems were published in 1870.

Nowadays a profusion of regulations controls the digging up of corpses and the opening of coffins. When it is suspected that the cause of death was unnatural, the coroner can have the body exhumed for further examination, or to test for poisons. Arsenic remains detectable for centuries. On rare occasion relatives want to rebury someone in a different location, and they must then obtain permission. There is still an obligation to rebury or properly dispose of human remains accidentally dug up. Old graveyards may be disturbed, for instance during excavation for new buildings. The bones are retrieved, placed in bags, then cremated or reburied. Firms with names like 'Necropolis Incorporated' specialize in this type of work.

Agreements reached in 1997 seem likely to lead to the opening of mass graves in Cyprus. The Turkish invasion in 1974 split the island into two parts, and it is estimated that about 1600 Greek Cypriots and 800 Turkish Cypriots who are missing lie buried in these graves. Returning the bodies to their families will be painful but is expected to help improve relations between the two communities.

Funerals

Funeral practices have varied greatly over the centuries. They are of immense interest and can be thought of as chapters in religious, social and economic history. In ancient Egypt, as elsewhere, the size of the funeral procession depended on the wealth and status of the deceased. Most burial places were on the west bank of the Nile, but nearly all the people lived on the other side, so a funeral procession had to make a river crossing, and there are many illustrations of mummies and coffins on boats. Many of those in the procession would have been servants carrying flowers, food and drink for the funeral feast, and also the dead person's bed, clothes, jewellery and other possessions to be deposited in the tomb with the mummy. Often the mummy could be picked up from the embalmers on the west side, after the river crossing, en route to the tomb.

Funerals are not necessarily mournful occasions. The Romans arranged games, free feasts, even gladiatorial combats, at the funerals of prominent people; 120 men fought at the funeral of Publius Licinius in AD 183.

Wealthier, more important people have more sumptuous funerals, although there are exceptions. When Charles de Gaulle died in 1970 he was placed in a plain wooden coffin, made by the local carpenter, and buried in a small village cemetery with only his own family, friends and neighbours in attendance. This can be contrasted with the pomp and magnificence of royal funerals in Europe, or the great ceremonies that must have accompanied the burial of Egyptian kings. The labour and resources taken up in building the larger Egyptian pyramids were a major drain on the economy of the country.

Undertakers appeared in England in the seventeenth century. Before then the laying out and preparation of the body had been the responsibility of the family, and was usually done by females. The first undertaker was William Boyce, who opened his shop near Newgate in London in 1675. His trade card announced that 'You may be furnished with all sorts and sizes of coffins and shrouds ready made. And all other conveniences belonging to funerals'. Towards the end of the nineteenth century, when undertakers were providing much more than the mere disposal of the dead, they began to be known as 'funeral directors', and they now belong to the British Institute of Funeral Directors. It is a well-regulated profession, with a National Association of Funeral Directors and its own Code of Practice, as well as a Funeral Standards Council which has apponted a Funeral Ombudsman.

Grand funerals for grand people have always been expensive. The elaborate funeral procession for Sir Philip Sidney on 16 February 1587 was recorded in an engraving by Johann Theodor de Bry. The engraved plates were printed in the form of a roll 38 feet long illustrating the 700 participants in their different costumes. The funeral was so expensive it was said to have bankrupted its sponsor. Spectacular processions helped to sustain a great man's memory, although in this case the motives may have been partly political. By the beginning of the seventeenth century in England, funerals for the wealthy had become expensive displays of private wealth. There were special arrangements for the procession, for the clothes to be worn and for heraldic items such as standards and armour, organized by the College of Arms. In addition to the bearers of the coffin there were the pall-bearers, who carried the heavy canopy or pall which lay over the coffin. There were banquets, mourners and other attendants, candles and other accoutrements to be organized and paid for. Elaborate instructions were often included in the wills of the wealthy, both for the funeral and for the details of the tomb. Sir Francis Page, who died in 1741, went so far as to leave money for his tomb to be regularly dusted. Provision for events after death can not have been more precise, or more cruelly calculated to have effect, than in the case of the nineteenth-century Frenchman who committed suicide after being betrayed by his mistress. He ordered that the fat from his body be converted into a candle, and that the lighted candle be carried to her, together with a letter saying that he had burned for her during his life and would continue to do so after his death. The fire of his passion would be illustrated by the light of the candle by which she read his letter.

The Puritans campaigned against the unseemly expense and pomp of funerals, and between the beginning of the seventeenth century and the middle of the eighteenth, cheaper nocturnal funerals became common. They took place between 9.00 p.m. and midnight and expenses were reduced to attendants and candles.

In Victorian England the funeral came to be a more important ritual for ordinary people. The rich would have a funeral that reflected their social and financial status, but the poor wanted above all a decent, respectable funeral. Death in the workhouse was the ultimate horror. But funerals cost money. After the Anatomy Act (1831) the various burial societies expanded, providing a personal insurance towards the cost of burial, with weekly contributions of from a farthing to threepence. By 1842 the Manchester Unity of Oddfellows had 220,000 members, many of them from the labouring classes. The ancient Romans had also had burial clubs through which contributions were paid towards funeral costs, and which also organized social activities.

In the UK in 1996 the total cost of a funeral plus cremation was about £1,000. A whole body burial is more expensive because of extras for the gravediggers' fee, the church service, and a larger space in the churchyard or municipal cemetery. The average price for the cheapest plot in a cemetery (1997) is £500, including digging. You can buy the 'grave right' of undisturbed burial for up to about fifty years – it used to be in perpetuity. But many churchyards have room only for caskets or urns containing the ashes. It is slightly cheaper if the ashes are scattered loose into a shallow hole. Woodland burials (see the end of this chapter) are the least expensive, and they are eco-friendly.

Is the funeral ceremony doomed to extinction? Probably not, because there is a need for such rituals implanted deep in the human psyche. So many societies have had them. We no longer have old-style puberty initiation rites, the times of sowing and harvesting concern only a handful of farmers and local communities, and christenings with religious overtones are becoming less common. But we still have weddings, we have ceremonies for handing out degrees and diplomas, and we have funerals. These rituals mark the stages of life and strengthen social bonds. Funerals, or assemblies to think about a dead person and celebrate their life, will always have a special role. To see the corpse, especially after the embalmer has been at work, to grieve, to meditate at the sight of it, is often to feel more comfortable about the death. Laying out the corpse so that it looked presentable, washing it, wrapping it in a winding sheet or shroud, was in itself a therapeutic ritual; it also ensured that the corpse was in a suitable state for viewing by kindred and friends. Viewing the corpse, as discussed in chapter 5, used to be a widespread custom in England, though it is now less common. Also, since at least the fourteenth century, there had been the custom of watching the corpse day and night during the period between death and burial. This would have served the purpose of keeping corpse robbers away, but it had been an established ritual long before the days of dissection. It was a ceremony, and one watcher was enough. With the advent of mortuaries, where the corpse could be stored before burial, the watching vigils became less common. Funeral wakes had a social function and took place on the eve of the funeral. Together, the watching and the wakes gave opportunities for thoughts about the dead person and the expression of grief, and also for purely social activities, joking, eating, drinking.

The subject of funerals has been dealt with extensively in other books and articles listed in the references at the end of this book. An up-to-date guide to funerals in the UK, including do-it-yourself funerals, is given in *The New Natural Death Handbook* (edited by N. Albery et al., 1997). For the unbeliever, an excellent booklet entitled *Funerals without God* can be obtained from the British Humanist Association, 14 Lamb's Conduit

Passage, London WC1R 4RH. It gives practical advice about dignified, non-religious funerals. It is necessary to make these arrangements before death rather than merely mentioning them in a will, which may not be read until after the funeral has taken place.

Burials at Sea

Burial at sea seems at first sight the simplest and cleanest method. The body is exposed to rapid and efficient destruction by a variety of invertebrate and vertebrate feeders, including sharks in warm seas. It is spared the slower, odoriferous putrefaction of the buried corpse. But this brings its own problems. There is nothing left for the bereaved, not even ashes, and there is nothing to mark the place.

In the past, burial at sea was quite common. It was unavoidable in the days of sailing ships because of the decomposition of the untreated, unrefrigerated corpse. The diary of the Revd A. Fraser of Cawdor gives an eyewitness account of the burial of soldiers who died on board ship while being invalided home from the Crimean War in the mid-nineteenth century: 'A shutter [wooden screen] was laid over the ship's side. On this the corpse was laid sewed up in a blanket, within which at the feet a cannon ball was rolled up – the whole was covered over during the religious service with an old flag. At the words "we consign his body to the deep" two men raised up the end of the shutter, when another laid hold of the flag, and the body shot off.'

The sea is a natural and convenient resting place for certain seafaring peoples and for those who live on small islands. Bodies, weighted down with stones, can be consigned to the deep waters. Sometimes there is a greater significance in sea burial. Norsemen and some Pacific islanders believed that when a chief died he sailed away and would one day return. Accordingly, the body was placed in a vessel, in the case of Pacific islanders a canoe, and sent off on its journey. The Norsemen sometimes included bitumen or tar in the package and set fire to it at launching, so that the ceremony was really a combined cremation and sea burial.

A tradition of ship burials developed in Scandinavia in the second and third centuries AD, and many of them have been excavated in Norway, on the Baltic island of Bornholm, and in places where the Vikings settled. The great ship burials were reserved for people of importance, but many ordinary folk were buried in small boats. Thousands of ship graves have been discovered. In the great ship burials, however, the ship was positioned in the earth, often some distance from the sea. For example, at Sutton Hoo in Suffolk, England, a royal Anglo-Saxon ship-burial took place in AD 655–70. The kingdom had already been converted to Christianity, but the custom had been introduced from South Jutland two centuries earlier. When it was excavated in 1939 it was found to be a large wooden seagoing ship and 26 metres long, with places for thirty-eight rowers. Yet the site was about a kilometre from the nearest river. The ship contained valuable buckles, gold clasps, and thirty-seven gold coins. Surprisingly, there were no signs of human remains.

The explanation for the elaborate ship burials in the earth is not entirely clear. It does seem reasonable, from a purely practical point of view, that such an important man with such a rich collection of artefacts should be interred in the ground rather than cast adrift in the sea where all would be sunk, lost, perhaps plundered. Indeed, some ship burials seem best regarded as burials in which the coffin happened to be a seagoing object. At Loose Howe, near Rosedale Head, Yorkshire, there is a burial consisting of a few dug-out canoes 9 feet long, laid together to form a coffin and containing human remains. This dates at about 1500 BC. The so-called 'cavalry' burials in Denmark perhaps come into the same category. High-ranking males were buried with stirrups, spurs, weapons and sometimes a horse, while females were placed in carriages serving as coffins.

Those who go down with a ship when it sinks have a watery grave, and the fate of the corpses is much the same as in those who are buried at sea or who drown after falling overboard. The body undergoes the natural process of marine decomposition. Things are different when the body is more or less sealed in the ship and inaccessible to scavenging fish, shrimps and other creatures – in sunken submarines, for example. The ocean floors are littered with submarines deposited during the First and Second World Wars. Each one contains an entire crew, serves as an aquatic tomb, and is officially classed as a war grave. For instance the German U-boat UB30, identified by a diver in 1993, lies 160 feet below the surface off the coast of Whitby, North Yorkshire. It was sunk in 1918 and acts as a tomb for the twenty-two sailors inside it. Divers can now plunder such wrecks for souvenirs, the modern aquatic equivalent of the grave robbers of the Egyptian pyramids. When the wreck of the *Titanic* was located at the bottom of the Atlantic in 1992 there was a strong feeling that divers were disturbing a peaceful grave.

In the UK there are complex rules about sea burial. The corpse cannot be deposited wherever you like; for one thing it might be trawled up by fishermen. A licence nust be obtained from the Ministry of Agriculture Fisheries and Food and from the Fisheries Inspector. There must be no embalming, the body must be weighted down with at least 100 kg of metal or concrete, and a steel band is required round the coffin in case it bursts open when it hits the water. Burials at sea are quite expensive.

Tombs and Monuments

Whether the corpse is buried in a coffin, shrouded, or just as ashes, there is generally something to mark the spot. This was no more than a wooden board ('leaping board') in old English churchyards, tombstones and headstones being rarities before the late seventeenth century.

Tombs and monuments reached their peak, as far as size and complexity goes, in ancient Egypt. Initially, in the first and second dynasties (3000–2680 BC), the tombs were underground with unimpressive structures above ground. Later (in the fourth dynasty) there was a low building with a rectangular ground plan and a flat roof, called a *mastaba* (= 'bench' in Arabic). The burial chamber moved deeper and deeper and the superstructures bencame more elaborate, full of rooms and corridors. The first pyramid, that of King Djoser (2650 BC) was a step pyramid, and was built at Saqqara. It has six steps and rises to a height of 60 metres, and was enclosed within a wall, together with its complex of funerary buildings. A hundred years later (2550 BC) the first true pyramid, with smooth sides, was built: that of King Sneferu. The ones of worldwide fame are the three great pyramids of Giza, a few miles from Cairo on the west bank of the Nile (see box). Pyramids remained the burial place of Egypt's kings for nearly 1,000 years. Surrounding the pyramid would be a number of buildings, including the pyramid temple. Their complex architecture, with sealed passages and false doors, arose partly in response to the threat of tomb robbers (see chapter 9). Pyramids were abandoned in the eighteenth dynasty (1567–1320 BC). The Egyptians had been building them for an astonishing 3,500 years, which reminds us of the extraordinary stability of their society.

The next type of royal tomb was a rock tomb, less conspicuous, and cut into rock. King Tuthmosis 1 (1520 BC) was the first to be buried in this way. Rock tombs had staircases, corridors and chambers, and ran for hundreds of feet into the rock. The associated above-ground temples were built miles away to avoid giving a clue as to the burial site.

People from Sudan conquered Egypt in the seventh century BC, and started to build their own pyramids in Sudan. The last of their pyramids, dating from the third and fourth centuries AD, are very small erections of brick.

Pyramidal structures are also seen in Mexico. The earth pyramids of the sun and moon are in Teocuitarlan, and although they are little more than great mounds of earth, the sun pyramid measures 700 feet on one side of the base, and rises to about 200 feet. They were probably crowned with temples, and were places for worship and sacrifice rather than funerary buildings.

The Babylonians made large temple-towers of clay, bricks and mortar, called ziggurats. These had up to seven terraces, with stairways between them and trees and plants on the upper terraces, watered by special water-raising devices. The best-preserved, at Ur, is 70 feet high; its base measures 200 × 150 feet. The best known, however, is the Tower of Babel. These, too, were not true burial places.

The Pyramids of Egypt

In Egypt today there are fifty or sixty pyramids still standing. Those of greatest fame are the three pyramids of Giza, a few miles from Cairo on the west of the Nile. The largest is that of Khufu or Cheops, son of Seneferu. Each side of the base measures 455 feet and the pyramid is oriented according to the points of the compass. It was originally 481 feet high, covered an area of 13 acres, and took 22 years to build. The core consists of more than two million blocks of stone, each weighing about two tons. This was faced with blocks of fine limestone, but the latter have been mostly removed, leaving the great steps formed by the underlying stone blocks.

The second pyramid is of Khufu's sucessor Chephren, and is slightly smaller, and the third, of Men-kau-Ra, is smaller still. In a nearby mortuary temple, where the royal corpse underwent ceremonies of purification, are five boat-shaped pits which held wooden boats needed by the king on his journey to the sky.

Why did the ancient Egytians build pyramids? It had something to do with their wish for permanent and impressive tombs. It also reflected the stability of their society, the supremacy of the rulers, the regularity of the Nile water, the flatness and fertility of the land, and the idea that this pattern of life was pre-ordained by the gods. They resisted change. There have been suggestions that the sides of the pyramid symbolized the rays of the sun. The Egyptian word for pyramid probably means 'place of ascent', up which the dead king could travel on his journey to the sun-god in the sky.

The Romans made a great cult of the dead, and set much store by tombs, monuments and funeral ceremonies. Money was left in wills to pay for stone inscriptions or for the eating and drinking rituals of the funeral. Over the centuries a great variety of tombs were built. Some were underground, cut into the earth or rock, into hillsides or cliffs. They might be covered with a tumulus (mound) of earth, up to 50 metres in diameter, true Roman barrows. Others were built above the ground, using masonry blocks. Near cities the number of individual tombs was so large that there were rows of them, arranged in streets, with squares, and forming a necropolis or 'city of the dead'. A large necropolis runs for 70 metres under the nave of St Peter's Basilica at the Vatican in Rome.

Great burial mounds were also constructed by less developed societies such as those of the neolithic and early Bronze Age in western Europe. These were hills of earth or stones (barrows, cairns, tumuli, schists) and are of immense interest to archaeologists. At least 10,000 megalithic tombs of this type have been identified in Sweden, Denmark and France. Comparable edifices were built by the inhabitants of North America 2,000–3,000 years ago. *Megalithic tombs* were communal burial sites, used over many generations. The organized labour needed for their construction shows that they played an important part in social life, and although their exact function is not known we should perhaps regard them as monuments to a lost faith, equivalent to the churches of Christendom. Like churches, they may have served religious or magical purposes as well as the accommodation of the dead.

Silbury Hill, the mystery mound

Silbury Hill, in Wiltshire, England, is the largest prehistoric mound in Europe. It stands 39.5 metres (130 feet) high, and is surrounded by a 6 metre (20 foot) deep ditch. The people who built this colossal conical structure, more than 4,000 years ago, had to move an estimated 350,000 cubic metres of chalk, using only stone or bone (shoulder blade) shovels and pickaxes made of antlers. This means about 18 million man hours of work, or 700 men working for 10 years. So far no corpses or burial chambers have been found, and the part it played in human life remains a mystery.

The nearby West Kennet Long Barrow, however, which is 100 metres long, with stone chambers in the first 12 metres, contains skeletal remains of forty bodies. Evidently it was only the bones of important people that were buried here, probably after the flesh had rotted away, and to the accompaniment of unknown rituals. This long barrow dates from around 3700 BC and burials here ceased about 2000 BC, when they were abandoned in favour of smaller round barrows.

Barrows (from an old English word *beorg*, meaning a hill or mound), and their shapes, sizes and meaning have always fascinated archaeologists. There are 30,000–40,000 barrows in England alone. But it can be argued that they are *not* basically burial grounds.

An Assortment of Strange Tombs

In prehistoric times the people of Malta were greatly preoccupied with death. Dotted over the island are twenty groups of huge stone temples, some of them connected with ancient burial sites. One of these sites, excavated in 1902 at Hal Saflieni, near Palwa, Malta, appears to be a temple for the dead. It consists of an underground burial complex, cut into the stony ground, with thirty-two chambers containing thousands of human bones. About 6,000–7,000 individuals are buried there, and it dates from 3500–2500 BC.

On the neighbouring island of Gozo is another burial ground, this time on the site of a megalithic enclosure, the Brochtorff Circle. Here the burial chambers have been constructed in a series of natural caves. As so often happens in burial sites, decaying bodies were removed from their initial positions to make room for fresh corpses, and the disarticulated parts placed in mass burial pits. These contain thousands of bones, some of them sorted and stacked so that skulls are in one place, femurs in another, and so on. Presumably special rituals were carried out at the time of burial and during the later rearrangement of the dead. We are reminded once again that we know almost nothing about the religious life and beliefs of prehistoric people.

Among the grave goods in these Maltese burial sites are clay and stone figurines of obese people, usually females. Anthroplogists have suggested that they are examples of a widespread cult of fertility. The figurines are more elaborate and more numerous than elsewhere and it seems that in prehistoric Malta the worship of corpulent images blossomed into a consuming passion.

The Merina people of Madagascar put a great deal of effort into their tombs. The very structure of this traditional society, with its systems of interlocking groups of individuals, is permanently fixed by being built into the countryside in the form of tombs. The tombs are massive buildings, partly underground, made of heavy, expensive blocks of stone. Each village has five or six tombs, and people spend five to ten times as much on their tombs as on their houses. Even the children talk about their tombs.

The largest tomb in the world is the Mount Li tomb for Zheng, the first Emperor of China. It is near Xianyang and was made in 221 BC. It contains an army of about 8,000 life-size terracotta soldiers. A more recent tomb, in Okinawa, Japan, contains about 18,000 Second World War dead.

Cemeteries, Catacombs and the Columbarium

A cemetery is a place of collective burial, and comes from a Greek word
meaning a dormitory or resting place. Although occasional cemeteries
date from the neolithic or early Bronze Age, they are generally a feature
of settled, well-populated communities. The Egyptian, Chinese and
Jewish peoples established cemeteries outside the walls of cities, partly
perhaps because of odours and hygiene. The Romans buried their dead
(as corpses or ashes) along the sides of roads leading away from towns
and cities, as for instance along the Via Appia Antica outside Rome or
the Street of the Tombs, Pompeii, and cremations were not permitted in
the cities. This was for hygienic and aesthetic reasons. Plutarch says that
Lysurgus the Spartan ordered the dead to be buried, wrapped in red cloth,
inside the city, so that Spartan youths became accustomed to the sight of
death and corpses. Catacombs were subterranean cemeteries, often cut
into rock. They were generally reserved for members of a certain religion,
such as the Jewish and the Christian catacombs under the Via Appia,
Rome. The catacomb of San Giovanni, Syracuse, is large and laid out like
a real city of the dead, with streets at three levels.

After the rise of Christianity many wished to be buried inside a church.
St Neot's church in Cornwall, England is said to have had 548 bodies
buried beneath its floors between 1606 and 1708. In seventeenth- and
eighteenth-century England family vaults became popular. These were
in essence brick-lined graves but were often large, holding up to fifty
coffins on shelves. They functioned as dynastic burial chambers.
Between 1729 and 1865 Christchurch in Spitalfields, London received
a total of about 1,000 coffins, placed in vaults. Tombstones were often
used to pave the church, which in some cases became little more than a
cemetery. It was not long before churches became too crowded for all but
the wealthy or the distinguished. Burials now overflowed into the
churchyard, which became a cemetery. Burial inside the church
(intramural burial) was finally prohibited in the City of London from
1850.

Before the late seventeenth century very few of the graves in
churchyards had tombstones and the position of the graves was not
marked. After this time an increasing number of noblemen and finally
common people had tombstones, and because of the increase in
population and the inexorable growth of cities and towns the churchyards, too, were eventually inadequate. For instance, the old churchyard
at St Andrews, Widford, Herts, was less than half an acre in size but had

received bodies continuously for 900 years. It became more and more crowded until finally in 1903 it closed, replete, with a total of at least 5,000 corpses.

Gardens came to be regarded as the ideal final resting place. Flowers have been placed on corpses since the time of the Neanderthals. Garden burials were suggested in the seventeenth century and early Quakers were often buried in their gardens. Wordsworth thought of gardens and rural scenery as appropriate places for graves, and in 1820 noted that urban cemeteries were inadequate spiritually. In the same year the first garden cemetery was opened in Liverpool.

Until late in the nineteenth century, cemeteries were peaceful places that could induce interesting thoughts about life and death (unless one was there, stricken with grief, to visit a loved one's grave). The poet Shelley wrote: 'the cemetery is an open space among the ruins, covered in winter with violets and daisies. It might make one in love with death, to think that one should be buried in so sweet a place.'

In France in the Middle Ages, church cemeteries were public places, and provided useful open areas in crowded towns and cities. They were places for strolling, meeting, playing games or selling goods. In other words, there was close and frequent contact between the living and the dead. In eighteenth-century England, too, cemeteries were public thoroughfares, places where people hung out their washing or emptied their chamber pots, and where domestic animals could graze, the parson having the rights of pasture. A fuller account of the burial problems of English and French cities is contained in the books by Claire Gittings, Philippe Aries, and Ruth Richardson listed in the references at the end of this book.

City Burials: Problems with Smells and Space

Aries describes how, in 1763, faced with complaints about smells and the perceived threat of epidemics, the Parisian authorities came up with a radical decree. All the existing church cemeteries in the city were to be closed down and eight municipal cemeteries were to be built on the outskirts of Paris. This promised to have an important effect on attitudes to death and religion, burial then becoming a municipal rather than a religious affair. The proposed cemeteries, moreover, were too far away for the traditional procession to the church and graveside. The clergy therefore opposed the decree, which was not put into practice for many years. With the long journey to the cemetery, funeral processions sometimes became so disorganized and profane that the French actor

Brunet remarked 'God, if I had to be buried like that, I'd rather not die at all.'

One of the largest cemeteries in Paris, Les Innocents, was closed by the police in 1780. All the corpses were removed and the site was eventually transformed into a public park. Digging up and removing the bodies, the earth and the bones was done by night with torches and bonfires, and took two winters and an autumn. There were eighty vaults and fifty common graves to be opened, containing more than 20,000 bodies. Over a thousand carts were used to carry these remains to the quarries of Paris, which now served as underground cemeteries. Later, Napoleon commissioned his chief engineer to arrange the bones artistically and open the series of tunnels to the public. They are called the Catacombs and the bones from several million anonymous corpses, in orderly arrays, are still visible. Notwithstanding these heroic relocations, the further growth of the city in the nineteenth century engulfed even the new cemeteries so that the dead were once again in the midst of the living. The available space was used up faster because of the universal desire for impressive and lasting tombs.

In England, municipal cemeteries were unknown before the 1820s, but then the pressure of space in town and city churchyards made them necessary. London was the first. Highgate Cemetery was opened in 1839, and soon there were others. The founder and architect of Highgate Cemetery, Stephen Geary, had already designed London's first gin palace and also the colossal statue of George III standing near that intersection of six roads which came to be known as King's Cross. The 17-acre plot had been purchased for £3,500, and there were two acres of unconsecrated ground for dissenters and agnostics. At least 166,000 people have been buried there. Local authorities throughout the rest of England were allowed to provide cemeteries in the Cemetery Clauses Act of 1847. Burial was a serious matter and attracted a great deal of legislation. In England between 1852 and 1906 there were no fewer than fourteen Burial Acts, stipulating who could be buried where, what rites should be performed and by whom, what monuments could be erected, and so on. These Acts were replaced, simplified and updated in 1974 in a Local Authorities Cemetery Order.

Given the shortage of space, could the ground in graveyards be used more efficiently? Family plots are currently allowed to take no more than four corpses, and moves are afoot to change the rules and allow multiple occupancy of graves by unrelated people. After seventy-five years skeletal remains would be reburied at a greater depth and fresh coffins placed above them. For those who will always prefer to be buried rather than cremated, there are now moves in the UK to arrange for the re-use of old graves.

What happened to graveyards in towns and cities, once they closed? The church itself could, if necessary, be pulled down and the site rebuilt, but there were restrictions about what could be done with graveyards. Although they could be converted into open spaces for the use of the public, legal cases arose over the construction of urinals, bandstands, bus shelters and electricity transformers. Matters were made easier by the Act of 1971, which allowed tombstones to be removed and human remains dug up and reburied elsewhere, as long as various permissions and approvals were obtained.

A similar movement from city churchyards to out-of-town cemeteries took place in the USA. The New York Board of Health recomended in 1806 that the city burial places be converted into public parks, and laws were passed and finally enforced in 1822. Burial in New York City was prohibited and cemeteries established at Mount Auburn, Laurel Hill and Greenwood.

One of the most beautiful cemeteries in the world is the Protestant Cemetery in Rome. Burials used to take place at night to avoid hostility from Catholics, and about 4,000 people, mostly English and German, and including many artists and poets, are buried here. In 1821 the English poet John Keats lay dying of tuberculosis in his apartment in Rome, and asked his friend Severn to go and look at the place where he would be buried. On hearing that violets, daises and anemones grew wild on the graves he was content and said that he 'already felt the flowers growing over him'. He died, aged twenty-six, only four months after arriving in Rome. Shelley's grave is also there, and his tombstone bears three lines from the song of Ariel in Shakespeare's *The Tempest*:

> Nothing of him that doth fade,
> But doth suffer a sea-change
> Into something rich and strange.

The story of Shelley's ashes and heart is outlined in chapter 7: the lines are grimly appropriate.

The Tomb-Cult and Columbaria

It was in the cemeteries of the nineteenth century that the tomb-cult of the Europeans and Americans reached its peak, with magnificent marble or stone memorials to the famous and the wealthy. In modern cemeteries or in the modern part of old cemeteries the dead are accommodated with greater economy of space. In many European cemeteries three or four vertical layers of coffin containers are stacked up on the side of the path, giving the appearance of small apartment blocks. A building displaying these walls of human remains is called a columbarium because it looks like a pigeon house or dovecote (from Latin *columba*= pigeon). The Roman columbarium was a subterranean sepulchre with niches in its walls for cinerary urns. Notable columbaria include the Columbari di Vigna Codini, on the outskirts of Rome, and a large modern one in the Père-Lachaise Cemetery in Paris. The latter has two floors above ground and two below, with 25,400 spaces for urns. Cremated remains are of course easier to accommodate than whole bodies, but cremation is still not popular in France, where only 2 per cent of the dead are cremated, compared with about 70 per cent in Britain. Père-Lachaise cemetery was opened in 1804, and many famous people, including Balzac, Molière, La Fontaine, Proust, Chopin and Oscar Wilde, were buried or reburied there.

Modern Cemeteries

Cemeteries are now becoming more pleasing places for repose and remembrance. Pioneering steps were taken by the Americans, as set out by Jessica Mitford in her book *The American Way of Death* (1963). Cemeteries are highly profitable enterprizes in the USA, although most are now organized as non-profit corporations. They supply morticians, caskets that can be chosen in special 'casket selection rooms', viewing of embalmed corpses in 'slumber rooms', any type of memorial service, flowers and music, future care of the burial lot, and so on. The funeral director (undertaker) provides a complete service, steering relatives through the unfamiliar procedures with tactful efficiency. One funeral home in Louisiana has a drive-in chapel of rest where viewing is made easy. When required, the dead person in an open casket rises up into a plate glass container, illuminated by blue argon lights. Staff must be properly trained, and the University of California runs one-year courses in 'mortuary science', followed by a two-year apprenticeship before an 'embalmer's licence' is granted.

Very successful campaigns of selling cemetery places to people before death ('pre-need' sales) took place in California in the 1960s. They were said to have sold enough places for the next 100 years, and now three-quarters of US funerals are paid for before death! The dead are converted into beautiful memories. A host of euphemisms protect relatives from the realities of death and bodily corruption. The funeral home is a 'chapel', the morgue a 'preparation room', death is 'deanimation', the dead the 'deceased', the body the 'remains'. There are plenty of tombs and statues, the cheapest memorial is a small (12 × 24 inch, bronze or plastic) grave marker, flush with the ground. In so-called 'lawn cemeteries', which are a feature of California, the metal grave markers are arranged regularly on the landscape in the midst of grass lawns. The largest of the five 'Forest Lawns' near Los Angeles is the one in Glendale. The Forest Lawn founder, Dr Hubert Eaton, envisaged it as 'a great park, devoid of misshapen monuments and other signs of earthly death, but filled with towering trees, sweeping lawns, splashing fountains, beautiful statuary and memorial architecture'. And that is how it is. Everything necessary is in the park, including mortuary, cemetery, crematorium, mausoleum, churches and flower shop. Works of art abound, including sculptures and huge paintings. The one of the Crucifixion, the world's largest religious painting, measures 195 feet long and 45 feet high; there is also the Resurrection (70 feet long and 51 feet high), and a stained-glass version of Leonardo da Vinci's *Last Supper* (30 feet long and 15 feet high). All this is set in a garden landscape.

Different religions have always had their own cemeteries; in 1996 Ajax, an Amsterdam football club, opened a special cemetery for dead fans. It is in the form of a miniature football field on which people can scatter the ashes of the departed fan.

Cemeteries have been getting bigger and bigger. Brookwood, Surrey (500 acres and nearly a quarter of a million bodies) was established in 1854 and is the largest in area in the UK. It used to have its own private railway, the Brookwood Necropolis Railway, with a station in London plus two stations in the cemetery, one for Anglicans and the other for Roman Catholics, dissenters and Jews. The charge for carrying coffins from London to Brookwood was between sixpence and one shilling a mile (single fare only). The City of London cemetery, Wanstead, is the largest in Europe in terms of number of burials, with 500,000 – many of them reburied from City of London churchyards that were cleared out during Victorian times.

Where land is scarce, coffins or urns have to be vertically stacked. In Shenhu, eastern China, where 60,000 graves have been dug up since 1988 to reclaim farmland, a seven-storey crypt was being built in 1993 to accommodate remains after they have been cremated. The largest cemetery in the world, in terms of the numbers of bodies, is Piskrevskoe in Leningrad. It is only 26 hectares but is the resting place of more than 470,000 victims of the German army's siege of Leningrad in 1941–4. The largest cemetery in the world in terms of area is the Rookwood Necropolis, New South Wales, Australia (see box).

Rookwood Necropolis

Show me the way a nation disposes of its dead and I will measure . . . the level at which their society exists.

William Ewart Gladstone, English statesman (1809–98)

Rookwood occupies 283 hectares (692 acres) in inner western Sydney, Australia. It contains 600,00 graves and 200,000 crematorium niches. With nearly 100 on-site staff and employees it is operated on a commercial basis by a public company, and is an excellent example of a thriving, forward-looking semi-urban cemetery that is trying to reach a suitable relationship with the community it serves.

It was originally dedicated in 1867, with its own railway, and space was allocated according to the religious populations in the 1861 census. Today Australia, with 17.5 million people, has no fewer than 270 different ethnic groups and about eighty different religions. This includes many types of Christians, Buddhists, Hindus, Muslims, and the largest Zoroastrian (Parsee) community in the world outside India. As a sign of the times, those professing no religion are now the third largest group (12.9 per cent), whereas they were only 0.3 per cent in 1947.

Rookwood takes about 3,000 burials and 3,000 cremations each year, and at this rate it will be full by the year 2020. There is a tendency towards the lawn cemetery idea, where rows of flat metal plates lie in acres of beautifully kept grass. But clearly you cannot preserve all burial areas indefinitely. It is hoped to introduce a renewable tenure system to replace the perpetual burial rights seen in most cemeteries in Australia and the USA. After fifty years relatives are often non-existent or uninterested, and it is proposed that graves be re-used after this period unless the tenure is renewed.

Rookwood is unique because of its plan for preserving the indigenous vegetation. Many native plants that have disappeared from the nearby suburbs have survived in the cemetery. After commissioning a botanical survey it was found that of the 225 endemic plant species in the cemetery, fifty-three were rare and two (shrubs) were both rare and endangered. A special programme of conservation has been developed, so that burials can be accommodated without losing native plants.

At Rookwood, moreover, they have the idea that the cemetery should be a place for the community and for ordinary people, not just for mourners. Although it is primarily a place of solitude, reverence and relaxation, it can also be used for walking, photography, drawing, painting, botanizing and bird-watching. Visitors (about a million a year) are made welcome.

An old cemetery in central Tokyo has a three-storeyed building with six to eight corridors on each floor, lined by lockers. In the lockers are 9,000 ceremonial urns containing the ashes of cremated citizens. Ashes are also buried in the crowded grounds of the cemetery, which has roads running through it. All future cemeteries in this city are to be multi-storey. The idea is not new: the 2,000-year-old burial chamber at Silwan outside Jerusalem has five storeys. The city of Genoa in Italy, faced with a chronic shortage of accommodation for the dead, plans to build a 130 foot high, ten-storey cemetery. There will be three lifts and space for at least 10,000 permanent 'guests'. At present the tallest cemetery in the world is the Memorial Necropole Ecumenicia, Santos, near Sao Paulo, Brazil. It is ten storeys high and permanently illuminated.

Personalizing the Memory

Arrangements in cemeteries reflect the universal wish to be remembered, as well as the survivors' need to have a defined place where the dead can be visited and remembered. The Romans kept masks of their ancestors at home and wore them at family funerals. We may wish to remember the dead but no longer want to see skulls or other parts of the corpse on display. Perhaps a plaque is enough. In Catholic cemeteries in Europe there are photographs on tombs and coffin containers to act as striking personal reminders of the deceased.

The ancient Egyptians often fitted an idealized mask to the mummy to show a dignified face to eternity. But during the Roman occupation some of the mummies from Greco-Egyptian families bore actual portraits of the dead person. These magnificent portraits were probably made soon after death and as they look at you they make personal, immediate contact across nearly 2,000 years. On one of these mummies someone wrote roughly in ink the poignant message 'Farewell. May you rest in peace.' The Romans introduced portrait masks made of plaster moulds, with individual features shown by painting them and glass or stone inlay for the eyes.

Long ago wax masks or wooden effigies of various kings of England were displayed in Westminster Abbey after their death. They were placed on the coffin during the often lengthy period before burial; it could take weeks to arrange the necessary ceremonies and processions. Edward III (d. 1377) and Queen Elizabeth 1 (d. 1603) had wooden effigies, and those of William III (d. 1702) and Queen Mary II (d. 1694) were made of wax. The wife of the late President Marcos of the Philippines had thirty-two wax dummies made of him, and gave one to her daughter as a wedding present.

Plaster masks can be made from the living or the dead face, and in the eighteenth and nineteenth centuries many famous and infamous people had their features memorialized in this way. Beethoven's death mask expresses his strong personality, and, together with the painted portrait, rescues his features from oblivion. Death masks, however, might also express the unnatural serenity of death itself or the equally unchar-acteristic violent spasms of the dying person. Masks taken during life did not have this disadvantage, although the procedure could be something of an ordeal. The subject had to be equipped with special breathing tubes before the plaster was applied. A very realistic, Madame Tussaud-type image was made of Edmund Sheffield, second Duke of Buckingham, who died while touring Europe in October 1735. The corpse was preserved and a replica made using wax, leather and wood. The Duke (or rather his funeral effigy) now lies in Westminster Abbey, looking good.

More permanent memorial sculptures of the dead person were made in wood, iron or stone. The body had died and decayed but the individual was still visible, to be remembered. The figure was generally incorpo-rated into an impressive monument. Brass commemorative plaques were popular in the sixteenth and seventeenth centuries.

Empty Tombs

Tombs and monuments are sometimes built with no human remains in them. General Gordon was killed in 1885 by the Mahdi's forces beseiging Khartoum. His body was never recovered and St Paul's Cathedral, London, contains only his recumbent effigy. Cenotaphs commemorate the dead in wars but contain no bodies; they are empty tombs (Latin: *ceno* = empty + *taphos* = tomb). The Tomb of the Unknown Warrior in Westminster Abbey, however, contains the corpse of an unknown, unidentified British soldier brought back to England after the First World War and buried, together with soil from France, as an emblem of all the unidentified dead of that terrible conflict.

The rune-stones in Jutland, dating from 800–1000 AD, were memor-ials to the dead, but often referred to property rights and had no graves associated with them. The hero-stones and sati-stones of India are further examples of monuments without bodies. Some of the largest Egyptian pyramids, such as that of Cheops discussed above, contain no royal corpse, and no remains have yet been recovered from Silbury Hill, as we have seen.

One of the world's greatest burial grounds is in Bahrain, an Arab island state in the Persian Gulf. It consists of several thousand sepulchral tumuli (burial mounds) containing tombs of limestone, and dates from the third millennium BC. There are so many burial mounds that someone suggested that Bahrain was an 'Isle of the Dead', receiving all the corpses from ancient Mesopotamia. Until recently one of the mysteries of Arabian archaeology was that most of the tombs appeared to be empty, but now this is thought to be due for the most part to poor preservation of bones over thousands of years.

Green Burials

England today, like many other crowded countries, is experiencing severe pressure on cemetery space. Many people are beginning to believe that rather than having an eventually neglected headstone in a cemetery, rather than placing the body in a wooden coffin at cost to the trees, rather than cremating it and causing air pollution, the loved one should be buried in a 'woodland cemetery'. This would be especially appropriate for people who were green-minded, who loved gardening and wildlife. Farmers, landowners and wildlife trusts are now establishing woodland burial sites where families can bury their loved ones in their own way, more simply and less expensively, maybe in a cardboard box with flowers and with a tree planted at the spot. At present there are more than eighty different green burial sites in the UK, and details are listed in *The New Natural Death Handbook* by N. Albery et al. (1997).

Can you be buried in your own back garden? The answer is yes, as long as it doesn't cause a public nuisance or contaminate water supplies. But family disagreements, problems with neighbours and the effects on the resale value of the house make it an unusual burial place.

7

Exposure and Cremation

Man is a noble animal, splendid in ashes, and pompous in the grave.
'Hydrotaphia' (urn burial), by Sir Thomas Browne (1605–82)

Exposure to the Elements

It may seem uncaring to expose the corpse, above ground and without burial, to the natural processes of decomposition and destruction. This means maggots, carnivores and vultures. But it is the accepted practice in some parts of the world. It is such a striking contrast to burial that it seems to suggest totally different beliefs about death and the human body. The corpse undergoes a more violent and piecemeal destruction as it is torn and eaten by the scavengers. Cremation has its own violence, but the body at least remains in position as it is destroyed. On the other hand, exposure does leave the bones behind, which are in some ways preferable to a mere handful of ashes.

Leaving the dead body above ground (so-called 'walk away' burial) was common in prehistoric times and was not directly connected with religious beliefs. The body was merely discarded, apparently with no special rituals or respect for the dead. It was convenient, and if placed in a tree or on a rock the body was less likely to be taken by animals. Perhaps it was done partly because of the horror of the corpse or fear of its spirit. Sometimes the dying person was abandoned and left alone to die. The word 'exposure', however, refers to a regular method for disposal of corpses that is a key feature of religious practice. Exposure in remote areas was practised in Tibet, possibly due to practical considerations such as a shortage of wood for burning or the difficulty of digging graves. The Zoroastrians (members of the ancient Persian religion) exposed their dead on hills, mountains or specially constructed towers, at some distance from the living. They regarded the corpse as unclean and thought that it would contaminate the pure elements of earth, fire and water. It was not therefore to be buried or burnt, although the Zoroastrians were fire worshippers.

The Parsees and the Vultures

In the region of Gujarat on the west coast of India, the Parsee (= Persian) descendants of the Zoroastrians expose corpses on circular stone buildings, the 'Towers of Silence' or *dakhma*. These are generally on a hill and with no roof, so as to give free access to vultures. The Towers of Silence outside Bombay were 25 feet high, about 45 feet in diameter and divided at the top into seventy-two receptacles arranged round a central pit in three concentric circles, one for men, one for women and one for children. The body, covered with a white cloth, was laid on a stone ledge and left to the waiting vultures. Within a day the flesh had been stripped from the bones, which were later dropped or swept into the central pit to mingle with the bones of others. To save having to do this, the corpses can be placed on an iron grating so that the bones fall automatically into the underlying pit. Today, 90,000 of the world's 155,000 Parsee live in and around Bombay, India, but most of them no longer expose their dead in the old way.

The Kaffir Kalash, a religious group in the North-west Frontier Province of Pakistan, expose their corpses in coffins with the lid open.

The body is returned to the food chain even more rapidly after preliminary butchering. On a mountain crag near the Ganden monastery, Llasa, Tibet, the bodies of the deceased are hacked to pieces by special people ('body breakers') before being fed to the vultures. It is a gruesome ritual, but for Tibetan Buddhists it is a religious act.

Exposure in trees was an alternative, and was popular among certain tribes in India, Bali, Australia and North America. The 'skeletonized corpse' could be recovered later and buried. Aboriginal people of Arnhem Land, Australia, used this method to obtain the bones of special individuals so that they could be painted and then placed in log coffins.

Sharks, Hyenas and Turtles

In different parts of the world, different types of scavengers are available. For instance, in the Solomon Islands dead bodies were laid on reefs to be eaten by sharks; something between sea burial and exposure. The Kikuyu in East Africa used to drag the bodies of certain people into the bush for the hyenas, or block up the door of the house and make a hole at the back so that hyenas could come and remove the corpse.

Exposure sometimes goes hand in hand with cremation. It has been an age-old practice to throw human bodies, often half burnt, into the River Ganges. Although thirty-two new electric crematoria are being set up along the banks of the river, Hindus still regard the Ganges as a sacred destination, with power to free the dying from the cycle of rebirth. Ashes from burning *ghats* (steps or landing places by the river) are sprinkled into the river, along with whole bodies or partly burnt bodies. More than 45,000 corpses a year are cast into this heavily polluted watercourse. At one time it contained freshwater dolphins, crocodiles, turtles and hundreds of species of fish, but they all disappeared as the river was overwhelmed by factory effluents, sewage and corpses. Floating corpses are common. But a solution is at hand: 18,000 meat-eating turtles have been released into the river, and will help dispose of the heavy burden of human flesh. In other parts of India crocodiles, bred on farms, are released into rivers for the same purpose.

Cremation

Contemplating the dreadful transformation wrought by death on a loved person, the putrefaction and decay of once-living flesh, it seems little wonder that cremation is a popular alternative. Sir Thomas Browne put it well in the seventeenth century: 'To be gnawed out of our graves, to have our sculs made drinking-bowls, and our bones turned into Pipes, to delight and sport our enemies, are Tragicall abominations escaped in burning burials.' He also noted that 'burnt Reliques lye not in fear of worm', although with characteristic honesty he adds that ' 'tis not easy to find . . . wormes in graves.' Browne had an enquiring mind and was critical of Christian beliefs. His most celebrated book, *Religio Medici*, was prohibited by the Catholic church in 1645.

There are many points in favour of cremation:

1   It goes some way towards solving the problem of crowded cemeteries. Cemeteries cover more than 16,000 acres of England and Wales, and in 1989 it was calculated that cremation was saving 200 acres each year.
2   If there is no life after death, no hellfire and no physical resurrection of the body – ideas that are being accepted by an increasing number of people – cremation becomes a more acceptable, more hygienic method of disposal.

3   It is convenient, and has the advantage that it converts the fresh corpse to ashes within the hour, without going through the stages of putrefaction and decay.

4   Relatives can dispose of ashes more freely than in the case of bodies or bones. They can be scattered almost anywhere, given to friends, or kept in a box on the mantelpiece. Sometimes it is wise, from a political point of view, to scatter ashes. When undesirable leaders or war criminals die or are executed, scattering their ashes means that there is no shrine to act as a focus for their followers.

5   Sir Henry Thompson, surgeon to Queen Victoria and one of the first enthusiasts, pointed out that the judicious use of cremated remains could save the country a vast amount of money in imported fertilizers.

6   Last but not least, cremation is cheaper because there is no need to buy a grave and a headstone.

There are also, of course, arguments against:

1   The occasional individual who feared being buried alive has the equally unpleasant fear of being burned alive.

2   It deprives the family of a traditional funeral and burial, with the opportunity to declare their social and financial status as well as their affection.

3   The consignment to the flames and fire was an unpleasant, if unconscious reminder of the fires of hell. Funeral directors still have to be careful; they do not talk about ovens, a word that calls to mind a Nazi extermination camp.

4   The original anxieties of the Christian (particularly the Catholic) church were understandable. While traditional burial focused on the body and its resurrection, cremation called attention to the complete destruction of the body and the uncertain fate of the soul. It was not obvious to ordinary people that God could arrange resurrection as easily from the ashes as from the rotting corpse. The clergy's misgivings about cremation were doubtless increased by the fact that many freethinkers supported it, not only because it improved sanitary practice but also because it went with altered attitudes to the dead and to religion. The Pope's ban on cremation was not lifted until 5 July 1963.

From a practical point of view, bodies do not burn all that well, and help is needed from wood, coke, oil, gas or electricity. The Greek philosopher

Heraclitus was said to have burnt with difficulty because he had dropsy (fluid retention). When it is done thoroughly, nothing is left but the ashes and small fragments of bone. Even the bones are consumed, as suggested by the word 'bonfire,' which originally meant a bone-fire made from the bones of domestic animals.

Cremation dates back to prehistoric times and is seen in the 10,000-year-old burials in Franchthi Cave, Greece, in Irish Passage graves and among the original inhabitants of Australia some 30,000 years ago. It was common in ancient Greece, and in ancient Rome until the second century when for unknown reasons it gave way to burial. Caesar's corpse was cremated, and in the funeral procession, above the bier, was a wax model of the great man, showing his wounds, which could be turned in all directions for the benefit of spectators.

Cremation is the preferred method of disposal for Sikhs, Hindus and most Buddhists (the Buddha himself was cremated), and is universal in Japan's Shinto and Buddhist people. But cremation was unthinkable for the ancient Egyptians, who knew that they needed their bodies in the afterlife, and for Muslims it is unacceptable.

Hindu Cremation

Hinduism maintains that the individual soul is a mere particle of the all-pervasive cosmic principle, which is temporarily trapped in a human body. By burning the dead, the soul is released and at the same time the body, which is no longer needed, is purified by fire. In the traditional Hindu cremation ceremony a lighted torch is handed to the eldest son, who lights the pyre. While the body is burning the soul is thought by some to seek refuge in the head and is liberated when the head bursts in the intense heat. Others say the soul exits through the nose, eyes or mouth, and that the souls of the irretrievably wicked exit via the rectum. Then, after certain journeys, the soul enters another body, not necessarily human.

The cremation ceremony does not always go smoothly. For one thing, fuel is expensive and the corpse may be only partially burnt. To the Western spectator the ceremony can appear casual and poorly organized, as shown in this 1933 description of the burning *ghats* in Benares by Patrick Balfour, from *Grand Tour*, London (1934):

Through stagnant water, thick with scum and rotting flowers, we drifted towards the burning ghats, where a coil of smoke rose into the air from a mass of ashes no longer recognizable as a body. One pyre, neatly stacked in a rectangular pile, had just been lit and the corpse, swathed in white, protruded from the middle. An old man, surrounded with marigolds, sat cross-legged on the step above. Men were supporting him and rubbing him with oil and sand. He submitted limply to their ministrations, staring, wide-eyed, towards the sun.

'Why are they massaging him like that?' I asked the guide.

'Because he is dead.'

And then I saw them unfold him from his limp position and carry him towards the stack of wood. Yet he looked no more dead than many of the living around him. They put him face downwards on the pyre, turned his shaved head towards the river, piled wood on top of him and set it alight with brands of straw, pouring on him butter and flour and rice and sandalwood.

The ceremony was performed with dispatch and a good deal of chat, while uninterested onlookers talked among themselves. When I drifted back, some ten minutes later, the head was a charred bone, and a cow was placidly munching the marigold wreaths.

In the UK today, the corpses of Hindus are burnt in crematoria. Ashes are sometimes sent back to India to be scattered on the sacred river Ganges. The great Indian leader, Jawaharlal Nehru (1889–1964), arranged for his ashes to be scattered from a light aircraft over the fields of India.

Suttee

The practice of suttee or *sati* (from a Sanskrit word meaning 'a virtuous wife') was once prevalent among the Brahmins of India. When the husband died the widow, as long as she had no small children, would join her husband on the funeral pyre and perish with him. If there were many wives they would all do it; sixty-four women burned with Raja Ajit Singh in Jodhpur in 1780. It was technically voluntary, but those who abstained were disgraced for life. Many parts of India have *sati* stones, the earliest one dating back to AD 510. They commemorate the wives of heroes who had immolated themselves in this way. Later, between 1680 and 1830, suttee became less of a heroic act. Current laws had given inheritance to widows and there were cases of compulsion, and of escape and rescue. Suttee was finally prohibited by the British in 1829. The exact details varied; the following eyewitness account, recorded by Jean-Baptiste Tavernier in 1650 (*Travels in India*, V. Ball, 1989), appears in the *Faber Book of Reportage*:

On the margin of a river or a tank, a kind of small hut, about twelve feet square, is built of reeds of all kinds and faggots, with which some pots of oil and other drugs are placed in order to make it burn quickly. The woman is seated in a half-reclining position in the middle of the hut, her head reposes on a kind of pillow of wood, and she rests her back against a post, to which she is tied by her waist by one of the Brahmans, for fear she should escape on feeling the flame. In this position she holds the dead body of her husband on her knees, chewing betel all the time; and after having been about half an hour in this condition, the Brahman who has been by her side in the hut goes outside, and she calls out to the priests to apply the fire; this the Brahmans, and the relatives and friends of the woman who are present immediately do, throwing into the fire some pots of oil, so that the woman may suffer less by being quickly consumed. After the bodies have been reduced to ashes, the Brahmans take whatever is found in the way of melted gold, silver, tin or copper, derived from the bracelets, ear-rings, and rings which the woman had on; this belongs to them by right, as I have said.

Burning of Witches, Heretics and Martyrs

In suttee a live human being is burnt, but more or less voluntarily. Burning alive as a punishment and as a public example was quite common in Europe a few hundred years ago.

Belief in sorcerers and witches is common in many societies. It is part of a general belief in magical practices, not all of them harmful, and the subject has been studied with enthusiasm by anthropologists. In sixteenth- and seventeenth-century Europe the idea of the malevolent witch was widely accepted, and accusations of witchcraft were common. Those accused included priests, monks, nuns, magistrates and mayors; the charge was not directed only at old women. It has been suggested that the 'epidemic' of witch-hunting in Europe was related to the demoralization, social disorganization and panic that followed the Black Death.

In England witchcraft trials go back to the Middle Ages. The first recorded was that of Dame Alice Kyteler in 1324. Punishments, prescribed by law, ranged from being placed in the stocks to being burnt alive. In Essex, more than 1,200 cases were brought to trial between 1560 and 1680. Altogether hundreds of thousands of 'witches' were killed, many by burning, in between the fifteenth and seventeenth centuries. In one of the last episodes of witch-mania, the witch trials in Salem, near Boston, USA in 1692, no one was burned.

Each of the thirty-one people tried (including six men) received the death sentence, but most were hanged. The last witch (a woman) to be burnt alive in England was in Sunderland in 1722.

The Medieval Roman Catholic Church was relentless in its pursuit of heretics and dissidents. The Inquisition was set up after the Cathari and Waldenses heresies, and torture was permitted by Pope Innocent IV in 1252. As indicated by the word (from Latin, *inquiro* = enquire into), the inquisitors did not wait for complaints but actively sought out heretics and invited reports and accusations. At first, in northern Italy and southern France, the Inquisition took a less severe form. Punishments varied in severity but the death penalty was uncommon. Between 1308 and 1322 the inquisitor Bernard Gui heard 930 cases in Toulouse, France. There were 139 acquittals, 300 received religious penances and forty-two were con-demmed to death. The regime became harsher, however, after Pope Sixtus IV authorized the Spanish Inquisition in 1478. Persecution continued and reached a peak in Spain, France and Germany in the sixteenth and seventeenth centuries. Victims were tortured and finally executed, often by burning. When Urbain Grandier was burnt in Loudun, France in 1634, after being accused of sorcery and witchcraft and tortured, his shirt was impregnated with sulphur to assist combustion. The ecclesiastical agents carried out the trials but the actual burning was done by civil authorities. One of the most 'successful' of the perpetrators of the holy terror was the Dominican and Grand Inquisitor Tomas Torquemada. He became a symbol of its worst features. At least 2,000 were burned at the stake during his term of office. Another memorable figure was King Philip II of Spain (1556–98). He personally attended an 'auto-da-fe' (act of faith) to watch the burning of twelve heretics, and during his reign 220 Protestants were burned at the stake. Hundreds of thousands of people were tortured and killed during the period of the Spanish Inquisition. It was not a good time to be alive if you were of an independent frame of mind.

A few of the martyrs and saints were burned alive; they gave their bodies to the flames for their faith. But the numbers are small compared with the multitude of witches and heretics who suffered this terrible fate. Joan of Arc, regarded as a saint and national leader by her supporters, but as a witch or political threat by her enemies, was burned at the stake in Rouen in 1431 at the age of nineteen.

Burning of War Dead

As a practical measure after battles, corpses were sometimes disposed of by incineration. In Rivas, Nicaragua in 1855, the bodies of twelve

The day a city died; the story of Pompeii

On the morning of 24 August AD 79, the people of Pompeii, at the foot of Mt Vesuvius in the Bay of Naples, felt a violent tremor, heard a clap of thunder, and saw flashes of fire. At 10 a.m. a cloud in the shape of a pine tree appeared at the summit of the mountain as Vesuvius burst open.

White-hot stones began to fall, then ash, blinding the citizens and filling their mouths and lungs. It happened fast, with no time to escape. Hot, toxic gases swept down and people died within minutes, their lungs overwhelmed. There was no lava, but the ash continued to rain down, covering the bodies, blocking doorways, reaching up to the level of house windows. Before long it had covered the houses themselves and by 1 p.m. the entire city was buried under several metres of ash and pumice. Very few escaped, and Pompeii, with its population of thousands (nearly half of them slaves) had disappeared from the face of the earth. The inhabitants, struck down and killed in the midst of their everyday activities, were faithfully preserved in the volcanic debris. Pompeii, on that warm summer morning, had been stopped in its tracks for posterity.

There had been an earthquake seventeen years earlier, with damage to the city, but the volcano itself had been in repose for at least a thousand years. In AD 79 the eruption was totally unexpected. Since then, however, there have been many more eruptions, with Vesuvius never entirely quiescent.

The buried city, covered over with earth, vineyards and mulberries, was not discovered until 1594, and serious excavations began in the eighteenth century. The bodies in the solidified ash could be cast in plaster by pouring liquid plaster into a small hole until the entire cavity was filled. Today, visitors to Pompeii can see the everyday details of Roman life. There are the houses, the graffiti, a sign at the entrance to a house saying 'Beware of the dog', freshly baked loaves in the baker's oven, the gladiator's barracks – the visitor feels he has come face to face with antiquity. Poignant scenes from that fateful morning are also preserved, such as a family clustered under a roof to escape the hail of burning stones. One of the women has fallen to her knees holding a cloth over her mouth as protection against the deadly fumes and ash, and her husband, who died nearby, holds their child's hand.

Pompeii has fascinated generations of writers, including Goethe, Stendhal, Mark Twain and Charles Dickens. Pliny the Younger gave a first-hand account of the eruption in his letters to Tacitus, the Roman historian. He was visiting his uncle, Pliny the Elder, who was commander of the Roman fleet in the Bay of Naples, and who lost his life by suffocation while trying to get a clearer view of the eruption and helping the stricken villagers on the slopes of the volcano.

officers and 100 men of the American Government Force were burnt after they had been killed in a battle with insurgents. The burning of Allied corpses after the Battle of Sedan in the First World War, in which more than 40,000 died, is described in chapter 2. The ancient Greeks burnt their dead after battles in distant lands; according to the Roman historian Pliny, this was because the enemy dug up the corpses if they were buried.

The Rise of Cremation in Europe

Jews and Christians were originally opposed to cremation, and in 789 Charlemagne ruled that those practising cremation should be put to death. Burial was the rule. How could resurrection take place if there was no body to be resurrected? Not only the clergy but also the general public were against it. Cremation seemed too much like a pagan ritual, akin to sacrifice. There were ominous similarities, perhaps, between the flames applied to the corpse and the everlasting flames of hell. Public opposition was strong and continued into the nineteenth and even the twentieth century. A few well-known people were enthusiastic, including Millais, Trollope, Spencer and H.G. Wells, but nearly everyone else looked upon it with horror. The police were worried that with forged death certificates murders could be concealed by cremating the victim.

Edward John Trelawny records the cremation of the poet Shelley on the beach near Leghorn, Italy, on 15 August 1822. Shelley had been drowned on 8 July and been given a preliminary burial.

> The lonely and grand scenery that surrounded us so exactly harmonised with Shelley's genius, that I could imagine his spirit soaring over us . . . We had to cut a trench thirty yards in length, in the line of the sticks, to ascertain the exact spot, and it was nearly an hour before we came on the grave . . . the work went on silently . . . We were startled and drawn together by a dull hollow sound that followed the blow of a mattock; the iron had struck a skull, and the body was soon uncovered. Lime had been strewn on it; this, or decomposition, had the effect of staining it a dark and ghastly indigo colour . . . the corpse was moved entire into the furnace . . . the heat from the sun and the fire was so intense that the atmosphere was tremulous and wavy. The corpse fell open and the heart was layed bare. The frontal part of the skull, where it had been struck with the mattock, fell off; and, as the back of the head rested on the red-hot bottom bars of the furnace, the brains literally seethed, bubbled, and boiled as in a cauldron, for a very long time. Byron could not face this scene . . . Leigh Hunt remained in the carriage . . . The only portions that were not consumed were some fragments of bones, the jaw, and the skull, but what surprised us all was that the heart remained entire. In snatching this relic from the fiery furnace, my hand was severely burnt.

Shelley's ashes are buried in the Protestant Cemetery in Rome; the heart was taken back to England and buried separately.

The eventual acceptance of cremation was largely due to changing beliefs about God and the afterlife, and the distancing of the living from the dead, but its widespread use was made possible by technical advances. More efficient methods for incineration meant that it was clean and reliable, and could be used on a large scale. The first designs for crematorium furnaces were produced in the 1870s. Sir Henry Thompson, surgeon to Queen Victoria, had visited a model crematorium at the Vienna Exposition in 1873, and had come away greatly impressed with the hygiene and convenience of this method of disposal. He and others founded the Cremation Society in 1874. They bought some land in Woking in 1878, built a furnace, and the next year successfully tried out cremation on a horse. But the authorities and the general public were not interested. When in 1882 a Captain Hanham in Dorset asked the Society to cremate his wife and his mother, both of whom had asked that this be done, the Home Secretary refused to give permission. Captain Hanham therefore had a furnace built on his own land and carried out the cremation. The Home Office took no action against him. Another who took matters into his own hands was Dr Price of Llantrissant. In 1884, when he was in his eighty-fourth year, he cremated his infant son in the open air, wrapping the body in napkins, placing it over a large cask of paraffin and setting it alight. A crowd had gathered and were so incensed that they would have harmed him had it not been for the presence of police. Price stood trial for this, but was acquitted. Cremation was judged not to be illegal as long as it caused no public nuisance.

In 1885 the Woking crematorium began to operate, and the first cremation, of a Mrs Pickersgill, took place on 26 March. The public continued to offer fierce opposition, and police protection was sometimes needed. It was not until 1902 that cremation was officially recognized in a Cremation Act, with rules and regulations about the siting and the operation of crematoria. Attitudes began to change, but only very slowly. In 1910 cremation was deemed necessary for burial in St Paul's Cathedral, but as late as 1990 the Free Presbyterian Church in Scotland opposed the building of a crematorium in Inverness. The Roman Catholic Church did not give its approval to cremation until 1963, and priests were not permitted to perform ceremonies in crematoria until 1966. Now, however, cremation is the accepted method for disposal in the UK. The big change came after the Second World War. Presumably this was due to the world upheaval and renewed mass killing in wartime, the feeling in the post-war period that big changes were taking place, and the weakening of belief in traditional religion and its rituals. In 1938 a mere 3 per cent of corpses were cremated, and the figure was still only 7.8 per cent in 1945; yet by 1991 it stood at 69.9 per cent. The number of crematoria rose accordingly, increasing fourfold between 1950 and 1973, and at present in the UK there are about 230.

Cremation is almost as popular in Switzerland (57 per cent of corpses cremated in 1988), less so in The Netherlands (43 per cent), and even less in Canada (31 per cent) and the USA (15 per cent). It is still uncommon in strictly Catholic countries, the figures for Italy and Spain being 1 per cent and 11 per cent.

Why was cremation accepted less unwillingly in the UK than in many other countries? One interesting suggestion made by Douglas Davies is that the British familiarity with Hindu cremation in India had an influence. In the early stages however, cremation progressed more rapidly in Germany than in England; the first crematorium opened in Gotha in 1872 and twenty-one were in operation by 1911.

When people are accustomed to burial the prospect of cremation may elicit extreme responses. In Haian county in Jiangsu province, China, officials recently decreed that anyone dying after 1 April 1993 would be cremated. The *Independent* reported that dozens of elderly people rushed to commit suicide before this date so as to avoid cremation.

Modern Crematoria

Cremators (cremulators) operate at 800–1000°C, and it takes about an hour for the process to be completed (longer for thin corpses). The aim is for complete incineration, so that invisible gases, not smoke, are discharged from the chimneys.

There have been concerns that harmful quantities of mercury from dental fillings could escape with the other gases, mercury being a volatile substance. Tests show that all the mercury is indeed released and, allowing for five amalgam fillings per head in the 70 per cent of us who still have their own teeth, about 11 kg of mercury would exit each year from an average-sized crematorium doing 3,500–4,000 cremations. Evidently this matter merits attention; perhaps mercury (and silver) could be recovered during the incineration process.

Heart pacemakers are removed from the body if they contain batteries because the latter explode during cremation and can damage the lining of the cremation chamber. Personal possessions such as books, leather jackets or crash helmets are vetoed because of smoke emission rules, so the dead have to make the journey on their own.

During busy periods more than one coffin is processed at a time, but the remains are kept strictly separate. Initially they contain small fragments of incinerated bone, together with metallic objects from rings, tooth fillings, artificial hips or coffin nails. The metallic objects are removed with an electromagnet and the rest pulverized so that it is truly in the form of fine ashes. In Britain about 40,000 hip replacement operations are performed each year, and there are metal residues in up to three-quarters of all cremations. Unexpected metal items have included coins, surgical forceps and scissors, a micrometer and a ring cutter!

Crematoria keep books of remembrance, and some of them display the entries on each anniversary of the cremation. Relatives have the opportunity to buy trees, shrubs, birdbaths, sundials or seats, and place them, together with small memorial tablets, in churchyards, cremeteries, crematoria or other public places.

The Nikolo-Arkangelskoye cemetery in Moscow has seven giant twin cremators, with several 'Halls of Farewell' for atheists.

Crematoria vary in the services they offer. That moment when the coffin disappears behind the curtain may be charged with emotion, and people expect it to be a reverent, respectful occasion even when they have no belief in an afterlife. At times, with just fifteen or twenty minutes allowed between services, there can be a conveyor-belt feeling to the proceedings. But there is a Federation of British Cremation Authorities and a Crematorium Code of Practice to maintain standards. Practical details are set out most helpfully in *The New Natural Death Handbook*, published in 1997. This is an excellent work, written with compassion and wisdom. It contains a guide to the best cemeteries, crematoria and undertakers, as well as useful advice about care for the dying, the process of grief, wills and so on. The authors feel that in a country where death and its details tend to be taboo and people are not very good at dying or at managing deaths, there is a place for what they call 'midwives for the dying'. Another source of information for the 'purchaser' is contained in a special 1995 *Which?* Magazine publication called *What to do When Someone Dies*.

Disposal of Ashes

In the UK today there are no laws to regulate the disposal of cremated remains. After being pulverized to about 2 kg (5 lb) of fine, greyish ash they are given for disposal to the next of kin. Initially, the wish to keep the ashes in an identifiable site meant that they were placed in a small container and either buried or placed in a columbarium (see chapter 6), with a memorial tablet to mark the spot. But the constant pressure on cemetery and crematorium space, and changing attitudes to death and the body, has led to more and more people deciding to scatter the ashes. In the UK in 1986 57 per cent were scattered and another 15 per cent buried at the crematorium. Only 4 per cent were deposited in graves or niches in cemeteries. Relatives took 22 per cent, most to be scattered at some private site such as a garden or a favourite place in the country. A few were scattered according to special requests of the deceased, such as a football pitch or bowling green (the latter a scene, perhaps, of more recent activity), or at sea. A final 2 per cent were left unclaimed at the crematorium, where after a time, they were scattered. Graham Swift's beautifully written novel *Last Orders* centres on the scattering of ashes. It begins in a pub when the box containing them is placed on the bar, and ends movingly when they are scattered into the sea at Margate, Kent.

Immigrant communities with special customs for disposal are, it seems, being catered for. *The Times* reported in November 1993 that Leeds City Council planned to build a £12,000 platform over the river Aire from which Sikhs can cast the ashes of relatives. Sikhs believe that human ashes should be disposed of only in running water.

In the USA cremation is looked upon by funeral directors not merely as a method of disposal of the corpse but as a means of preserving the remains for what they call 'memorialization'. The ashes can be placed in an urn where they will receive perpetual care. Scattering of unpulverized ashes, in particular, is not recommended, and could have distressing consequences. Recognizable human fragments could lie about on the ground or wash ashore on beaches, upsetting strangers and causing great distress to the bereaved. Many Americans still prefer burial, and in some states it remains illegal to scatter ashes.

Some *unusual instances of disposal of ashes* include the following:

- A massive accumulation of ashes from the 100,000 who died in the 1945 atom bomb attack are buried at the Funeral Park in Hiroshima, Japan.
- The ashes of Beatrix Potter (1866–1943) are scattered in one of her fields at Near Sawrey, Lancashire. Her shepherd, Tom Storey, scattered them and had promised not to say exactly where. He died in 1988 having passed on the secret to his son, who then died suddenly in 1989, the secret dying with him.
- Artemesia, the wife of Mausolus, is said to have sprinkled his ashes in wine and drunk them, thus achieving complete unity with her husband.

- Mr Jeff Thorp's last wish was that his ashes should be discharged acoss the Cheshire Plain in twenty-eight giant fireworks. After clearance from the Civil Aviation Authority this was done, a shower of silver, red and green stars acompanying his ashes down the hillside. *The Times* reported that he departed, in August 1992, as he had wanted, 'with a real bang, colouring the sunset with his ashes'. More recently, in 1996 a man arranged for his ashes to be incorporated into a 5-foot-long rocket with his name on it, and hoped friends and family would come to the party with a bottle and watch him go out in a trail in the rocket. The firework company was supplying the rocket free, but some of those at the factory had asked to be excused from this particular product.
- The ashes of Timothy Leary, the 1960s guru, were sent into orbit on 21 April 1997, together with ashes from twenty-three other people. He had planned to have his body frozen in a cryonics laboratory, but changed his mind at the last minute. The Pegasus rocket with its human remains was launched from the Canary Islands, and the satellite will circle the earth for up to ten years before it re-enters the atmosphere and burns up.
- In Tibet, partly burnt bones were sometimes ground to a powder, mixed with clay and moulded in the form of a votive offering. This sounds like a good idea: you end up with a sculptured object that actually consists of the dead person. The object could also be a household item such as a cup, plate ('eating off uncle Jack'), or paperweight. You can even buy a garden gnome with a compartment for ashes! Talking of eating 'off' Uncle Jack, there is also the gruesome cannibalistic possibility of *eating* Uncle Jack by incorporating some of his ashes into the family meal. As a useful and longer-lasting reminder, they could be incorporated into an egg-timer – as in one case publicized in 1997, when a widow did precisely this, on her dying husband's own suggestion.
- Disposal of the bodies of large animals can be a problem: the first camel cremator is being built in Dubai by a Welsh incineration company.
- So that they could not be removed, the ashes of the writer D.H. Lawrence were mixed with a ton of concrete and incorporated into his shrine. The shrine is in the garden of the house his wife Freda built to protect him, near Taos, Santa Fe, and has a phoenix perched on its gable.

8

Unusual Methods of Disposal

Not everyone is burnt, buried or exposed. There are one or two additional ways of disposing of corpses, easier for the soft parts than for the bones. Bones tend to be left over and may be treated or used in different ways, as described in chapters 12 and 14.

Cannibalism

Cannibalism has occurred sporadically in most parts of the world, from prehistoric times up to the present. During the famine of 695–700 in England and Ireland, men ate each other just to survive. Cannibalism was commoner in parts of New Guinea, Australia, North and South America, and Polynesia. Accounts suggesting the regular practice of cannibalism come from tenth-century China, nineteenth-century Africa, and Sumatra before its colonization by the Dutch. In some of these places human flesh was reputed to be for sale in the marketplace.

Why Do People Eat One Another?

Cannibalism can arise from the wish to acquire the qualities of the thing that is eaten, for example by eating a valiant foe slain in battle. Another motive is to pay ritualistic respect to a dead relative. There is a curious logic to the idea that the most fitting burial place for the deceased is inside the body of their living relatives.

In one interesting instance ritualistic cannibalism led to unforeseen results. The Fore people in the highlands of Papua New Guinea used to eat their relatives when they died. Unfortunately a virus-like infectious agent had been present in the brain of an early ancestor and was transmitted through the population by cannibalism. Those who ate the brain (usually women) became infected, and between four and twenty years later developed a severe and in the end fatal neurological disease called *kuru*. The disease, spread by cannibalism, affected many of the Fore people: and altogether there were 3,700 cases. At one time as many as half the women in villages were ill with *kuru*. Cannibalism ceased in 1957 and no one born since then has developed the disease. The infectious agent that caused *kuru* is almost identical with the one that causes scrapie in sheep, BSE in cows and CJD in humans.

Cannibalism as a Way of Life

According to anthropologists, some cannibals eat human flesh simply because it is the best type of food: 'Man's flesh is best of all, and afterwards follows monkeys flesh.' This has been referred to as 'gourmet cannibalism'. In Melanesian pidgin language the word 'long-pig' refers to human flesh as an acceptable item of food.

In 1910 the anthropologist A. P. Rice described the cannibal feast of a Fijian chieftain. The men eaten had been captured in local raids.

> And such a feast! It consisted of 200 human bodies, 200 hogs, and 200 baskets of yams. The preparation of the human bodies and of the hogs was identical, and every member of the tribe had of course to partake of the two main dishes – he was not allowed to select one or the other. This was to ensure that the men did not glut themselves on the human bodies to such an extent that there was insufficient human flesh and others would therefore have to content themselves with mere hog flesh.

The number of people who were feasting is not stated, but it seems to have been a gargantuan meal. In Victorian England it was commonly held that savages were cannibals, and that this practice was only restricted by the spread of civilization. Cannibalism was what was expected of primitive people. One therefore has to look critically at many of the earlier travellers' tales: they sometimes saw cannibalism where none existed.

The topic exercises a morbid fascination. While it may be an exaggeration to say that certain tribes have a lust for human flesh, it seems that humans have no inherited disinclination to eat it. Just as we have to learn in infancy that coprophagy (eating faeces) is revolting, so we acquire but are not born with our deep distaste for human flesh. In any case, it would be the butchering and the cooking as much as the actual eating that would horrify us today. The taste of human flesh is said to be something between veal and pork. The Maoris of New Zealand used to be enthusiastic cannibals, as recorded in Captain Cook's journals;

they noted that black men had a better flavour than white men because there was much less salt in their diet.

Survival Cannibalism

For the most part, humans take to cannibalism only under unusual circumstances. When faced with death by starvation as a result of siege, famine, shipwreck or plane crash, there is repeated emergence of what may be called survival cannibalism. The choice is between cannibalism or death. But cannibalism as a way of life is rare.

In the seventeenth and eighteenth centuries, the days of sailing ships, when people were shipwrecked and starving it was acceptable to eat one of the party so that the rest might survive. Sometimes lots were drawn to determine who should be killed; often the cabin boy was the victim, perhaps because of his low rank rather than for the quality of his youthful flesh. A case that came to the English courts in 1884 was that of *R.v. Dudley and Stephens*. After shipwreck at sea and more than a week without food or water, the defendants killed and ate the cabin boy. They were convicted of murder and lost their appeal. Although you are allowed to commit certain crimes to save your life, murder is not one of them.

Other episodes are recorded in some detail. Robert Hughes describes a celebrated case in *The Fatal Shore* (1988). In 1822 eight convicts escaped from the dreaded prison settlement on the isolated west coast of Tasmania. After ten days in very rough country, hunger overcame their inhibitions and one of the party was killed with an axe as he lay asleep: 'Matthew Travers with a knife also came and cut his throat, and bled him; we then dragged him to a distance, and cut off his clothes, and tore out his inside, and cut off his head; then Matthew Travers and Greenhill put his heart and liver on the fire and eat it before it was right warm; they asked the rest would they have any, but they would not have any that night.'

The next morning they shared out the flesh and continued the trek. Two of them left the group and later on were found half-dead from exposure, but for the remaining five there was no other food and it was a question of who was next. Greenhill, the one with the axe, was at an advantage, and one by one the others were killed and eaten until only two were left, Greenhill and Pearce. They walked apart, watching each other. There was no question of sleep. Finally, one night towards dawn, Greenhill fell asleep and Pearce took the axe from under his head and killed him. Pearce, now on his own, walked on, carrying part of

Greenhill's arm and thigh, until, nearly seven weeks after the escape, he was found. He confessed the whole story to the authorities, but they did not believe it, and he was sent back to prison. He escaped once more, together with another convict, and was recaptured after five days, during which time he had killed and eaten his companion. After his trial the court ruled that he be hanged and, as the ultimate brand of infamy, that his body be 'disjointed' and delivered to the surgeons for dissection. His skull ended up in a glass cabinet in the Academy of Natural Sciences in Philadelphia.

The story of the *Mignonette*, a 33-ton vessel that sailed from Essex for Australia in 1884, provides another example. She was lost in a storm about 2,000 miles east of South America, and four survivors took to an open boat. They existed for twenty days with nothing except 2 lb of tinned turnip and no water, although they caught a small turtle on the fourth day. The entire turtle was eaten, including skin and bones; they drank their own urine and occasionally caught some rain. On the twenty-ninth day Richard Parker, a youth of seventeen or eighteen who was very ill, was killed by the others. His throat was cut with a penknife, and the other three then drank his blood, and cut out and ate his heart and liver. The body was dismembered and supplied them with food for another four or five days, when the decomposing remains were thrown overboard. The next day the survivors were rescued. On arriving back in England they said what they had done, and eventually two of them were sentenced to a year in prison.

The prospector Alfred G. Packer attempted to cross the San Juan mountains, Colorado, in February 1874 with a party of six others. Fifty-five days later he came back, and it turned out that he had killed and eaten the other five in a place called Dead Man's Gulch, near Lake City. Although he ended up in Gunnison County Jail he was not tried until 1886, when he claimed to have acted in self-defence ('eat or be eaten'). The judge gave him forty years (eight years for each victim) and he spent seventeen years in prison. During this time he served the community as a tourist attraction, being known as 'The Colorado Man-Eater' or 'The Great American Anthropophagian'. There seems no doubt that he enjoyed his notoriety.

Survival cannibalism continues to this day, as in the case of the sixteen young Uruguayans wrecked in a plane high in the Andes in 1972. They survived for seventy days by eating the flesh from fellow passengers who had been killed in the crash.

Cannibalism based on eccentricity must be rare. Another American, John Johnson (1820–1900) was known as 'the Crow Killer' because he killed Crow Indians and then ate their livers. He was not driven to this by hunger or gluttony, but by the fact that the Crows had killed and scalped his wife. The outrageous cannibalistic suggestions put forward by Jonathan Swift in his satire 'A Modest Proposal' are referred to in chapter 13.

The Acid Bath

The acid bath is hardly a common method of disposal. Its fascination is that it leaves almost no trace of the corpse, and that it was used by a murderer in a famous case in England. In 1949 John George Haigh shot Mrs Durand-Deacon, and, after drinking some of her blood, placed the body in a 40 gallon steel tank of concentrated sulphuric acid. She was quite fat, and three days later he skimmed off the fat and added more sulphuric acid. After four days he poured the contents of the tank into the ground. The body had been completely dissolved, with no recognizable remains; only a greasy, granular mass on the ground, spread over an area 6 × 4ft. Haigh was a suspect, but there was no body. Unfortunately for him, the sharp-eyed forensic pathologist found a human gallstone among the debris. It was covered with a fatty substance and had not been dissolved. On further careful searching a hairpin, a few tiny pieces of bone and dental fragments were recovered. Haigh was convicted and condemned to death. He had been in too much of a hurry. If he had left the body in acid for a whole month nothing identifiable would have been left behind.

The Compostorium

In a compostorium the buried corpse decomposes by natural means and its components are eventually reutilized by living things. Mr Bloom in James Joyce's *Ulysses*, meditating on burial, is aware of this potential for recycling. 'Every man his price. Well-preserved fat corpse gentleman, epicure, invaluable for fruit garden. A bargain. Buy carcass of William Wilkinson, auditor and accountant, lately deceased, three pounds, thirteen and six. With thanks.'

It is not surprising that in an environment-conscious age there have been suggestions that the natural process of decomposition be encouraged and accelerated. Back in the 1870s Dr Francis Seymour Haden believed that earth should be allowed more ready access to the corpse. He advocated burial in a perishable coffin made of paper pulp or cardboard with a wooden framework, and cloth round it. In 1899 a Dr Young suggested perishable coffins for paupers, but this was on the grounds that they would be cheaper. Needless to say, both of these suggestions were opposed. People did not like the idea of a return to coffinless burial in what was little more than a shroud.

But the technology now exists for building an environment-friendly corpse disposal machine. It would be based on knowledge acquired from sewage and refuse disposal. After removal of bowels and preliminary treatment in a mechanical macerator, the human fragments would be fed into giant fermentation tanks. Microbial digestion would yield methane to power the plant, and a product rich in nitrogen and phosphorus that could be conveniently recycled as fertilizer. Unfortunately the occasional proposals for a compostorium have not been taken very seriously. It nevertheless has certain attractions, and artificially accelerated microbial disposal of human bodies may yet have its day.

Eccentric Exits

Some ingenious individuals have made highly unusual and personal suggestions for corpse disposal. James O'Keilly of New York, the inventor of the original penny-in-the-slot machine, had an idea about corpse disposal in the stratosphere. The machine was called the Navohi, and it was egg-shaped and filled with gas. When the coffin was introduced into it, an automatic mechanism caused acid to flow over the corpse. As a result more gas was released, and finally the mixture was ignited, the entire Navohi roaring into the sky like a rocket.

9

Embalming and Mummification

So Joseph died, being an hundred and ten years old; and they embalmed him, and he was put in a coffin in Egypt.

Genesis 50:26

It is the fear of being forgotten that makes people want gravestones, monuments and other memorials. These act as reminders, and can also be thought of as attempts to maintain the shape of a former life. Behind these impulses lies the alarming thought that one day one's body, albeit lifeless, is going to decompose, disappear and be physically destroyed. Some people believe that the body is needed in the afterlife or on the day of resurrection, but others worry more about the permanent disappearance of the only visible and tangible relic of the self. Even if we no longer believe in life after death, it may be difficult to view with equanimity the complete decay and dissolution of our bodies. The great British geneticist and writer J. B. S. Haldane (1892–1964) said that he would have no more concern about the disposal of his corpse than about the disposal of an old boot. Such ruthlessly logical people are rare.

Accordingly, a variety of methods have been devised for preventing or delaying this dissolution. It is not surprising to find that the methods for preserving dead bodies are similar to those used to preserve meats and fish. Bodies can be dried in the sun, salted, smoked, cured by impregnation with preservatives or frozen. These treatments prevent the growth of the microbes that cause decay. Sugary solutions can also be used because they too inhibit the growth of many microbes, which is why jams are preserved. The body of Alexander the Great was said to have been preserved in honey and then exhibited in a glass container. However, there is no requirement for edibility in the case of corpses, so some of the methods go further than would be acceptable for meat and fish. Mummification is an example of one such process.

Embalming

The word 'embalm' means to impregnate with aromatic substances (balms). To embalm the body is to preserve it, at least for a month or so, so that the smell of decomposition is avoided and it remains a not unsightly object for relatives and friends. Embalming is not usually needed for legal or religious reasons but, in the terminology of American funeral directors, the common corpse is thereby transformed into a beautiful memory, suitable for viewing (although at a later stage it can begin to look like old shoe leather). From the eighteenth century, substances like alcohol, camphor, essential oils and saltpetre were used; in the modern process of embalming, the preservative fluids are introduced directly into the body through blood vessels. As well as preserving, this counteracts dehydration, which otherwise leads to loss of elasticity and wrinkling of skin, and the sinking of eyeballs into their sockets. Also, because most of the blood is replaced, the discoloration of the skin caused by blood collecting on the underside of the body (see chapter 5) is avoided.

In the UK embalming is not carried out as a matter of course because the custom of viewing the corpse is less common than in the USA. The American point of view is nicely put in Frederick and Strub's *Principles and Practice of Embalming*: 'A funeral service is a social function at which the deceased is the guest of honour and the centre of attention . . . A poorly prepared body in a beautiful casket is just as incongruous as a young lady appearing at a party in a costly gown and with her hair in curlers.'

As a practical procedure present-day embalming entails the following steps:

1   The blood is drained from the body. A tube is inserted into the heart or into the large veins near the heart, and blood is withdrawn into a suction apparatus.

2   Tubes are then inserted and tied into the axillary (armpit) artery, and 1–2 gallons of embalming fluid are introduced under pressure. If the corpse is not to be disposed of for a month or two the carotid (neck), femoral (thigh) and brachial (arm) arteries are also intubated. Blood and blood-rich embalming fluid is drained off from the veins until most of the blood has been replaced. The fluid is also introduced into the main body cavities of the chest and abdomen.

Embalming fluid consists mainly of:

- formaldehyde to preserve ('fix') tissues;
- glycerin to counteract dehydration;
- borax to help keep the blood liquid so that it can be readily drained off;
- phenol, potassium nitrate and acetate as disinfectants;
- dyes (saffranin, methyl red) to restore a lifelike tint to the skin;
- water.

3 There may still be parts of the body that have not filled out with the embalming fluid, so more of it is injected directly into tissues such as the skin, massaging and kneading the body until it assumes a lifelike appearance.

4 All joints should be moved to counteract rigor mortis (which subsides between sixteen and forty-eight hours after death; see chapter 5).

5 The mouth is kept shut. This is done by passing a needle and thread up behind the upper lip, across the nose via the nostrils, and down behind the upper lip to the lower lip where it is tied. Cotton wool is placed between the gums if there are no dentures, and also to fill out the cheeks.

6 The face is shaved if necessary, the hair washed and combed, and the nails cleaned.

7 Vaseline or cream may be used to anoint the body and cosmetics are then applied. These include a foundation cream, with rouge for the cheek bones (deep pink for blondes, clear red for brunettes, brown for the rugged face). Vaseline is put on the eyelashes and eyebrows (the eyes being kept closed), and lipstick applied as required.

Special techniques are needed when the body is emaciated, diseased or injured. An interesting extra stage in the process, which gives a very lifelike appearance, is the infiltration of the body with wax. Eva Peron is said to have been beautifully preserved for posterity in this way, and so has the head of Tollund Man, one of the bog-bodies (see chapter 10).

The ancient Egyptians were pioneers in embalming. In the case of the wealthy this included filling out the breasts and refashioning nipples (if female), painting the face, applying a wig and painting the nails with henna. The heart was left in the chest because it was recognized as the source of life but the brain, an organ thought to do no more than produce the mucus that runs out through the nose, receives almost no mention in medical papyruses. The Egyptians were more interested in permanent preservation than in making a lifelike corpse. Mummification included embalming but was a more radical technique for preservation and is described below.

Embalming was practised in fourteenth- and fifteenth-century England, although it was not quite the same as the modern method. First, soft organs from the abdomen and chest were removed and buried, either at the place of death, or separately with the rest of the body. The chest and abdominal cavities were then sluiced out with disinfecting and aromatic fluids, and the outside of the body treated with ointment or paste containing preservatives and spices. Clearly the process needed care, experience and the necessary equipment. The following, therefore, was no more than a romantic fantasy:

> For my Embalming (Sweetest) there will be
> No Spices wanting, when I'm laid by thee.
>   'To Anthea: Now is the Time'
>   by Robert Herrick (1591–1674)

When Henry VIII died in 1547 a group of surgeons, apothecaries and wax chandlers were summoned. The corpse was spurged (washed), cleansed (bowels emptied and rectum plugged) and disembowelled: the removed viscera were placed in a lead box, to be buried in St George's Chapel at Windsor. They played no part in the official funeral rites but were reunited with the rest of the corpse in the coffin nineteen days later. The body was then seared (major blood vessels cauterized to prevent leakage of blood) and embalmed. It was dressed with spices, wrapped in layers of wax-impregnated cloth, and covered with a linen square.

Queen Katherine de Valois, the wife of Henry V, who died in 1437 and was buried in Westminster Abbey, was also embalmed. Samuel Pepys reported seeing the corpse on 23, February 1668, and 'did kiss her mouth, reflecting upon that I did kiss a Queen, and this was my birthday, thirty six years old, that I did kiss a Queen'

Another royal corpse to be embalmed was that of Charles I. After he was beheaded on 30 January 1649, his remains were removed in a coffin to the Palace of Whitehall where the parliamentary surgeon sewed the head back on to the body and embalmed it all. Charles was buried at St George's Chapel, Windsor, seven days after the execution, burial in Westminster Abbey having been refused. The coffin was subsequently lost, and when it was rediscovered in 1813 the Royal Surgeon, Sir Henry Halford, carried out a post-mortem examination. He took the opportunity to remove the fourth cervical vertebra at the site of the axe cut, and used it as a salt holder at dinner parties. The bone was eventually returned to Charles I's coffin.

When Sir Walter Raleigh was beheaded in 1618, his wife had the body buried, and the head embalmed and placed in a red leather bag. She kept it by her side until she died, twenty-nine years later.

Embalming was uncommon in eighteenth-century England and almost disappeared in the nineteenth. However, Martin van Butchell (1735–1812) gives a very practical, matter-of-fact account of the careful preservation of the corpse of his thirty-six-year-old wife. He was an eccentric pupil of the surgeon and anatomist John Hunter, and a successful dentist and eminent truss-maker. In his Memorandum he notes:

> 14 January 1775: At half past two this morning my wife died. At eight this morning the statuary took off her face in plaister (a death mask). At half past two this afternoon Mr Cruikshank injected at the crural arteries 5 pints of Oil of Turpentine mixed with Venice Turpentine and Vermillion . . .
>
> 15 January 1775: At nine this morning Dr Hunter and Mr Cruikshanks began to embalm the body of my wife.

The process took a month to complete before the body was eventually sewn up. The orifices were then filled with camphor and the body washed, dried and finally rubbed with fragrant oils by her husband.

She was then placed on exhibition in a glass-topped container in their drawing room, and soon became one of the sights of London. On 21 October 1775 van Butchell was forced to announce that viewing hours would have to be restricted. But his actions were not merely those of an eccentric. He had good reason to keep his wife preserved and visible because her marriage settlement had stated that he would have control of her fortune as long as she remained above ground! His new wife, understandably, did not like the exhibit and he therefore presented it to The Royal College of Surgeons.

C. Cobbe saw the body in 1857 and commented that it was 'a wretched mockery of a once lovely woman . . . with its shrunken and rotten-looking bust, its hideous mahogany-coloured face, and its remarkably fine set of teeth. Between the feet are the remains of a green parrot . . . it still retains its plumage; it is a far less repulsive-looking object than the larger biped.' The body lasted until the Second World War, when it was destroyed in an air raid.

Embalming Modern Despots

Lenin was embalmed and has stayed more or less presentable in the Kremlin mausoleum for the past seventy years. He is open to public viewing, and twice a week his face is treated with an embalming ointment. As a national monument he gets the best attention, and every year or two is taken to a special laboratory, given a two-week-long bath, an injection, and a new suit and tie. Stalin is also embalmed, as is the former Czechoslovak leader, Klement Gottwald. The Russians at the Centre for Biological Structures are experts and have similarly preserved Ho Chi Minh who lies in state in Hanoi, and Kim IL Sung, the North Korean dictator.

Fortunately, these embalmed gentlemen are little more than effigies, on view for posterity but dead beyond recall. Cryopreservation (see chapter 10) would have at least raised the threat of a comeback.

Mummification

When we think of mummification we think of the trussed corpses of ancient Egypt, and this is the type of mummy descibed here. But the word also has a wider meaning and may refer to any well-preserved dead body, including the bog men and the frozen corpses described in chapter 10.

In ancient Egypt the physical preservation of the body was a central feature of funerary practices. Existence after death was not disembodied, and one had to be prepared to participate physically in the life hereafter. Indeed, the spirit could not continue to exist on its own if the body had disappeared. It had nowhere else to go. It was essential to make sure you had your body, more or less intact, your name, and a magical or real supply of food and drink. Hence the incorruptible mummy in the tomb, inscribed with texts that include the owner's name, and some food and drink nearby.

The deities associated with death and burial included – Osiris, supreme God of the Dead; Nut, the Sky Goddess, associated with coffins; and Anubis, the jackal-headed Embalmer God. At one time Anubis had been the chief deity to whom prayers at a funeral were made, but Osiris overtook him in importance and Anubis then became the guardian of the underworld, guiding the newly arrived dead to the hall of judgment. Here he helped the scribe Thoth in weighing the heart of the deceased against the feather of truth before presenting the dead soul to Osiris. Was the deceased worthy to enter the Field of Reeds (the Egyptian equivalent to the Elysian Fields)? A monster was waiting nearby, ready to devour the heart if it was found unworthy.

The Egyptian Book of the Dead is a collection of magic spells, written on papyrus, and buried with the dead. It has about 167 chapters and includes legends such as the weighing of the heart; knowledge of the magic spells was thought to be essential for happiness after death. It is easy to think of ancient Egypt, the land of the Pharaohs, as a land of tombs and mummies. But they were also great builders of cities, several of which had up to 80,000 inhabitants, and they were pioneers of the large-scale use of stone in architecture. They constructed dams. They invented writing ink and the first paper-like material (papyrus), and produced the world's first novels. In other words, the ancient Egyptians were not totally obsessed with death and burial.

To convert a corpse into a mummy the key requirement is to remove water. The hot, dry sands of Egypt acted as a powerful desiccating agent, so that bodies were well preserved even when buried untreated, uncoffined, in simple shallow graves. Corpses, wrapped in skins or matting, were buried in this way in the pre-dynastic period (before 3100 BC). Cultivated land was always needed, so the graves were at the edge of the desert. They were sometimes lined with matting, boards or bricks, but it was a simple grave, not a tomb. This was at all times the burial method for poorer people who could not afford coffins, embalming, mummification or tombs. The final appearance of a corpse buried in this way is that of a skeleton covered with tightly stretched skin, giving off a sound like a drum when flicked with a fingernail. Probably some hair is still present on the head, but the flesh and soft parts have shrunk away to nothing.

Towards the end of the pre-dynastic period, more than 5,000 years ago, the Egyptians began to enclose the body in a container, some kind of coffin, tomb or chamber, before burial, and now, without the hot dry sand, the natural process of putrefaction took place. Attempts to preserve the body by wrapping it in resin-soaked linen before placing it in the coffin failed because it still underwent decay, and all that was left in the end was a hollow shell of bandages containing the bones. More extensive treatment was needed. It was largely in response to the threat of putrefaction that the technique of mummification was developed. But it was not until the New Kingdom (about 1560 BC) that the basic requirements were understood, and the skill of the embalmers reached its peak after about 1100 BC. Even then there were many poorly prepared, poorly preserved mummies.

Mummification the Ancient Egyptian Way

An early method was to fill embalmed bodies with molten resin or pitch, but this produced a black, brittle and inflammable corpse. The Arab word for bitumen or pitch is *mummiya*, and this is the origin of the word 'mummy'. In Persia there was a 'Mummy Mountain' that oozed a black, bitumenous substance said to have medicinal and preservative properties. It is not clear how often bitumen was used, but petrochemical techniques show that it is present in mummies from about 1200 BC; it came mainly from the Dead Sea and from Iraq. Eventually a satisfactory procedure was worked out, and this was used from the early fourth dynasty (2600 BC) for nearly 3000 years, until it almost disappeared with the rise of Christianity, and was finally eliminated after the Arab invasion in AD 641. Such procedures were abhorrent to Muslims. The quality of mummification varied over the centuries but at its best (and most expensive) it is a triumph of ancient technology. In a successful mummy we can still see the actual features of someone who lived fifteen centuries before the birth of Christ. We can also see whether men were circumcised (most were). Occasionally clear fingerprints can be taken, which somehow give the mummy a true individual humanity and bridge the gap of thousands of years.

Mummification is based on the use of natron, a mixture of salts including sodium chloride (which we use as table salt), found in large amounts at the edges of lakes north-west of Cairo. This substance absorbs water so that the corpse is dried out, and at the same time acts as a mild antiseptic. For a while it was used dissolved in water, and the mother of King Cheops was discovered in her tomb, inside a calcite chest, still soaking in a weak solution of natron. But this method meant that a large container was required to soak the body in, and it was much easier to use solid crystals of natron.

The full process, which includes embalming, is as follows. Herodotus gave an eyewitness description of it after his visit to Egypt in about 450 BC, and the picture has been filled in by examination of mummies and by experiments in laboratories. The entire process, from death to burial, took an average of seventy days. The exact technique varied in different periods; in earlier times no attempt was made to remove the brain, and sometimes mummies, instead of being eviscerated, were injected with resin via the anus.

1   The brains were drawn out of the skull by means of an iron hook inserted through the nostrils. The fragments were thrown away

and the inside of the skull rinsed out with fluid. The function of the brain was not understood: it was considered merely the organ that produced mucus, and was not thought worthy of preservation.

2   An incision was made in the flank with an obsidian knife, and the internal organs extracted. Bowels and lungs were taken out (the latter via the diaphragm after cutting the windpipe), but the heart was left behind. The heart was thought to be the seat of the intelligence, and would be weighed in front of Osiris, God of the Dead, to decide whether the person deserved to enter the afterlife. The kidneys, perhaps because they were not recognized, were also often left behind. The Fallopian tubes, uterus and ovaries were removed, but not the penis or testicles. The chest and abdominal cavities were rinsed with palm wine and spices, packed with temporary materials, and the skin incision sewn up.

3   The body was then covered with natron, and left, on a sloping stone slab, for about six weeks.

4   After washing and thorough drying, the chest and abdomen were repacked with linen cloth and bags of resin, sawdust and natron. Resin-impregnated cloth was put into the skull.

5   The skin was by now very shrivelled and stiff, and the face and limbs were shrunken and distorted so that the person was almost unrecognizable. The skin was treated by massaging into it a lotion consisting of juniper oil, beeswax, natron, spices and wine. From twenty-first dynasty (1085 BC) linen pads, mud, sand or sawdust were packed under the skin as filling material to restore outlines. This was done on the face, under the skin of the arms, legs, back and neck, and sometimes, in the case of women, the breasts. Cosmetics such as rouge were applied as necessary. The mummy began to look more lifelike. In later periods the whole surface of the body was painted with an ochre-gum mixture, red for men and yellow for women.

6   The edges of the skin incision in the flank were drawn together and covered with a plate of gold foil or wax. The incision was not usually sewn up.

7   The eyes, sunken as a result of the dehydration, were covered with resin-soaked linen pads and the eyelids drawn over them. In later periods artificial eyes were used, with obsidian pupils and whites made of alabaster. Nails, often loosened after the natron treatment, were tied on with thread.

8   Jewellery, or other adornments such as gold leaf on the face and

chest, were added at this stage. King Psusennes (twenty-first dynasty) probably holds the record: when he was ready for bandaging he had twenty-two bangles on his arms and twenty-seven rings on fingers and thumbs.

9   Finally, the body was carefully, thoroughly, wrapped in gum-impregnated linen bandages. Roles of bandage were up to 15 metres long and a single mummy might be enveloped in 37 square metres of bandaging, arranged in intricate geometrical patterns. It was an elaborate process, taking about fifteen days. When it was finished, the corpse looked like our idea of a mummy.

The completed mummy wore a mask on the head and shoulders, consisting of plaster-stiffened linen with a painted face. It was handed over to relatives, placed in an anthropoid (human-shaped) coffin, and taken to the burial chamber. Each mummy was labelled round the neck with a wooden tablet to avoid confusion. Mummies were always being moved around the country, and in later times less important folk were mummified and buried without coffins in communal tomb-pits.

A cheaper, economy class mummification was carried out by injecting cedar oil into the anus with a syringe and plugging this orifice, or by removing the intestines via the anus, and leaving the body under natron for many days. The dehydration was less thorough, and oils, resins and wrappings were used more sparingly than in good-quality mummies. When a poor-quality mummy is unwrapped the skin is not well preserved and the limbs and ears easily fall off.

The Egyptian embalmers and mummifiers formed a powerful and highly organized profession, with priestly titles and numerous assistants. Their workshops must have been hot and evil-smelling. They would have been crowded with corpses during wars or plagues, and there was not always time to make good-quality mummies. A corpse would decompose to an unpleasant extent during a long journey by river from the place of death. Flies, beetles and other insects abounded. The pupae of flies can be seen on some mummies, embedded in the sticky resins on the body and between the layers of bandage. This means that eggs had been laid on the corpse and that the maggots had fed. In one instance a dead mouse and a lizard had been incorporated into the wrappings. It was a messy job, and the wrappings have stains from resin-coated fingers.

In that hot climate corpses soon rotted. Herodotus, in his account of the mummification procedure, says that corpses of women who were especially beautiful or famous, or corpses of the wives of important men, were kept for three to four days before being handed over to the embalmers. This was done, he suggested, to prevent the embalmers violating the corpse.

Sometimes a mummy contains bones from different corpses, or there are bits missing. This could mean that the mortuary workers had been careless, or it could mean that on the occasions when bodies accumulated (wars, famines, plagues) they rotted before they could be properly processed, or that parts were missing because of death following acccidents. The mummy of King Sequenre (1500 BC) has a hole in the temple exposing the brain, his cheekbone and nose are smashed, and there is a stab wound below one ear. He was killed in battle, biting through his tongue as he died, and his body was rotting by the time he arrived at the embalmers.

Unwrapping a Mummy

In nineteenth-century England, imported mummies were unwrapped not only for scientific study but also publicly, in front of morbidly interested crowds, often at public lectures. Napoleon's Egyptian expedition and the deciphering of the hieroglyphic script had generated great interest in ancient Egypt. The craze for unwrapping mummies waned after 1850, and since 1900 only four have been unwrapped under controlled conditions in the British Isles. The last one was the mummy of Horemkenesi (see below). Attitudes have changed, and now we are aware that the remains of the dead should be treated with greater respect. Even an archaeological dig can be regarded as an unseemly disturbance of an ancient graveyard.

President Sadat of Egypt decreed in 1980 that the public would no longer be able to view the mummy collection in Cairo Museum. 'Egyptian Kings are not to be made a scene of,' he said. It was, however, reopened in 1992, and since 1997 eleven royal mummies from the Valley of the Kings are on display. These include the six-toed Seti I (nineteenth dynasty) and Segnere II (seventeenth dynasty). The latter has an injury to his skull, having died a violent death in battle. They are all inside special showcases with regulated temperature, humidity and oxygen levels. It is hoped that having them permanently displayed will be good for the tourist industry. But some people still have reservations about displaying royal corpses in a way that would have been so distressing to them had they known what was to happen. For scientific study, various non-invasive methods are now available and there is less need to unwrap mummies.

The story of Horemkenesi, a well-studied mummy, is described in J.H.

Taylor's *Unwrapping a Mummy* (1995). Horemkenesi lived in a mud-brick house near the Valley of the Kings on the west bank of the Nile. He was one of the lower-ranking clergy, responsible for local rituals and also for supervising the workmen who were building and decorating tombs. He was literate, although 99 per cent of the population were not, and he died in about 1040–50 BC, aged fifty to sixty years. At that time the prosperity of the New Kingdom, with its healthy economy and stable central government, was beginning to wane. Food shortages were frequent, and state revenue had fallen when the Nubian gold mines became exhausted.

Horemkenesi's coffin and mummy were found in 1905 and brought to England. The mummy was unwrapped in 1981 because it was in such a poor state of preservation. This was partly due to the warm and humid conditions in the Bristol Museum during the hot summer of 1976. Hopes for its survival were poor and it was decided to make a very careful, thorough job of the unwrapping, using all the latest technology. The team consisted of a pathologist, an archaeologist, an anatomist, a dentist, a radiographer, a textile expert, an entomologist and a chemist. It was like an archaeological rescue dig on a small scale.

The first layers were removed on 1 April, and the team took two weeks to remove them all. (In Cairo in 1886 the bandages enclosing King Rameses II had been removed in ten to fifteen minutes.) The entire procedure was transmitted to the public on closed-circuit television, an echo of the public unwrappings of the nineteenth century (discussed above). As was often the case, the linen (made from flax) of Horemkenesi's bandages was not new, but had come from domestic items such as sheets and shirts. (One mummy was wrapped in bandages made from the sail of a boat.)

Horemkenesi had already been dead a few days when he was mummified, and in that hot climate the body must by then have been in an advanced state of decomposition. It had been attacked by insects, and their larvae were busily devouring the corpse. Numerous carrion beetles (*Dermestes*) and larvae were trapped among the wrappings as well as inside the body. They preserve well, and had died without being able to get out. One place on the neck contained a group of forty-nine beetles. Any regular maggots (fly larvae) would have been eaten by the beetle larvae. The internal organs were therefore more or less non-existent.

Horemkenesi was short (180 cm), and somewhat obese as judged by the skin folds. Priests had to be clean, and Horemkenesi would have been expected to bathe twice a day. He was clean-shaven and his head was shaved, head lice being common. His nose had been disfigured by some earlier accident, and this would have given him trouble breathing. The spine showed osteoarthritis and spondylitis, which is quite common in Egyptian mummies. Life in ancient Egypt was based on the water and soil of the Nile – wet soils and stagnant sheets of water that were fine breeding grounds for the larvae of the schistosomiasis parasites that penetrate the intact skin, and for the mosquitoes that carry malaria. Although none of the organs remained, modern techniques enabled the scientists to test for the proteins (antigens) of schistosomiasis and malaria. Horemkenesi suffered from both. His teeth were heavily worn; three were missing due to root abscesses, and at the time of death he had two other root abscesses. He must have suffered greatly from toothache. Two other teeth showed caries. Presumably his diet contained enough sugar from honey. His face, built up from the skull (see chapter 11), showed a crooked nose and blocked nostril. Because the organs are missing we cannot be sure why he died, but a stroke or heart attack is a possibility. There were no signs of violence.

Unwrapping and dissecting a mummy like this is now a rare event. With CT scans and other non-destructive techniques there is almost no need to remove the bandages.

What Happened to the Internal Organs

The internal organs received special treatment. They were dried out with natron, treated with sweet-smelling ointments, coated with molten resin and wrapped in linen bandages. Separate packages were made for different organs, as each was the concern of a different son of the God Horus. These four sons were guardians of the entrails:

- Imsety, portrayed as a human head, took care of the liver;
- Hapy, portrayed as a baboon's head, the lungs;
- Duamutef, a jackal's head, the stomach;
- Qebhsenuff, a falcon's head, the intestines.

The four packages were placed in separate boxes, in miniature coffins, or in 'canopic' jars or chests. These jars, initially made of carved stone, were decorated and had human-headed stoppers. In later periods (twenty-first dynasty) the organ packages were returned to the body cavities of the mummy, although the empty canopic jars were retained as part of the burial equipment.

In earlier times, before the invention of mummification, the corpse was placed on its side in a hunched-up position and coffins, accordingly, were square in shape. But the body had to be set in an extended position to give access to the abdomen during embalming and mummification; as a result coffins became rectangular, full-length. They were richly decorated, and at one time had two eyes painted on the side so that the mummy, lying on its side, could look out at the world. There was also a vogue for inner and outer coffins, or even a series of coffins, each fitting neatly inside the next like a set of Russian dolls.

The Funeral and the Ceremonies

To understand the Egyptian obsession with the dead we have to remember that more was involved than simply the physical objects we see – the tombs, the mummies and the tomb 'goods'. There was also a complex set of rituals and magic spells that accompanied burials and continued after death and entombment. A rich man might have a large procession of servants, relatives, officials and hired mourning women, the corpse being pulled on a sledge by oxen to the tomb. During the funeral the all-important ritual of Opening the Mouth was carried out, and for this the coffin had to be in an upright position. The idea was to revive the mummified body so that the spirit could dwell in it again. While the ritual words were being read out, incense was burned, water poured over the coffin, and with a variety of instruments the priest touched the eyes, ears, nose and mouth of the face painted on the coffin. This magically restored the use of the mummy's faculties, so that it could pass into the next world breathing, seeing and hearing. Nothing short of a time machine will bring to life for us these ancient and holy ceremonies.

The Constant Battle with Tomb Robbers

Decorating the mummy with expensive materials and providing it with the objects, often made of gold, needed in the next life, was an open invitation to tomb robbers. They were a prominent feature of life in ancient Egypt, a constant threat to the security of the dead, and a constant challenge to tomb builders.

The only royal tomb to survive more or less undisturbed was that of Tutankhamun – an unimportant king from the eighteenth dynasty (1567–1320 BC) who died when he was eighteen. The tomb is a small one, but is famous because it survived intact for us to see. The entrance was covered by rock chippings and soil from the diggings of Rameses

VI's larger tomb (1156 BC) nearby. It had nevertheless been entered by thieves, but was sealed again by the officials of Rameses IX. It then remained unknown until 17 February 1923 when the archaeologist Howard Carter entered it. What he saw has become one of the wonders of the world.

The royal coffin was made of gold; the mummy had a mask of beaten gold and a magnificent gold collar. There was a golden dagger, rings, necklaces, amulets, and a throne covered with gold. Many other treasures filled the different chambers, including chariots, statues, couches, magic oars, furniture, shrines overlaid with gold and a silver trumpet. There was nourishment in the form of 116 baskets of fruit, forty jars of wine, boxes of roast duck; and, of course, a canopic chest containing the royal viscera. No wonder these tombs were robbed. Some of the greater kings had had even more magnificent tombs. Threats to robbers were carved on the walls, but were no deterrent. The 'Curse of Tutankhamun', a favourite for horror films, was a journalistic invention, and seems to have had little effect on Howard Carter, who lived on for another sixteen years, dying at the age of sixty-five.

There could be rich pickings also in the *mastabas* (see chapter 6) of the royal and the wealthy. One royal tomb had twenty different chambers, containing vast amounts of equipment, in the superstructures above the ground, and in two others there were no fewer than forty-five chambers above ground level. These *mastabas* were so vulnerable that they were later replaced by smaller underground storerooms, cut into the rock.

As well as many valuable items, most tombs contained food. One funerary meal found in a tomb at Saqqara consisted of:

    a loaf of bread
    porridge from ground barley
    a cooked fish
    a pigeon stew
    a cooked quail
    two cooked kidneys
    ribs and legs of beef
    stewed fruit
    small cakes with honey and cheese
    jars of wine

Certainly not the meal of a poor man! Obviously it would be a tremendous burden to keep the mummy supplied on a daily basis, and later, from about 1567 BC, things were simplified. Pictures or models of food and other offerings were placed in the tomb, not for decoration but because they were thought to provide the real things by magical means.

The vast majority of tombs and graves, large and small, were plundered in antiquity, in spite of precautions and deterrent devices. The sloping entrance of a *mastaba* would be closed by a heavy limestone block acting as a portcullis, or the entrance would be at the bottom of a deep vertical shaft filled with rubble. A huge plug of granite might block the shaft. But the robbers were expert tunnellers, and would burrow round such barriers. The entrance to the tomb might be concealed, or a number of fake entrances placed round it to make it harder. In the royal tombs there were concealed passages, closed by sliding trapdoors of stone, sometimes locked in position by metal bolts. The blocking stones could weigh as much as 20–40 tons. There were false passages and ingenious contraptions that can only be explained in diagrams. But the tomb robbers kept coming, and got round nearly all these defences. Tomb robbing was well documented, and official commissions were set up to do something about it. Suspects could be beaten and tortured, and the guilty killed by impaling on a stake, but even these severe penalties did not stamp it out. The next anti-robber strategy was to choose a lonely valley (the Valley of the Kings) for burials, with entrances to these rock tombs kept small and inconspicuous; but the robbers got there too.

In the twenty-first dynasty (1085–945 BC) the priests, seeing the futility of guarding tombs safely, decided to move all the royal mummies to secret hiding places. Valuables that had escaped the attention of tomb robbers were removed, and mummies were rewrapped and reburied elsewhere. The new accommodation was more modest than in the original lavishly furnished tombs, but some of them remained undisturbed until 1875. Two famous caches of 'royal' corpses were discovered, one of them containing 153 mummies.

The final stage in the constant battle with tomb robbers was reached in the sixth century BC, when additional security devices were employed. For example, the burial chamber of one wealthy individual was at the base of a 30-metre-deep pit in solid rock, and covered by a stone roof. This main shaft was full of sand, and the burial chamber could be reached only by a narrower vertical shaft next to it, with a short horizontal passage connecting the two. The stone roof covering the burial chamber had a hole in it, blocked by a pottery jar, and when the burial cermony was completed the last man out broke the jar, allowing sand to fill the main shaft. He then made his way out via the narrow shaft, which was later filled in with sand. This ingenious arrangement meant that anyone entering the burial chamber from the small shaft would immediately encounter a torrent of sand. The only way in was to excavate the whole of the main shaft.

It was the Egyptians' belief that the body must be provided for that made the whole burial system so vulnerable, so attractive to robbers. The burials of the Coptic Christians, from the third century AD, remained undisturbed because grave goods were not required by their religion.

It is interesting that there are still expert tunnellers in twentieth-century Egypt. In 1924 thieves burrowed into the antiquity store of the Metropolitan Museum of Art in Thebes, but did not find anything of value.

Mummification of Animals

The ancient Egyptians regarded certain animals as sacred, and there was a flourishing trade in embalmed or mummified animals. There are mummies of snakes, fish, gazelles, cats, falcons, crocodiles, dogs, even beetles. They worshipped the animals associated with specific gods, and these animals were actually present, alive, in some of the temples. The god Thoth, for instance, was represented by an ibis, the god Apis by a bull. Other sacred animals included falcons, baboons, dogs, cats, rams and snakes. At a later stage of Egyptian civilization they were all mummified, and in the tombs there are whole galleries containing the mummified remains of a given animal: a gallery of baboons, a gallery of dogs, and so on.

At Saqqara, from about 600 BC, there was a special animal necropolis, containing, among other things, about half a million mummified birds. This large total reminds us how long-lasting these beliefs were, with the steady addition of specimens over the centuries. Saqqara was the site of the ibis cult, and the temple had up to 60,000 living birds, with an average rate of burial of 10,000 birds a year. A little arithmetic shows that the birds could not have died of old age, and the burial was evidently a mass burial, almost certainly accompanied by sacrificial ceremonies. The number of mummified cats was such that in the nineteenth century hundreds of tons of them were sold and shipped to Liverpool, England, to be converted to fertilizer.

How Many Mummies Are There?

Mummification was at first reserved for pharaohs, their families, and nobles, but it later became widely sought after so that almost everyone ended up as a mummy. Burial was cheaper if the mummy was in a pottery coffin instead of a wooden one, and the poorest burials were in communal underground chambers. These chambers were stacked with pitch-soaked mummies. It has been estimated that there are about 50 million mummies buried in Egypt. They are uncovered whenever new roads are built, and although hundreds of thousands have been destroyed or exported, most of them are still underground. Of those discovered, thousands found their way to museums throughout the world. In addition, mummies were at one time thought to have great healing powers, and by the sixteenth century 'extract of mummy' was sold in most apothecaries' shops in Europe. It was used to treat wounds and was also taken internally. Francis I of France always travelled with a supply of mummy mixed with pulverized rhubarb in case he fell ill or was wounded. When demand exceeded supply, fake mummy preparations soon became available: the corpses of executed criminals were treated with pitch to look like mummies.

Up until the twentieth century, too, mummies were used to manufacture bituminous paint.

Mummies in Other Parts of the World

Mummies are not exclusive to Egypt. In ancient times the inhabitants of the Canary Islands, the Torres Strait Islanders, and the Peruvians practised a similar embalming and preservation of their dead. Some of the details are so similiar that it has been suggested that the practice spread to the Canaries and Torres Strait from Egypt. The sepulchral caves of the Canary Islands at the time of the Spanish conquest in 1402 contained thousands of mummies. Embalming was carried out by a professional class of morticians. After removal of the bowels and the brain, the body was dried in the sun for a few weeks and then treated with grease and wrapped in a covering of sheep leather. The Torres Straits Islanders removed the viscera and brain, dried the corpse in an upright position on a wooden frame, and then painted it. The tongue, palms, soles and nails were stripped off and presented to the spouse. During the Inca Empire (1471–1534), the corpses of kings and high-ranking individuals were eviscerated, treated with Peruvian balsam and other substances, and wrapped in a sitting position with the head over the knees. Smoke-curing was used as a method of mummification in parts of Australia, New Guinea and South America. In southern Panama the body of a dead chief was slung in a hammock over a slow-burning fire, and in this way preserved.

Natural Mummification

Strictly speaking, mummification means the purposeful preservation of the dead. But mummies can be formed naturally when the corpse remains in a dry, protected environment such as a cave or rock shelter. Of forty-two mummies of this type from the south-western USA, about half were found to have head lice. The town of Guanajuato, Mexico, at 2,000 metres altitude, provides another example. Guanajuato not only has the most elegant, marble-lined public lavatories in Mexico, but also, nearby, the Valenciana silver mine, once the richest in the world; and in the cemetery of the Church of La Valenciana are rows of mummified corpses arranged along the walls of the vaults.

The mummies of Chinchorro, in coastal Chile, are recently described examples. About a hundred men, women and children of all ages, even foetuses, have been discovered. They are 5,000–6,000 years old, and have been defleshed, smeared with clay, bound in fabric and tied with cord before burial. Natural mummies are also found in Alaska, where freezing accompanies the drying action. In the Aleutian Islands, on the other hand, the climate is damper, and mummification, as carried out 200–300 years ago, needed artificial assistance. The body was eviscerated, stuffed with dry grass, and dried as far as possible in air before being wrapped in seal or otter skins and placed on a platform in a burial cave. Mummies from Greenland include those of six women and two children buried 500 years ago.

High humidity is also a feature of the climate in Japan, but mummification fitted Buddhist principles, and here we see something extraordinary that can be called 'self-mummification'. It was practised by priests, mostly old men, and consisted of self-starvation plus contemplation, during which time the body gradually dried out and eventually died. The process was sometimes assisted at the post-mortem stage by embalming. By self-mummification the holy state of *nyujo* was attained. Nineteen of these Japanese mummies have been described. In Tibet, numerous holy persons were mummified, including most Dalai Lamas, and some were later encased in gold or silver shrines to protect them.

Drying, plus treatment with herbs and vinegar, has preserved to a varying extent the bodies of 8,000 men, women and children in the Capuchin catacombs in Palermo, Sicily. Some of them are monks but most are not and they present an eerie appearance, standing or lying in rows on either side of the corridor, dressed in their best clothes.

Shunken Heads

The head is a particularly potent part of the corpse. Skulls or heads of the loved or respected have been kept for remembrance, and those of criminals or the vanquished displayed in public places. We have already noted that Sir Walter Raleigh's widow kept his embalmed head in a leather bag until she died. In ancient Assyria it was common practice to remove the heads of the defeated enemy. In the twentieth century this practice was still carried out by certain tribes in Africa, Indonesia, South America and the Oceanic Islands. The heads were dried, smoked and skinned, and the skulls, often painted, kept as trophies.

The Jibaro people in the high Amazon river went further. They were warriors and were skilled in the preparation of shrunken heads or 'Tsantsas'. First, the skin was separated from the skull via a midline incision in the scalp, using a sharp bamboo knife, shell, or flint stone, and with careful preservation of the eyelids, lips, nose and ears. The skin was then put in a bowl containing various plant extracts, including tannin, to shrink and preserve it. The mouth, eyes and neck opening were sewn up with plant fibres so that the head was now a sort of sack, and into this hot sand was poured. Surplus tissue under the skin was burnt away, and further shrinking took place. The face was finally treated with oils and fats. The head, with its long hair (not cut during life in this people), ended up about the size of a clenched fist. But the Jibaro were not mere bloodthirsty savages. The process involved a good deal of ritual, reflecting their beliefs about the soul, its place in the head, and the qualities of the dead person.

10

Freezing and Other Methods of Preservation

This is the Hour of Lead –
Remembered, if outlived,
As Freezing persons, recollect the Snow –
First – Chill – then Stupor – then the letting go.

From 'After great pain, a formal feeling comes',
by Emily Dickinson (1830–86)

Natural Freezing

Freezing is a reliable method for preservation, but the actual temperature matters, and it must be maintained. Thawing and then refreezing damages tissues.

It is not often that corpses are accidentally frozen and remain frozen, but it must occasionally take place in polar and mountainous regions. Just as it has been calculated that thousands of mammoth lie frozen beneath the Arctic ice (their ivory tusks worth millions), so there must be human corpses awaiting exhumation and study. Mt Everest, for instance, has endured more than 233 expeditions since its peak was first climbed in 1953. No fewer than 932 people have now reached the top, queues sometimes forming for the fixed rope sections of the ascent, and since 1920 a total of 156 have died en route. In addition to about fifty tons of rubbish strewn near the summit, plus the wreckage of a crashed helicopter, there are the corpses of twenty people.

Bodies frozen and dried under natural conditions are often very well preserved, and in recent years there have been some striking examples.

Frozen Children

In the 1950s, high up in the mountains of Chile, the remains of a frozen boy were discovered. It was an Inca site and, at 15,000 feet, was remote and inaccessible. It seems to have been a place where the Incas sacrificed their children to the gods. The boy, whose skin, hair and nails were intact, was kept in a freezer in the Natural History Museum in Santiago. With him was a small pouch containing his baby teeth and nail clippings. His feet were calloused and swollen, his fingers frostbitten, and it is thought that he, together with his proud parents (it was an honour to have your child selected) and others, marched to the mountain top in a ritual procession. Here he was killed and buried. His face looks peaceful.

But the frozen bodies of other children, victims of the same *capacocha* sacrifices, tell of a more violent end. Another boy was found on the tallest mountain top in South America in 1985. He was eight or nine years old. The head was damaged by exposure, but there is evidence for the cruel details of his last moments. He had been forced to drink something containing a red dye; his vomit can be seen on the teeth and his body and clothes are red, coloured by his vomit and diarrhoea. It seems that he suddenly realized he was about to be killed and was terrified. Then in 1996 a frozen girl was found under similar circumstances in the mountains of Peru. CAT scans of her skull showed a skull fracture, evidence that she was killed by a hard blow on the side of the head.

Human sacrifice may seem to us an unbelievably strange and cruel practice, but in those distant days it was an accepted method for placating the gods and thus helping the community. It was the only way of doing something about life's threats and uncertainties (see also chapter 3).

The Iceman: An Unexpected Corpse in the Alps

An ice-mummy from the Alps made the headlines in 1991, and tells a much more ancient story. Two hikers found a man's frozen body by a rock in the mountains, 10,000 feet up, at the border of Austria and Italy. It was not buried, and freak weather conditions had revealed it. Many bodies are lost in the Alps and it was not clear that there was anything special about this one. It was removed to Innsbruck University and, because a fungus had begun to grow on the body, it was placed in a freezer. A copper axe with a handle of yew wood found nearby was discovered to be at least 4,000 years old, and radiocarbon dating put the body as 5,300 years old – the late Stone Age, long before Stonehenge was built and before the pyramids were raised in Egypt.

The man was about forty-five years old, dressed in finely stitched leather clothes, and his shoes were stuffed with grass. He had a 6-foot-long bow, a quiverful of arrows, a belt, and a backpack containing fruit and tools made of metal that had been mined 300–400 km away. The stomach contained his last meal of meat and roughly milled corn. There were strange tattoo lines on his lips, his leg and ankle, and on his back: fifty-seven tattoos altogether. Could these have been done as treatment for his aching joints? We know that primitive people apply tattoos to painful spots as well using them for adornment. X-rays showed heavily used joints with arthritic changes, and there was extensive calcification (hardening) of the arteries. There was evidence of chilblains on the little toe.

Was he a man who was performing some ritual? Was he a hunter? The most likely bet is that he was a shepherd and was taking flocks of sheep or goats up the valley in the spring. He lost his way in a snowstorm and as the temperature fell he sat down to make a fire. It grew dark, the snowstorm continued; then he ran out of wood, fell asleep, and died of hypothermia. His body was dried by winds, presumably protected from scavenging animals and birds by snow, mummified, and eventually frozen under several metres of ice and snow.

He is the first prehistoric human to die unexpectedly and be preserved intact with his everyday clothes and tools. Investigations continue, with careful detective work by scientists. His hair contained copper and arsenic, probably acquired during the smelting of copper ore, which produces arsenic vapour. Hundreds of gallons of snow from the site where he was found have been melted and filtered, and are being analysed to find more clues. There was some dispute as to whether it is an Austrian or Italian matter, but the actual site of discovery appears to have been 92 metres inside Italy.

An Ice-Maiden

In the late 1980s a Russian archaeology team was exploring burial mounds high in the Altai mountains of southern Siberia. The ground was permafrost, which never completely thaws, and as they dug into the hard earth on a windswept plateau they came upon a coffin. It was secured with copper nails, and seemed to contain a large piece of ice. They slowly thawed it and what was revealed was the heavily tattooed body of a woman. The tattoos had been made with bone needles and soot and were spectacular, consisting of fantastic deer-like creatures. She was a tall young woman (1.7 metres) and had died about 2,500 years ago.

Removed from its ice cocoon the body began to decay, and it was flown back to the Institute of Archaeology in Akedemgorodok. Subsequent study showed that she had been embalmed, and many of the organs had been removed, including the womb and the eyeballs. Her skull was stuffed with fur from a pine-marten and her body with peat and bark. She was twenty-five years old, and when her features were built up from the skull (see chapter 11) they were seen to be Mongoloid. She was the first woman to be found buried on her own in this way, and had died of natural causes. Her metre-high head-dress had taken up a third of the coffin and she wore a necklace of wooden camels, coated in gold leaf. Her dress, woven from sheep's wool and camel hair, was in three colours and tied at the waist by a braided cord.

Six horses had been killed by an axe and were buried nearby, with ornate harnesses. The scientists knew she had died in June, although the exact year is unknown, because of an amazing piece of detective work. There is a type of fly that lays its eggs on the skin of a horse. The horse licks the eggs off and swallows them, and they hatch inside the horse. Larvae of the fly were found in the stomach of one of the horses, and since the larvae are known to be in horses until the second half of June, this gives a fairly precise dating for the burial.

The ice-maiden seems to have been an important person, possibly a storyteller who memorized the history and myths of that ancient nomadic society. The body is now in Moscow, where further studies are being done. Some of the local people from the place where she was discovered are angry that her tomb has been disturbed, and think it was barbaric to melt her out and take her away. They hope she will be reburied in the mountains.

Cryopreservation: Suspended Animation

Cells can readily be stored at the temperature of liquid nitrogen $(-180°C)$, and when they are thawed many of them are still alive and can divide. To minimize damage to cells, freezing should be rapid and thawing slow. The trick is to avoid ice crystals forming inside cells, because the crystals pierce vital membranes and kill the cell. Ice crystals between cells do much less damage.

What are the prospects for suspended animation of humans by deep freezing? So far there is no evidence at all that whole human bodies can be deep frozen and later thawed out, yet remain alive. A few frogs and reptiles can survive after having their bodies partly frozen, and they do this by producing their own anti-freeze compound. The idea of preser-

ving humans was proposed as long ago as 1861 by the French writer
Edmond About, who suggested that even though people were 'declared
incurable by the ignorant scientists of the nineteeth century' they 'could
wait peacefully in the bottom of a box until the doctors had found
remedies for their ills'.

Children's fairy tales and mythology have many stories of long sleeps.
In one of Washington Irving's stories (1819), Rip van Winkle takes a
magic drink and falls into a deep sleep which lasts twenty years. But he
ages as he sleeps and wakes up to find he is an old man; his wife is dead
and buried, and America has become independent. The Sleeping Beauty
is more fortunate, and retains her youthful charm during seven years of
slumber in readiness for the awakening by her Prince. In these stories
sleep is induced by magic, but nowadays we are familiar with the idea of
whole body freezing because in science fiction stories it is the way
astronauts survive immensely long interstellar journeys. On the other
hand, a lot of people associate freezers with dead meat.

A handful of wealthy individuals have already chosen the deep freeze
rather than burial or cremation. There is always the possibility, they
would argue, that science might one day thaw them out alive and
perhaps deal with their terminal disease. They have high hopes of
medical scientists, who one day will surely be able to accomplish
anything. Death, they think, will become an unacceptable imposition
on the human race. Most scientists, however, regard the idea as
ridiculous, pointing out that trying to bring a frozen body back to life
is a hopeless undertaking, like trying to turn a hamburger back into a
cow. Although we can freeze cells, many of them die during thawing;
and while we can freeze very early embryos consisting of a small number
of cells, we cannot freeze whole organs, let alone whole animals or
people. Liquid helium, at a temperature of $-270°C$, might possibly be
better than liquid nitrogen, but it would be a lot more expensive.

The American Cryonics Society arranges to freeze people in liquid
nitrogen when they are at the point of death. An anticoagulant is first
given and the moribund patient, on a life-support machine, is perfused
with preservative and cooled. The preservative contains antifreeze
agents such as glycerol to stop ice crystals forming. The frozen bodies
are stored in large aluminium tanks, usually four bodies in each
container. At present this treatment costs at least $120,000 dollars
and about thirty bodies have been cryonically suspended in this way,
the first in 1967. Pets can be preserved, too. It is less expensive ($40,000)
to cryopreserve the head by itself, and many would in any case prefer to
be supplied with a new body at the time of resuscitation! Many hundreds

of living clients have signed up for preservation with the Alcor Life Extension Foundation in the USA, and this company is represented in Britain. There are reports of unscrupulous operators allowing 'old' bodies to thaw.

In September 1990 the *Independent* reported that enquiries about freezing and cloning were said to have been made on behalf of the Iraqi tyrant Saddam Hussein. Ferdinand Marcos, the head of state in the Philippines, was placed in a refrigerated crypt when he died. When the government cut off power because of the family's unpaid electicity bills his widow, Imelda Marcos, demanded a hero's burial.

Interesting legal and religious problems arise when we take cryopreservation seriously. If you can freeze people at the point of death, could you not have it done well before you are dead? Would that be suicide? How about the death certificate? Does a frozen corpse have legal rights? If the freezer is turned off because of failure to pay the premium, does this amount to the death penalty for failure to pay? What happens to the soul during freezing? The gravest objections to the idea arise when we think of a world filled to capacity with the thawed remnants of previous generations. Does anybody truly want this? Do we in any case want to live for ever (see chapter 4)?

It will probably turn out that cryopreservation ceases to be an issue. If scientists learn how to create an entire individual from the DNA in single cells (see the section on cloning below), then because we already can preserve cells indefinitely, we already have the tools for conferring immortality. Replicas of a person could be produced in the laboratory at will: replicas of docile workers or of impossibly beautiful and talented archetypes. For most people this is the ultimate fear, which is why cloning experiments on humans are banned. But we have to remember that immortality by cloning would be much less satisfactory for the individual. Although the replicas would be genetically identical, the conscious self and the memories of the original person would have died, obliterated beyond recall when the body died. The individuals created in the identical bodies as they grew and matured would be different. At least the freezer might keep that sense of personal identity intact.

The Bog Men

The bog men are a series of at least 2,000 bodies found, more or less well preserved, in the peat-bogs of Denmark, The Netherlands, northern Germany and the UK. The first discovery to be reported in detail was in Denmark in 1773; then came a more thoroughly documented report from County Down, Ireland in 1781. Bog men date from 100 BC to AD 500 and are preserved in a remarkable state. The humic and tannic acids in the peat-bogs halt the process of decay and preserve and tan the body, while gradually dissolving the bones. The skin is perfectly preserved. The hair is still present, reddish in colour due to the action of the bog acids, and the intestinal contents can often be analysed. Some of the bog bodies looked so recently dead that the local police were called in. But the bodies have a collapsed, shapeless appearance because the supporting bones have been partly dissolved away.

Tollund Man was found by farmers digging for peat in Tollund Fen, Denmark, in 1950. The body lay 8–9 feet below the surface, naked except for a pointed cap made of skin, and a belt round the waist. Round the neck was a plaited leather thong rope. He had been strangled, in 210 BC. The head was the best-preserved ancient head from any part of the world. It was clean-shaven with close-cropped hair, and the face was astonishingly peaceful and composed in view of the manner of death. The heart, lungs, liver and alimentary canal were well preserved and the stomach contained the remains of a vegetable gruel he had eaten twelve to twenty-four hours before he died.

Grabaulle Man was discovered by peat-cutters in another fen, only 11 miles away, in 1952. He had a less well preserved face but beautiful hands and feet from which finger and sole prints could be taken. His throat had been cut, from ear to ear, in 310 AD. X-rays showed that he had also been hit by a blow to the right temple. Three teeth had been lost during life, two more attacked by caries, and the dental appearance indicated an age in the late thirties. The conservator at the local museum kept the body for eighteen months in increasing concentrations of tannic acid, which completed the tanning process and enabled the body to be exhibited dry, with the help of regular applications of linseed oil.

Another bog body was that of a young girl of fourteen. She appears to have been led naked out to the bog, with a woven woollen bandage round the eyes and an ox-hide collar round the neck, and then drowned, some time in the first century AD.

At least forty-one bog people have been found in England and Wales. Lindow Man was discovered in Cheshire, England in 1984. He had been felled with an axe, garrotted with a thin cord and his throat cut, in about 300 BC.

The real mystery lies in the manner of death, because many of the bog men, women and children had been strangled, hung or had their throats cut. Some seem to have been of noble status because their hands show no signs of hard manual labour. This was an age of taboos and magic, and the killings were probably some sort of ritual sacrifice. Bodies were not generally buried in bogs. Analysis of grains and flower seeds in stomachs indicate death just before spring, a time when such a sacrifice might ensure good luck and fertility for the peasant community during the coming year.

Alcohol, Formaldehyde and Other Preservatives

A corpse thoroughly impregnated with alcohol or formaldehyde will be preserved, but these substances penetrate tissues quite slowly. Hence it is generally small objects such as a worm, a mouse or a human finger that are particularly well preserved after immersion in these preservatives. In September 1840 Mr Gladstone, the British Prime Minister, was injured in a shooting accident. One of his fingers had to be amputated and was placed in spirits where it was preserved and buried with him when he died more than fifty years later.

Although an intact human corpse can also be preserved by immersion in a bath of alcohol or formaldehyde, preliminary decomposition of the viscera is likely to take place unless they are removed, or the preserving fluid is introduced directly into the abdomen.

Alcohol

Suzanne, the wife of Jacques Necker, a Swiss financier and statesman, was celebrated for her beauty, wit and learning. She was afraid of being buried alive and also wanted to maintain communication with her beloved husband after death. Hoping that her features would be preserved for him, she arranged for her body to be placed in a tank of alcohol in a mausoleum on their estate on the shores of Lake Geneva. She died in 1794, and the mausoleum was opened ten years later when he died. The black marble basin was found to be still half full of alcohol, and the body preserved and recognizable.

When Lord Nelson was killed at the Battle of Trafalgar (21 October 1805), he had asked Captain Hardy not to have him 'thrown overboard' (buried at sea). His body is said to have been placed in a barrel of rum on board ship and carried back to England. It was buried in St Paul's Cathedral nearly three months later.

The American inventor and scientist Benjamin Franklin (1706–90) thought it would be a good idea to invent a method for embalming drowned people. He added that as far as he was concerned, 'having a very ardent desire to see and observe the state of America a hundred years hence, I should prefer to any ordinary death, being immersed in a cask of Madeira wine, with a few friends, til that time.'

Formaldehyde

Formaldehyde is used for preserving anatomical specimens, and human corpses for dissection. It dries and hardens tissues and is present in embalming fluids. It is a powerful preservative but was not discovered until the late nineteenth century. Formic ('methanoic') acid is the weapon used by stinging ants (from the Latin word for an ant, *formica*) and stinging nettles. The aldehyde is formed when methyl alcohol loses hydrogen (*alcohol dehyd*rogenatum). Formaldehyde deodorizes but leaves its own characteristic smell.

Mercury Salts

The body of the Marquise of Tai, who lived in Hunan, China 2,100 years ago, is very well preserved inside a series of coffins. He had been immersed in a solution of mercury salts, and the coffins hermetically sealed. A large gallstone obstructs his bile duct and the coronary arteries are atheromatous. He also had pulmonary tuberculosis, various worm infestations and a poorly treated old facture of the forearm, but death seems to have been due to biliary colic which led to a heart attack.

Silicone

At the Technical Museum in Mannheim, Germany, you can see bodies that have been preserved by injecting silicone, after sucking out subcutaneous fat. This gaves a dry, flexible, odourless, more touchable corpse.

Eccentric Methods of Preservation
Jeremy Bentham's Self-image

Jeremy Bentham, the English philosopher, died in 1832. As a courageous example to others in a period when hanged murderers were the only legal source of bodies for dissection (see chapter 11), he arranged for his body to be publicly dissected. The skeleton was then to be reassembled in a seated position, clothed, and placed in a glass display cabinet to form his 'auto-icon', his actual self-image. The head was to be removed, dried and preserved – he had been carrying a suitable pair of eyeballs in his pocket for the last twenty years.

Most of this was done. The bones were wired together and padded out under the clothes. Unfortunately the head was not well preserved, so a substitute wax model was made and placed on the shoulders while the real head was put inside the thorax. Jeremy Bentham wanted this type of preservation to be used also for other famous people. His lifelike auto-icon is still to be seen at University College, London.

An Eccentric Tramp

Things are not so easy to arrange these days. Mr Edward McKenzie was a tramp who lived in a barrel on a rubbish tip, and he died in 1984 at the age of seventy-two. His friend, a painter, wanted to embalm him, coat him in acrylic, and display him in his library as a sort of varnished paperweight. But, as The Times reported, Plymouth Council, worried about being seen to approve of such an unconventional proceedure, vetoed the proposal.

Electroplating the Dead

In 1891 Dr Varlot, a surgeon in a Paris hospital, invented a method for electroplating the dead. The body, after being made conductive, was immersed in a bath of copper sulphate and covered with a 1 mm layer of metallic copper. The outer shape of the cadaver, with its brilliant metallic finish, was preserved for eternity. The contents, however, would have consisted of the juices of putrefaction.

Preservation in Gold

Preservation in gold seems a fate suitable only for supremely important people. The body of Chih Hang, a Chinese Buddhist monk, was placed in an urn when he died. After five years the urn was opened and when his body was found to be still intact it was decided that his holiness had been established beyond reasonable doubt. The body was then gilded, and it remains on display in a pagoda in Taipei, Taiwan.

Slicing Up the Human Body

A truly permanent preservation of the structure of a human body can now be achieved by computer imaging. The National Library of Medicine in Washington, DC, in its Visible Human Project, has created a permanent image of the average human male and female, and this is described in chapter 11. The corpses have been reconstructed in the computer from thousands of very thin slices. Although the original corpses will have been destroyed, the exact anatomical details will have been recorded for posterity.

Preservation of DNA

Each human being has a unique set of genes (DNA) which, expressing themselves as the embryo develops in the womb, determine the exact shape and structure of the individual (as well as many psychological features). The environment (the womb and early life) is part of the actual equation, and to argue whether the genes or the environment matter most is in some ways like arguing which of the numbers 2 and 4 is the most important in reaching the product 8. To preserve the DNA would be, to a large extent, to preserve the actual person. Hence preserving a person's cultured cells in the deep freeze can be looked on as a form of bodily preservation.

To move the idea a giant step forward, could the total DNA of the individual be charted on a computer in terms of base sequences, and the DNA itself synthesized at any stage, perhaps a million years later? The difficulties with this type of unlimited immortality would probably be of a purely technical nature. Also, as pointed out earlier, it would be only the genes that had attained immortality, not the conscious, self-aware person, who would be extinguished when the original body died.

The US Human Genome Project, in which the whole of human DNA is spelt out, is now well under way. It is time-consuming and expensive, but by the year 2004 nearly all of the 90,000 genes that make a human being will have been located and defined. One of the payoffs from this extraordinary achievement, other than the unique gift to human biology, will be that harmful genes will be pinpointed. As a result we will be able to understand and thus do something about many unpleasant diseases. Like many other scientific advances, this will bring its own problems. For instance, if insurance companies get hold of people's DNA profile they will be able to charge high premiums to or exclude those possessing genes that lead to life-threatening diseases. The Human Genome Project has been devoting about 3 per cent of its yearly budget to research into the ethical, legal and social implications of these advances in human genetics. In terms of the amount of money involved this is probably the largest philosophy research programme ever undertaken.

The smallpox virus offers an example of such an approach to an *individual* organism. Smallpox virus now no longer exists anywhere in the world except in two refrigerators in the USA and in Russia. There are unanswerable arguments for the destruction of these last physical survivors of a microbial species still capable of causing a terrible human disease. The date of destruction has at last been agreed upon. But the DNA sequence of the virus is known, so that theoretically the monster could be recreated at a later date.

Cloning

Cloning is one of those words that makes people apprehensive. It calls up the image of the mad scientist and the evil despot. Genetic engineering, irradiation, and nuclear are similar words. The trouble with the latter word is that people still associate it with the dreaded nuclear weapons, and transfer their fears to other concepts like nuclear power (although not so far to the nuclear family). Nuclear power stations have had their difficulties, but in the end are destined to be the main source of world energy.

Let us allay our fears of cloning by understanding it. A clone is a group of cells or organisms with identical genes, and to clone a piece of DNA is to produce identical copies of it. It is not too difficult to do this in the laboratory using molecular biological methods, and in 1991 geneticists in the USA were given permission to clone individual genes from the remains of Abraham Lincoln. But to clone an entire organism is not so easy, because the DNA must be in a suitable place (the nucleus of a fertilized egg) if it is to create the new individual by supervising the development of the embryo. And the embryo, in the case of mammals, must be correctly installed inside the uterus. At present it cannot all be done in the test tube.

In 1997 a sheep called Dolly was cloned, and she attracted worldwide attention. This was done by taking nuclear DNA from unspecialized body cells (fibroblasts) of an adult sheep X, and transferring it into unfertilized eggs (oocytes) from which the nucleus had been removed. The eggs, containing DNA from sheep X, were then introduced into the uterus of another sheep where they developed and grew into embryos, one of which was born. This lamb (Dolly) had only the genes (DNA) from sheep X, and was therefore an exact genetic replica (clone): as identical as an identical twin. In actual fact, a very small piece of the cell's DNA is kept outside the nucleus, in the mitochondria of the cell, and because this was not transferred, Dolly was not quite 100 per cent identical with sheep X.

The news of her cloning caused a great shaking of heads. What if people were cloned? Where are the scientists leading us? Will they eventually be able to recreate dead people? Surely cloning is morally repugnant . . . The religious problems, needless to say, are formidable. What happens to the soul? Does it reside in the fertilized egg, or does it only take up residence when the foetus looks like a human at six or seven weeks of gestation, as argued by Thomas Aquinas? Does the sperm or the egg have a soul? Hundreds of years ago the great microscopist Anton van Leeuwenhoek (1632–1723) suggested that a human sperm contained a tiny replica of a human being called a homunculus.

Most people, including scientists, feel that cloning, like some of the other biological wonders of today, needs to be regulated. There must be some control over the fiddling about with human genes. The first reaction of politicians was to ban human cloning. US President Clinton, with his National Bioethics Advisory Committee of eighteen experts, has recommended that cloning of human beings be made a crime in the USA. Clinton says that while 'there is nothing inherently immoral or wrong with these new techniques if used for other things, [cloning] has the potential to threaten the sacred family bonds at the very core of our ideals and our society'. The European Parliament also wants to ban it. The Vatican has called for a worldwide ban on theological grounds, pointing out that the creation of human life outside marriage is against God's plan. But it is impossible to turn the clock back. The technique will undoubtedly lead to some useful applications, for instance the preservation of endangered or rare species of animals, and it will advance medical research by telling us more about the interaction of the cell's nucleus with the cytoplasm.

In 1997 American scientists cloned two rhesus monkeys, using a different method. They produced a monkey embryo by means of *in vitro* fertilization, and when the embryo had developed to the stage of eight cells, the cells were separated from one another and transferred into the womb of a female chimpanzee. This is cloning on a very small scale. Two of the cells developed and resulted in live births. Cloned cows are in the offing and human cloning will possibly be with us before long.

It can be argued that there is nothing new about the prospect of someone having thousands of children. The unbelievable production line in the testis means that sperm are always available, and the children could be produced by the artificial (or theoretically the actual) insemination of thousands of females. It is said that some of the early African chieftains had entire regiments in their army consisting of their own children. But feeling about clones depends on the fact that they are identical. Knowing that identical twins are clones does not help. It is the threat of large numbers. Identical genes, of course, do not mean identical people, because the environment always make a substantial contribution. The human brain, for instance, undergoes a four-fold increase in size after birth, which gives an amazing plasticity during infancy and childhood. Different environmental stimuli give a different end result, a different person. You need more than Mozart's genes to make a Mozart, and without his father's music lessons and influence, without the music environment of eighteenth-century Austria, his genius might never have flowered. Nevertheless clones are rubber-stamped replicas, the sort that have been portrayed in science fiction stories as awesome armies of some malevolent Big Brother.

PART III

The Use and Abuse of Corpses

11

The Body and the Laboratory: Dissection

Dead bodies have their uses, not just when they are freshly dead but even after decaying, being buried, reduced to a pile of bones. They can also be abused, maltreated. This and the following two chapters deal systematically with the different ways in which human corpses can be used or abused.

Who Owns a Dead Body?

What does the law say? It turns out that in English law, no one owns it. In other words, no one can be arrested for stealing a body, a fact that led to difficulties in the prosecution of the 'body snatchers', as described later in this chapter. In the USA and in Canada, in contrast, the next of kin do have some property rights in the corpse.

The odd English law is said to date from the case of a Mr Haynes (1614), who was accused of stealing the shrouds (burial sheets) from a corpse. In the judgment it was stated that the corpse could not own the sheets, and this has been misinterpreted as meaning that the corpse itself could not be owned. Thus, in 1856 a son removed his mother's remains from a graveyard, but because the corpse was no one's property he could only be charged for trespass in the graveyard.

In a case heard in 1996, a person had died with a brain tumour, and the brain was wanted by the son as evidence in litigation against the local health authority for negligent treatment. But part of the brain had by then been taken (at the request of the coroner) and immersed in formaldehyde for histological study; the rest had been discarded after a storage period. The court ruled that the son had no rights of ownership over the corpse, but if it had undergone a preservation process requiring human skill it would then have become an item of property. Mere immersion in formaldehyde was not enough, but presumably mummification would have been. Exactly who owns a skilfully prepared human skeleton is still not clear.

The law, with its need for definitions, often makes macabre distinctions. A corpse, it seems, unlike a urine sample or a lock of hair, is not an item of property and cannot be stolen!

Post-mortem Examination

On the walls of some post-mortem (autopsy) rooms is displayed the following message (in Latin): 'This is the place where death rejoices to come to the aid of life!' A post-mortem is the classical method for discovering the exact cause of death. It is carried out to satisfy the coroner that death was not due to foul play, and in former times was done for the interest and education of physicians, who thereby learnt about the disease processes and the organ pathology that lay behind the final illness. Much of our present knowledge of human diseases has come from the examination of corpses. Over the past 150 years pathologists have been able to provide us with a rational, scientific explanation of most human diseases.

Post-Mortems in History

A thousand years ago no one would have dreamt of cutting up a corpse and examining it. Death had probably been due to 'worms' or other noxious agents and it was best not to let them out. It was in any case an affront to the dead person and a threat to the peace of their soul. But Pope Sextus IV (1471–84) gave official Catholic sanction for autopsies by issuing a bill that allowed students at Bologna and Padua medical schools to study corpses. This was confirmed by Pope Clement VII (1523–34), and at times it proved useful from a purely religious point of view. In one case a woman had given birth to female twins joined from the abdomen to the neck, and the local priest had carried out a double baptism. The church requested an autopsy in order to decide whether the 'monster' was really two individuals rather than one, that is to say whether it had two souls rather than one. The autopsy confirmed that there were indeed two sets of organs and therefore two souls. Nevertheless, in those days autopsies were rare. A memorable one was reported by Theophilus Bonetus (1620–89). He examined the corpse of a seven-year-old girl, and nicely described the kidney changes that follow an attack of acute nephritis. This disease was not at that time even recognized, let alone understood.

A more dramatic autopsy was recorded by Dr Boerhaave (1668–1738). The body was that of a nobleman, a self-indulgent man who enjoyed his food. On his last meal he had eaten the following:

> veal soup with fragrant herbs
> white cabbage, boiled with sheep
> calf sweetbreads lightly roasted
> thigh and breast of a little duck
> two larks
> apple compote and bread
> pears, grapes, sweetmeats
> beer, Moselle wine.

Shortly after the meal he felt an irritation in the abdomen, as if 'something had ruptured, was torn or dislocated'. It became very painful and he took medicines with drinks, plus emetics to make him vomit. But it got worse and he died. Boerhaave's autopsy revealed nothing abnormal in the abdomen, but when the chest was opened it was found to contain 10 liters of fluid, and a tear in the oesophagus (gullet). This was the first ever reported case of a ruptured oesophagus.

Autopsies remained uncommon until the nineteenth century. They were brought to their modern level by the great Rudolf Virchow (1821–1902), the founder of the discipline of pathology, with its systematic examination of organs and tissues and the use of the microscope. Baron Karl von Rokitansky (1804–78), a Czech who was professor of pathology at the University of Vienna, claimed to have performed 30,000 autopsies.

The Modern Post-Mortem

In the post-mortem room each of the main internal organs is displayed and examined by naked eye before and after slicing through it to reveal any internal abnormalities. Suspicious or abnormal tissues are then prepared for histological examination. This is done by embedding them in wax, cutting very thin (hundredths of an inch) slices or sections, staining these with coloured dyes, and mounting them on small sheets of glass (microscope slides). The pathologist then examines the slides under the microscope, reports on any diseased organs, and gives the probable cause of death.

To carry out a post-mortem (PM) examination, permission must be obtained from relatives, and nowadays this is not always easy. The relatives are distressed, and often have difficulty agreeing to such a request. They may fail to see why it is needed, may be worried about disfiguring the body, or may have have religious objections. Also, within the past thirty years the techniques for body imaging, (X-rays, CT scans, etc), together with other non-invasive diagnostic methods, have become so sophisticated that there is generally less mystery about the condition of the dead person's organs and tissues. Although the post mortem is the final arbiter, 'the ultimate medical consultation', the physician tends to be less interested in it unless the diagnosis has been doubtful or the disease is uncommon. As a result, there has been a steady fall in the number of post mortems conducted during the twentieth century.

The Benefits to Medicine from Examining the Corpse

Post-mortem examinations, however, are still vital for teaching medical students, training pathologists and studying certain diseases. A small number must continue to be carried out for these purposes. Furthermore, in at least one in ten cases the post-mortem shows that the pre-death diagnosis was not correct, and medicine learns from these errors.

As an example of learning about specific diseases, an important study on Alzheimer's disease (senile dementia) in the USA is proceeding with the help of nuns' brains. The Sisters of Notre Dame live in convents in different parts of the USA. They are college-educated and elderly, and 678 of them have agreed to participate in the study by undergoing mental tests and giving their brains for pathological examination when they die. They are a good group to focus on, because possible environmental effects on the disease are kept to a minimum. They all share a similar environment and food, and do not drink excessively or smoke. So far, from the study of 102 brains it has become clear that blockages in small blood vessels in strategic regions of the brain can trigger the onset of this type of dementia.

Post-mortem examinations are also an essential part of the work of a forensic pathologist. A person is found dead in water. Was it an ordinary drowning, was it drowning due to intoxication or illness, or was the person murdered and then thrown into the water afterwards? Was this wound on the head a result of him falling down after a heart attack, or did someone kill him by hitting him on the head? Are toxic substances present in the liver, kidney or brain? Many fascinating murder trials have hinged on expert evidence provided by the pathologist. This subject is dealt with in greater detail in chapter 14.

Not all studies of the dead depend on a full PM examination. For instance, heavy metals such as antimony, arsenic and thallium can be detected in hair or nail clippings within hours after they have been ingested. Even the ashes can be a source of evidence. In the Graham Young murder trial in Britain in 1971 thallium, a rat poison, was detected in the ashes of the cremated victim.

Using the Corpse for Divination

Divination means obtaining knowledge about the future by supernatural methods that are beyond the bounds of reason or science. In ancient days it referred to knowledge about the future as revealed by the gods, and often involved the examination of fresh corpses. The word is now used to describe, for example, a water diviner – someone who detects water underground by means of a stick held in the hand. Methods of divination, used by many primitive and many not-so-primitive peoples, include the examination of abdominal contents, usually of sacrificed animals.

The Future According to the Entrails and the Birds

The word haruspication comes from *hira* = entrails (Sanskrit) + *spic* = to behold (Latin). In ancient Rome this was of immense importance. An animal was slaughtered and the entrails carefully examined by the experts, who made decisions and predictions according to the findings. The shape of the liver or the scapula (shoulderblade), for instance, gave a great deal of information. The process was controlled by the Roman College of Augurs, a small group of about a dozen powerful people, who also interpreted any other significant signs, such as thunder and lightning, or the flight and song of certain birds, or the pecking behaviour of sacred chickens. Their function was not so much to tell the future as to find out whether the gods approved of a certain course of action. The pronouncements that were based on evidence from birds, especially their flight, were *auguria* (Latin *avis* = bird) or *auspicia* (Latin *avis* = bird + *specere* = to observe). Before a battle was fought, or before any important political activity, the experts had to be consulted and the signs had to be read. Was it a safe day or time, was it auspicious, did it *augur* well? Words such as inauguration derive from this origin.

Other methods of divination, perhaps easier to perform than haruspication but probably no more reliable, include:

> *Astrology.* This reached its peak in the seventeenth century, and it enjoys great popularity today, partly as a reaction against the no-nonsense, strictly logical doctrines of modern science.
>
> *The interpretation of dreams.* Since time immemorial this has been a favourite method for predicting the future.
>
> *Bibliomancy* (Greek *mantis* = a prophet). In this method a book is opened at any place, one or two lines are selected at random, and these are interpreted to give advice about the future.
>
> *Pyromancy.* This is the method of divination from the behaviour of fire, from the patterns in the flames and the burning wood.
>
> *Crystallomancy.* This involves gazing into a crystal ball.
>
> *Palmistry (chiromancy).* This is performed by examining the lines and creases in the palm of the hand (Greek *cheir* = hand).

In China for thousands of years people have consulted the *I Ching*, the Book of Changes, before taking an important action. This book interprets the hexagram made by the crossing of yarrow stalks.

The Study of Human History and Evolution

The crypt of Christchurch, Spitalfields, London, was recently excavated as part of a restoration study. Skeletons from more than a hundred people who died between 1729 and 1852 were examined for bone density. The necks of the femurs (thighbones) in females of different ages were compared with those of modern women, and it was found that thinning of bone begins at an earlier age now than it did 150–200 years ago. Bones laid down in early life are not so strong as they were in nineteenth-century people, and therefore post-menopausal weakening has a more marked effect in modern women. The most likely cause of this difference is not diet but the fact that the life of nineteenth-century women entailed more physical activity, such as walking, and this strengthens bones.

One of the conditions most commonly found when ancient bones are examined is osteoarthritis, due to the mechanical stresses on joints of a hard physical lifestyle. For instance, sixty skeletons from an Iron Age

(110–900 BC) tomb in Jordan showed marked osteoarthritis in knees, hands, and feet. A 1996 study of monks' skeletons stored in the crypt of St Stephen's Monastery in Jerusalem found that the skeletons, about 1,500 years old, nearly all show arthritis affecting the knees. This was probably due to the monks kneeling daily for long periods – an occupational disease; modern monks generally pray standing up.

Thousands of skulls, from the earliest periods of human history, have rounded holes in them. We know the holes were made during life because they have healed at the edges. They were either made with a drill-like instrument or, less dangerously, scraped out. It must have been a painful, bloody operation, and often fatal because of infection or bleeding. Why was it done? Societies have always had 'mad' people (suffering from conditions we know as, for example, epilepsy or schizophrenia), and it was probably done in the attempt to cure them, to 'let out the evil spirit'. One skull from Cuzco, Peru has seven trepanning holes.

Are We Descended from Neanderthal Man?

Fossil bones and teeth are classic sources of information about evolution. The ancestors of humans left few fossils, and the relationship between Neanderthal man (*Homo neanderthalis*) and *Homo sapiens*, for instance, had been uncertain. Do the two species have a separate lineage, different lines of descent? Neanderthals were similar to us, but more muscular and thickset, with prominent eyebrow ridges, low forehead and protruding jaws. They were human, had tools and used fire, but were they our actual ancestors? They died out 4,000 years ago.

It is an intriguing fact that Neanderthal man had a larger brain than ours. What does this mean? Were they cleverer than us? A study on body size was carried out on 163 sets of bones from early man ranging from 2 million to 10,000 years old. Instead of calculating body size from the size of the teeth or the eye socket as most people have done, the scientists used the width of the end of the femur (thighbone). The conclusion was that Neanderthal man was 30 per cent bigger than us; consequently, his brain size in relation to body size was 10 per cent smaller than ours. There were one or two other interesting conclusions. It appears that although our brain increased in size 600,000 years ago, *for the past 50,000 years the human species has been getting smaller*. Perhaps the shrinking (of body and brain) began when we abandoned hunting and gathering, and adapted to a life based on agriculture.

Recent studies of fossil bones from twenty-four individuals found in a chamber deep in a cave in the Sierra de Atapuerca in northern Spain have generated some more valuable information. These individuals lived 300,000 years ago and show many Neanderthal and at the same time a smaller number of *Homo sapiens* characteristics. Perhaps they were primitive Neanderthals, because the true Neanderthals did not appear until 120,000 years ago. So did Neanderthal man give rise to the modern European, or at least interbreed with *Homo sapiens*? Could modern man be descended from a variety of stocks?

In 1997 scientists managed to obtain *Neanderthal DNA from bones*. The bones were from the original specimen discovered in the Neander valley near Düsseldorf, Germany in 1857, stored since then in a Munich museum. The analysis was made using the very latest methods, and the result was clear. Neanderthal man was a separate evolutionary line from the one that gave rise to *Homo sapiens*, and did not interbreed with the latter. Their line diverged from the modern human one 550,000–700,000 years ago. This investigation was unusual, because DNA would not normally be expected to last for as long as 100,000 years. The Neanderthal bones, however, had been painted with varnish to preserve them (something that is not done nowadays) and this had protected the DNA from oxygen and water, which would otherwise have destroyed it.

The Kontiki expedition in the early 1950s, led by Thor Heyerdahl, showed that people from South America could have colonized Easter Island by travelling on a raft made of balsa wood. Heyerdahl excavated bones on the island, dating from AD 1100, and they were stored in the National Museum of Natural History in Santiago. Forty years later, in 1994, the DNA from the bones was analysed in the laboratory, and the conclusion was that although the present population is mostly Chilean, the original inhabitants were of Polynesian origin.

Life and Health in Ancient Egypt

Much has been learned from the examination of mummies. Large numbers have been studied, although often not very carefully or thoroughly. In nineteenth-century England, as we saw in chapter 9, mummies from Egypt were often unwrapped before audiences as a social occasion rather than for serious scientific study.

In the early 1900s at least 6000 mummies were rescued and autopsied during work on the Aswan Dam in Egypt. Fairly accurate dating of mummies is done by the radiocarbon method, using carbon obtained from skin collagen or from bone. A small proportion of carbon in all tissues is in the form of the isotope carbon-14. After death this isotope is progressively lost from tissues, and by measuring how much of it is left the time since death can be estimated. X-ray examination of mummies gives information about age, parity (child-bearing), state of health and the presence of diseases. This was first done in 1896, but more systematic studies have been carried out since the 1960s and 1970s, and a selection of mummies have been examined with computed tomography (CT). Mummies are often well enough preserved for histological examination of organ slices to be carried out. The Manchester University Museum in England has had an Egyptian Mummy Research Project since 1973, and has established a Mummy Tissue Bank and an International Mummy Database.

Blood vessels of mummies commonly show arteriosclerosis and atheroma (arterial walls thickened and/or degenerated). It is clear from the bones that many suffered from osteoarthritis, tuberculosis of the spine, middle ear infection or osteoporosis. Tuberculosis probably came from cattle, because it first appeared in humans after the domestication of these animals, spreading through the expanding urban populations. The ancient Egyptians seem to have had many of the ailments of the modern world. Syphilis, however, was absent, and so was rickets because the sun produced enough vitamin D in the skin. Cancer was not seen. There were birth defects such as achondroplasia (dwarfhood), hydro-cephaly, cleft palate, club foot and polydactyly. The twelve long toes of Seti I (1318–1304 BC) are visible in his mummy. X-ray examination of the tibial (shin) bone gives information about bodily stresses, and tells us that 30 per cent of mummies had suffered at some stage from disease or malnutrition. Pneumoconiosis (silica-containing lesions in the lung) occurred, due to inhalation of sand in desert storms, and also anthra-cosis (carbon particles in lungs) due to inhalation of smoke from indoor fires used in cooking. Parasites were common, including schistosomiasis (affecting the kidneys), liver fluke (liver), and tapeworm and roundworm infestations (intestines). Schistosomiasis was diagnosed in a 5,000-year-old mummy by detecting the parasite molecule (antigen) in extracts of skin.

For some reason ano-rectal problems were widespread. Kings and nobles were as susceptible as the common man to this condition, as well as to the other diseases – it is said that certain physicians were regarded as shepherds of the royal anus. Teeth were heavily worn down as a result of the coarse, grit-containing diet, which exposed dentine and pulp and led to *tooth abscesses* (as revealed, for example, by the examination of the mummy of Hortemkenesi described in chapter 9). A well-to-do

Egyptian lady from Thebes, mummified 3,000 years ago, underwent a full body CT scan in 1994; This showed an enormous dental abscess, and she had died of blood poisoning when the infection spread through the body. Samples of bread, the staple diet, taken from the tombs show that the flour contained many impurities, including sand, soil and grit from the grinding methods. Periodontal disease was common but not caries, the only source of sugar being from honey or ripe fruits. Caries did not become common until sugar was generally available in the sixteenth and seventeenth centuries (the first factory opened in the West Indies in 1641).

Building Faces from Skulls

This was first tried out on Neanderthal skulls in 1910 by the anatomist Solger. Another pioneer, Gerasimov, in Moscow, reconstructed the face of Ivan the Terrible from the skull. The methods for doing this have recently been greatly improved using more sophisticated techniques. In particular, Richard Nearn, Medical Artist in the Department of Surgery at the Manchester Royal Infirmary, has pioneered the reconstruction of facial features from the skull. The site and size of muscles can be judged from their attachments to the skull, and twenty-one small wooden pegs are put in position, projecting from the skull at different places to represent the position of the muscles and other soft tissues. A model of the face is then built up, using a 3-D image generated on a computer.

One use of this technique has been in bringing to life an individual's face for medico-legal purposes when only the skull is available. It has now been applied to mummies, notably to Natsef Amun, a priest in the Temple of Karmat more than 3,000 years ago. This mummy was discovered in 1823, eventually purchased by the Leeds Society Museum, and first unwrapped in 1828. Under the forty layers of linen wrapping the body smelt of cinnamon, cassia wood and myrrh. The skin was soft and greasy to touch and the fingernails well manicured and hennaed. A second, very through examination was carried out in 1989, using X-rays, CT scans, histology, serology and dental techniques. The reconstructed face provides a striking image of an individual who lived three millennia ago. It is impossible to be sure about details such as skin folds, spots, the shape of lips, tip of nose and ears, but the Manchester mummy team think that his close friends would have recognized him! The box shows how well the method works.

Mummy portraits

The archaeologist Finders Petrie found some skulls and mummy portraits in a burial pit near Cairo in 1888. Petrie was fascinated by skulls and kept more than fifty of them in a special part of his house, called the 'skullery'. He died in Jerusalem, after arranging for his head to be returned to England in a hat box.

A series of more than 200 mummy portraits from Roman Egypt was exhibited in 1997 at the British Museum, London. As some of these ancient Egyptians look you in the eye they give an intensely personal link with the distant past. One of the portraits is that of an Egyptian woman, and Richard Neave, who had not seen the portraits, was given her skull. The face he rebuilt from this skull bore a remarkable resemblance to the portrait.

Another face recreated from the skull, reported in the *Independent* in 1991, is that of a 2,200-year-old Etruscan aristocrat. She was about seventy years old, had lost at least twenty teeth, and had suffered many tooth abscesses. The modelled face is that of an old woman looking down at us, severely it seems, from the past.

A French anthropologist, Pierre-François Puech, has recently reconstructed the face of Mozart. The skull, recovered by one of the gravediggers who buried the composer in 1791, ended up in the Mozarteum in Salzburg. It has a healed fracture on the left temple; the eye sockets are small and the brow ridges weak; the front teeth stick out slightly, and marks between the teeth are due to frequent use of toothpicks. The reconstructed head bears a close resemblance to contemporary portraits.

DNA Messages from Human Remains

DNA can be recovered from dry bones, and has been studied in human remains up to 8,000 years old. The DNA is present only in fragments, but even a single molecule or very small piece can yield sufficient material for study after it has been copied many million times by the polymerase chain reaction. DNA was recovered recently from 100,000-year-old Neanderthal bones, as described above. Preserved human remains are generally suitable, although in the corpses of bog men (see chapter 10) all the DNA has been destroyed by the tannic acid in the peat.

Studies of mitochondrial DNA are valuable. Because mitochondrial DNA is present in the cytoplasm rather than the nucleus of cells it is not present in sperm, which has virtually no cytoplasm. Therefore it is transmitted only via the mother's eggs. There is not much mitochondrial DNA (enough for only thirty-seven genes) and the mitochondrial genes do not mix with the main gene bank in the nucleus. Mutations in mitochondrial DNA therefore give unique evidence about maternal lineages. Also, assuming that the mutations accumulate at a steady rate, analysis of this type of DNA can be used as sort of evolutionary clock. When we examine mitochondrial DNA from modern people and compare it with that from 3,000–8,000-year-old human remains in North and South America, we learn something about the earliest settlers of the Americas. Apparently the first humans to arrive in the Americas came from Asia, perhaps in four separate waves, 42,000–21,000 years ago. Similar studies from bones in Polynesia and Easter Island indicate that the Easter Islanders are Polynesian in origin, as mentioned above. Mitochondrial DNA gives us evidence that is totally independent of regular archaeology and anthropology.

Enthusiasts have gone further and suggested on the basis of similar studies that there was a single female ancestor of *Homo sapiens* ('Eve') who lived in Africa 100,000–200,000 years ago.

DNA technology is thus proving to be a powerful tool for the study of human remains. It provides eloquent and independent evidence about human evolution, and serves as a type of molecular archaeology.

Ethical Problems with Human Remains

The immediate, personal impact of the reconstructed face reminds us that a museum or tomb containing mummies is really a graveyard, not a mere spectacle for morbid interest. In their search for evidence about human origins, Victorian scientists were sometimes led by their enthusiasm into unacceptable actions. The last native inhabitant of Tasmania, William Lanner, died in 1869. The Tasmanians had evolved separately because they were isolated on their island, and they were a unique race of human beings. Dr Crowther, from the Royal College of Surgeons, dug up the body, took the head, and reburied the rest of it. Others took the hands and feet, and someone else the ears and nose. A Dr Stokell from the Royal Society of Tasmania is said to have made a tobacco pouch out of Lanner's skin. Not unexpectedly the last female Tasmanian, before dying in 1876, asked to be buried at sea. Nevertheless the Royal Society of Tasmania obtained her skeleton and displayed it in the Tasmanian museum until 1947. It was not until 1976, a hundred years after her death, that the skeleton was cremated and the ashes scattered at sea in accordance with her wishes.

The skeleton of Charles O'Brien (1761–83), known as the 'Irish Giant' (he was 7 feet 6 inches tall) was obtained for the Royal College of Surgeons, London, by the anatomist John Hunter. O'Brien, not wanting to become an anatomical specimen, had asked that his body be sealed in a lead coffin and buried at sea. Instead, Hunter bribed those in charge of the body, paying them £500, and after boiling the flesh from the bones in a huge vat, installed the skeleton in a glass case where it is still to be seen, an anatomical specimen, in the Royal College of Surgeons.

It is true that in the nineteenth century enthusiastic archaeologists and anthropologists, following the explorers, enriched the museums of the world with skulls and bones of peoples from distant lands. But collectors sometimes desecrated ancestral burial sites and on occasions received material from murdered natives. The phrenologists in particular, who thought they could judge mental faculties from skull measurements, had encouraged a veritable export of bones from Africa, Australia and the Americas. In recent years descendants of these people, distressed by the thought that their ancestors had become either objects for anatomical interest or mere curios for public display, asked that the bones be returned for proper disposal.

An estimated 2,000 skeletons, skulls and fragments of bones from Australian Aboriginals were held in European (mostly British) museums. Their descendants have campaigned to have these relics, some of which may have been from their grandfathers, returned, and in many cases this has been done. The tattooed heads of Maori warriors were prized specimens, and there are about 200 of them in British museums. Captain James Cook brought the first ones back to England and later on traders accepted them in exchange for muskets. Some of these remains are now being repatriated, with the help of the New Zealand government. The University of Edinburgh's 300–400 Aboriginal bones, and those from the Pitt-Rivers Museum in Oxford, have already been given back. Mr Burnum Burnum, a campaigner for Aborigine rights who offered his own head in exchange for that of the legendary Aboriginal warrior Pemulwy (sent to Britain in 1803), pointed out with justification that 'If I requested the loan of King George III's bones for display in Australia as a scientific souvenir, the British would be outraged.'

In Australia itself certain collections, such as the Murray Black with its 1,800 skeletons, and the Crowther collection, have been returned, but there was an understandable reluctance to hand over a collection of 400 Ice Age bodies, 10,000–15,000 years old, held in the Museum of Victoria. The bones would have been reburied, to decay, disappear and be lost to science. In this instance the case for scientific study seems to outweigh the demands of the descendants.

As an example of changing attitudes to human remains, the London auctioneer, Bonhams, had in 1988 planned to sell a preserved Maori head, valued at £10,000. But, following a public outcry, the owner, whose grandfather had brought the head to England 150 years earlier, agreed to give it back to New Zealand. Similar requests have been made for the return of the remains of Native Americans. Stanford University has agreed to hand over its collection of 400 to 3,000-year-old bones. On the other hand, the Smithsonian Institution in Washington has decided to keep most of its 18,650 Native American skeletons for research purposes, although in response to the Udall Congressional Bill about the return of human remains, hundreds have now been shipped to tribes in Alaska and Hawaii. At least twenty-two US states now have laws against disturbing Indian burial sites. It has to be accepted that the treatment of human remains is largely a religious and moral issue. Archaeologists have paid attention to these issues, and they generally earn the cooperation of indigenous communities when they explain to them why they need to study funerary remains, and provide them with clear reports afterwards.

Four Eskimos provide another examples of the increasing tendency to restore human remains to the place or country where they were collected. The polar explorer Robert Peary brought six (living) Greenland Eskimos, or Inuit, to New York in 1897. They were treated as live scientific exhibits and 30,000 people came on board the ship in New York harbour to see them. But after living for a few months in the American Museum of Natural History, New York, four of them had died of respiratory disease. The bodies were autopsied, the flesh removed, and the skeletons stored in boxes in the museum, where they stayed for a hundred years. Their remains were finally returned to Greenland in 1993 and buried in a casket under a mound of stones.

Emotional feelings about human remains extend to the ashes. There have been requests for the return to Russia of the ashes of the ballerina Anna Pavlova. She died in The Hague in 1931 and her body was brought back to London: she had lived for twenty years in Ivy House, Hampstead, and the white marble urn containing her ashes rests in Golders Green Cemetery. Perhaps more surprisingly, such feelings may apply also to tissue sections. Between 1940 and 1944 nearly 400 human brains from a Nazi 'euthanasia' centre in Brandenburg-Gorden, Germany, were sent to Berlin for study. Some of them ended up in the Max-Planck Institute for Brain Research in Frankfurt as fine slices of organs on about 10,000 glass slides. This embarrassing matter became known to the authorities, and the slides and other materials were cremated in 1989. When the

University of Tübingen found that they too had some tissue samples and skeletons from Nazi victims, they made arrangements to cremate them.

Dissection and the Training of Doctors

The dissection and study of normal human corpses has yielded a massive harvest of medical knowledge and has been a foundation stone in the training of physicians and surgeons. The fascinating history of dissection is charted in scholarly fashion in the book *Death, Dissection and the Destitute* by Ruth Richardson. Much of my account draws heavily on this classic text.

Until the Renaissance, dissection was thought to be unnecessary. Enough was known about human anatomy and all that was required was to study the descriptions given by authorities from ancient times. Medical texts were based on the writings of Galen in the second-century AD. During a dissection the learned professor read from Galen's text while an assistant pointed out the parts and a dissector did the actual cutting. But most of Galen's dissections had been of animals and he had transferred his findings by analogy to humans. The sternum, for instance, was said to have eight separate segments and the heart to have a bone (*os cordis*) in it. Because of errors like this it was difficult for medicine to progress until there was better information about the actual structure of the human body.

In the sixteenth century a few schools of anatomy were established in Europe, where careful dissection and observation was encouraged. Italy was foremost in this development, and in Padua, for instance, Vesalius made classical anatomical studies and carried out public dissections. He published his book on human anatomy in 1543 and recorded his observations in a series of 289 woodcuts. Gabriele Fallopius (1523–62), a pupil of Vesalius, discovered the Fallopian tubes leading from the ovary to the uterus, although it was another 200 years before it was demonstrated that the eggs were formed in the ovary, and he also described the clitoris and coined the word vagina.

The discovery of the circulation of the blood was just about the most important single step in the history of medicine. It depended on an accurate understanding of anatomy, and William Harvey (1578–1657), who made the discovery, had dissected the corpses of his father and his sister during his researches.

But it was often hard to get bodies that are suitable for dissection, and this problem has persisted until recent years. They have to be obtained soon enough after death, before the onset of putrefaction; and many people do not like the idea of their body being dissected after they have died.

Possible sources of fresh corpses have been the following:

- executed criminals;
- corpses dug up after burial;
- unclaimed bodies of the poor and dispossessed;
- unclaimed bodies of those dying in hospitals;
- bodies imported from other countries;
- bodies donated by owners before their death.

Executed Criminals

In former times, ordinary people regarded dissection as a terrible fate, a desecration of the body and a denial of a proper grave and burial. When James IV of Scotland granted the Edinburgh Guild of Surgeons and Barbers the bodies of certain executed criminals in 1506, and Henry VIII of England in 1540 gave the Companies of Barbers and Surgeons the right to take for dissection each year the corpses of four hanged felons, it was made clear that dissection was the ultimate punishment. The body was transferred straight from the gallows to the anatomists and was exhibited publicly after the dissection. Until 1832 the law used dissection as a final punishment, one that was added on to death at the gallows.

An Act of Parliament in 1752 gave judges the power to have the body of a convicted murderer dissected as an alternative to being gibbeted. The Act specifically prohibited the burial of the murderer. A gibbeted body was covered in tar after hanging, and suspended in an iron frame in a public place, the corpse gradually disintegrating as birds demolished it and fragments fell to the ground. Dissection was clearly a fate worse than gibbeting, worse than death. It was a systematic cutting to pieces of the corpse, a deliberate assault on the integrity of the body, and quite distinct from an ordinary post-mortem examination. How could the soul rest in peace after such indignities?

Corpses were needed by artists as well as anatomists. In 1775 William Hunter, Professor of Anatomy at the Royal College of Arts, obtained the corpses of eight hanged criminals from Tyburn, London. One of them was used to prepare an anatomical specimen. The corpse was flayed, that is to say the skin was removed to display the muscles, and the body was then set in a suitable posture before rigor mortis set in. Finally a mould was prepared, and the cast of this is still present in the Royal Academy of Arts. Benjamin Robert Haydon, who began as an art student at the Royal Academy in 1804, had opened his own school by 1815. He made his students draw from the actual corpse before starting on the living model. They had to bend over rotting corpses, dissecting for long hours.

The need for fresh bodies increased as the schools of anatomy expanded. In 1701 there were twenty-four lecturers in anatomy in London, many of them in private anatomy schools. Each year hundreds of bodies were needed and clearly the gallows were not providing enough. In twelve London anatomy schools in 1826, 701 students dissected a total of no fewer than 592 corpses. Clearly, corpses were in great demand, and had a monetary value. This led to unseemly competition and immoral practices. Surgeons would invite those in Newgate prison due to be hanged but not under sentence of dissection to barter their corpses for money. Bribes were then necessary for the executioner and for others. Public outrage at these practices ended in riots at the Tyburn gallows. Dissection as a punishment for murder was finally abolished in the Anatomy Act of 1832.

Corpses Dug Up after Burial: the Body Snatchers

An obvious source of bodies, for those who did not mind handling them, was recent burials. As noted at the beginning of this chapter the law said nothing about the ownership of a body, so legally a corpse did not count as property and it was not a crime to steal one. By 1795 there was a professional gang of fifteen body snatchers, known as 'resurrectionists', operating in London. They worked in thirty different burial grounds and supplied eight well-known surgeons. Other groups sprang up and gang warfare soon developed among the rivals. Cooperation from grave-diggers, sextons and undertakers, was achieved by bribes. Working in the dark or by lamplight, they developed the appropriate techniques. It was not necessary to dig a long pit, lift out the coffin and take the lid off. There were short cuts. The Borough Gang, who supplied Guy's Hospital, used the following method:

> First, any pebbles or other markers placed on the grave by suspicious relatives were removed and their position noted. They then dug down to uncover the head of the coffin, and prised it open. A rope was slipped round the corpse's neck, or a hook through the shroud, and the body pulled free. The shroud was returned to the coffin because its removal, unlike removal of the actual corpse, counted as theft. The corpse was

carried away in a sack as soon as the earth, together with any markers, had been tidily replaced.

Grave robbers generally removed the teeth before giving the corpse to the anatomists. Even if the body was beginning to putrefy and was unacceptable, the teeth still fetched a good price, and were bought by dentists to make dentures. Indeed, two of the resurrectionists had accompanied the armies in the French and Peninsular campaigns in the Napoleonic Wars, traversing the battlefields and extracting teeth from the mouths of the dead. But artificial teeth made of porcelain lasted longer than these real teeth, which were subject to decay.

The price of an adult corpse in the 1790s was about two guineas and one crown (less for children). By 1828 it had risen to at least eight guineas. Ruth Richardson traces the emergence of the corpse as a commodity and summarizes in these words:

> Corpses were brought and sold, they were touted, priced, haggled over, negotiated for, discussed in terms of supply and demand, delivered, imported, exported, transported. Human bodies were compressed into boxes, packed in sawdust, packed in hay, trussed up in sacks, roped up like hams, sewn in canvas, packed in cases, casks, barrels, crates and hampers; salted, pickled, or injected with preservative. They were carried in carts and wagons, in barrows and steamboats; manhandled, damaged in transit, and hidden under loads of vegetables. They were stored in cellars and on quays. Human bodies were dismembered and sold in pieces, or measured and sold by the inch.

Details of the activities of the body snatchers were brought to light in the 293-page report of a Select Committee on Anatomy set up in 1828 'to enquire into the manner of obtaining subjects for dissection in Schools of Anatomy and into the state of the law affecting the persons employed in obtaining and dissecting bodies'. A variety of strategies were used to foil the body snatchers. These included the anchoring of the corpse in the coffin with iron straps; the use of patent resurrection-proof iron coffins; placing iron cages over coffins (the 'mort-safe') or round coffins; and putting huge, heavy pieces of stone over the grave. Because the body snatchers needed fresh corpses, certain parishes had a 'dead house' where bodies could safely be stored for a few weeks before burial and undergo some decomposition. Watches were kept on the most recent graves and sometimes special resurrectionist observation towers were built in churchyards for these (nocturnal) vigils.

Feelings ran high. There were near riots when body snatchers were caught. It was suggested that each resurrrectionist should be 'coupled to a corpse and paraded through the streets until both live and dead bodies were amalgamated in putrefaction'. The official view, perhaps, was that although these practices were regrettable there was no alternative supply of bodies that was legally recognized.

Breaking point was reached in the Burke and Hare scandal. Burke and Hare were not resurrectionists; they had taken things one step further. Mrs Hare kept a cheap lodging house in Edinburgh. An old man died owing rent and they decided to sell the body, which they did for £7–10s. Later, another lodger fell ill, and this time, after plying him with whisky, they smothered him, and sold the body to the anatomy school for £10. A total of fifteen more people were murdered, all poor and friendless, twelve of them women. Together their bodies fetched nearly £150, which was a princely sum at a time when even skilled urban workers did not earn more than £75 a year. It seemed an easy source of money.

Then in 1828 Burke invited an old woman, Mary Docherty, home for a meal, killed her, and hid the body in bed straw. But the body was discovered, and in the trial Hare gave evidence against Burke, who was hanged and, appropriately, dissected. Hare was released. The dissection was witnessed by a select number of ticket holders, and the next day 30,000–40,000 people came to view the dissected body. It had been flayed, and the Royal College of Surgeons has a book bound with leather made from Burke's skin.

The actual receiver of the Burke and Hare corpses escaped justice. Dr Knox, who ran the anatomy school, was not called to give evidence, but must have known that the seventeen very fresh bodies he had received from Burke and Hare had not died naturally. A large crowd demonstrated outside Knox's house in Edinburgh, and throttled and hanged his effigy. The police charged the crowd, many were injured, and the windows of the Royal College of Surgeons were broken.

In 1830 two London 'burkers' were convicted, hung, and dissected. The words to burke, burking, a burker and burkophobia were added to the English language.

The terrifying details of body snatching are vividly described in a short story, 'The Body Snatcher', by R.L. Stevenson. The 'singular freshness' of the bodies made them ideal for dissection, but at the same time was a reminder of their sinister origin. In the story the bodies, in 'long and ghastly packages', are delivered to the teacher of anatomy. They include those of a freshly murdered young woman as well as one, of unexpected identity, newly dug from the grave.

The Burke and Hare affair had stirred the nation, and Parliament responded. The first Anatomy Bill was passed in 1831, and the Act became law in 1832. It was complicated, with twenty-one different clauses. Dissection was now legalized. The unclaimed dead could still be dissected, but dissection was no longer a punishment for murder. Anatomy inspectors were to be appointed to oversee the operation of the new law. There was, however, no legislation about the sale of corpses.

The latest Anatomy Act (1984) allows dead bodies to be used only for teaching and medical research, and under licence from the Department of Health. Even now the law occasionally comes into action. In April 1997, an artist and an ex-employee of the Royal College of Surgeons, London, were arrested on suspicion of using stolen parts of bodies and buried bodies without consent. The artist had made casts of various parts of the body, gilded them with gold or silver, and offered them for sale. He was charged with theft. Of course, art has always used the human body, but there have to be limits. For instance, in 1991 an artist and a gallery owner were convicted of displaying a freeze-dried aborted foetus as earrings on a sculpted head. There is a point at which public decency is outraged, or unacceptable distress or embarrassment caused to living people.

Grave robbing was by no means exclusive to England. In 1788 the black people living in New York petitioned the City Council to stop medical students removing their dead. The petition was ignored until the corpses of well-respected white citizens were snatched. In a subsequent riot, Columbia Medical School was ransacked, and by the 1880s seventeen states has passed anatomy laws. Corpses nevertheless remained in short supply and in the Baltimore medical schools in 1893 there were only forty-nine cadavers for 1,200 medical students. Grave robbing continued in certain states, and up until the 1920s Nashville's four medical schools are said to have been supplied from this source.

Unclaimed Bodies of the Poor and Disposessed

During the nineteenth century in England the body of a poor person was worth more dead than alive. In 1827, in 131 London parishes, 4,056 people died in workhouses. Of these, 736 were buried by friends, but more than 3,000 of the bodies were unclaimed. Many must have ended up on the dissection table. Workhouses were at this time very unhealthy places. Sanitation was non-existent, doctors were overworked and underpaid, nurses were often unpaid and unqualified, and beds were often shared. The poor would do anything to avoid death in the workhouse. Rather than face a pauper's funeral and the threat of dissection they would steal, starve, emigrate, take up prostitution, even commit suicide. Most people, including the anatomists, did not want to be dissected after death. But because dissection was necessary for training and for medical progress, people were resigned to the use of the pauper's corpse. After all, the workhouse pauper had already died, in a way, socially. The unclaimed dead could still be dissected under the Anatomy Act of 1831, but in the following decade there was a great growth in burial clubs and societies, through which poor people could save throughout life towards a burial grant and so avoid the pauper's grave and the anatomist's table.

In some countries the problem is with storage rather than availability. A report has draw attention to the shortage of bodies for dissection in medical schools in Bombay, India, in the 1990s. Each year about 500 unclaimed bodies are available, but the law requires them to be kept in storage for three days before they can be regarded as unclaimed. Many putrefy during this period because of inefficient air conditioners and the hot, humid climate.

Unclaimed Bodies Dying in Hospitals

There was a high risk of being dissected if you died in a nineteenth-century hospital. Patients were often easily deceived. At St Thomas' Hospital, London, very poor patients had to give one guinea when admitted so that they could be removed when cured or suitably buried if they died. However, if they died the body was quite likely to be sold to the dissecting room for four guineas.

Public feelings ran high at the time of the cholera epidemics, with rumours that cholera victims were being taken and used for experiments. In 1832 a child of three had died of cholera in the Cholera Hospital, Swan Street, Manchester. Her grandfather, on examining the coffin, was horrified to find that the head had been removed and replaced with a brick. In no time a crowd had gathered. The headless body was paraded round the streets and a rumour spread that the child had been murdered. Two thousand people rushed the hospital gates and attacked the building, breaking all the front windows and burning all the furniture they could lay their hands on. The wards were left untouched. After the police and troops had been called in, a courageous Catholic priest explained to the angry crowd that it was not the hospital doctor who had taken the head but an apothecary. The apothecary had run away and was never caught, but the head was recovered in his lodgings, sewn back on to the body, and the child then buried.

Bodies Imported from Other Countries

In the nineteenth century the importation of corpses from abroad (Ireland and the continent of Europe) was discussed as an alternative to body snatching. It never became an important source. Difficulties included the preservation of bodies en route, and the need to camouflage them because ships' crews hated 'abominable cargoes' of this sort.

Bodies Donated by Owners before Death

In Dublin in 1828, 400 people signed a pledge bequeathing their bodies for dissection. Many of the signatories were said to be connected with the medical profession, and although we do not know how many of them were eventually dissected, it was a spectacular attempt to set an example. In spite of this mass gesture it took over a hundred years for this volunteer source to provide more than a small proportion of the bodies needed. Even in 1934–5 only nine of 261 corpses dissected in Great Britain had been donated. The rest continued to be the bodies of paupers obtained from hospitals, asylums, prisons and, to a lesser extent, workhouses.

It was after the Second World War that the number of bequests showed a marked rise, and in the 1960s it accounted for 70–100 per cent of all bodies dissected. Of the 242 bodies dissected in 1969–70, 238 (98 per cent) were from bequests. The reasons for this change are complex, but it was a time of changing attitudes to religion, God and the afterlife, a growing disbelief in the spiritual significance of a corpse, and a more favourable view of science and medicine. There was an almost parallel rise in the incidence of cremation. Also, after 1949, government death grants and the National Health Service may have increased people's willingness to help with the training of doctors in this way, and today in the UK about 800 people a year donate their bodies to medical science.

The occasional enthusiastic donor stands out. A penny pamphlet published by Richard Carlisle in 1829 told the story of Pierre Baume, a French scholar. This man deplored the romanticism and wastefulness of regular burials. He was not content with writing in his will that his body was to be dissected. He also wanted his skin tanned and used to cover an armchair, either his skeleton presented to an anatomy class, or the skull presented to the London Phrenological Society and the other bones made into useful things like knife-handles, pin-cases, small boxes, buttons, etc.

Dissection

During dissection there is a piecemeal whittling away of bits of the body as organs and tissues are revealed. It is proper that rituals of respect for the dead should be observed, and many medical schools have brief remembrance services for the year's cadavers after the remains have been gathered prior to burial.

Not everyone in medicine enjoys dissection. The father of the great composer, Hector Berlioz (1803–69), was a physician, and he wanted his son to be one too. He forced him to attend medical school, saying he would lose his allowance if he didn't, and gave him an anatomy textbook with a promise to buy him a fine flute from Lyon if he applied himself to his anatomical studies. But Hector hated the dissection room, humming opera tunes while he dissected, and began to spend more and more of his time in the library of the Paris Conservatory of Music. He finally abandoned medical studies, at which point his father cut off his allowance. His son was thus launched, unrepentant, into the world of music in which he could fully express his romantic imagination.

Nowadays there is no shortage of bodies; more than enough come from bequests. As it happens there is at the same time less emphasis on dissection in medical schools. As generally practised it is not a particularly efficient method of learning anatomy, and although essential for specialist surgical training it is less necessary for the ordinary medical student.

Alternatives to Dissection

Several UK medical schools no longer dissect the human body. Instead, the students learn from models, permanent dissection specimens and bones. Teaching aids include computer programmes such as ADAM (Animated Dissection of Anatomy for Medicine), in which thousands of colour pictures take the student through forty layers of the human body; it also allows for interactive learning in which the user cuts through the body with an on-screen scalpel to display organs and tissues. ADAM's library contains various surgical proceedures that give simulated hands-on experience for students and junior hospital doctors. Another teaching aid, the Visible Human Project, is described below. Perhaps dissection is destined to become no more than a minor part of anatomical and surgical education.

Further Uses for Whole Corpses

There have been moves to use the freshly dead corpse not for dissection, but for training medical students and junior doctors in how to insert a breathing tube into the windpipe. It is more satisfactory than using mannequins (models) and certainly better than practising on live, anaesthetized patients. But people are not enthusiastic about this. It seems less necessary than dissection, and the British Medical Association and other European medical associations have ruled that it is unacceptable. As an alternative, a life-size human patient simulator has been created in the USA. It is a 'living doll': it has a beating heart, can talk and breathe, and produces blood, tears, and urine. Controlled by sophisticated electronics, it can be male or female, young or old, fit or unwell, even pregnant. It will be used to train doctors to handle medical crises in preparation for the real thing, the equivalent to a flight simulator for an airline pilot.

A possibly less objectionable, but still controversial use for corpses is in car crash tests. Researchers at Heidelberg University have used about 200 corpses for this purpose, with permission from relatives, and studies have also been carried out in other countries. The bodies are strapped into cars which are then deliberately crashed to discover the effects on passengers and drivers and the actual injuries suffered. Human corpses are said to give more reliable results than dummies, and have played a part in the development of air bags.

The Visible Human Project

The National Library of Medicine in Bethesda, Maryland, USA, has made a remarkable record of the human body. Two corpses, male and female, were obtained, the male being a thirty-nine-year-old killer who was executed by lethal injection in Texas, and the female a fifty-nine-year-old from Maryland who had bequeathed her body to science. The bodies were subjected to various scans (X-rays, CAT and magnetic resonance imaging) and then frozen in gelatin and sliced from head to toe at intervals of one millimetre (the male) or one third of a millimetre (the female). Each slice was photographed to give 1,800 (male) and 5,000 (female) cross-sectional pictures of the corpses, and these, together with the data from the body scans, were fed into a computer. Thus were produced databases on the two persons (the 'Visible Human Male' and the 'Visible Human Female') large enough to fill more than seventy compact discs. Using computer graphics, the information can be converted into three-dimensional anatomical models that can be rotated, cut open with an 'electronic scalpel' and examined, and then put back together again, as required. A similar computerized model has been made of a human embryo.

This archive of anatomical images of the Visible Human Male and the Visible Human Female can be browsed, and downloaded on to personal computers via the Internet. It will help, for instance, in training medical students and surgeons, as well as providing a complete visual record of the human body. It is one step towards realizing the fantasy of Isaac Asimov's 'Fantastic Voyage', in which a group of scientists are miniaturized and launched on a voyage of exploration by being injected into the bloodstream of a dying man. It is the anatomical equivalent of the Human Genome Project (see chapter 10).

12

Using Parts of the Body: Transplantation

The subject of transplantation sits naturally in a book about death, first because dead people supply organs and second because these organs prevent death in those that receive them. The human body is a wonderful, complex structure. Once dismantled, many useful things can be done with its component parts, either the living tissues or the dried, dead remains. As a student I was taught by the mathematical biologist J. B. S. Haldane (1892–1964). He once pointed out that even the ugliest human exteriors may contain the most beautiful viscera, and he said that occasionally on buses, to make up for the facial drabness of his fellow travellers, he dissected them in his imagination. This was many years before the birth of modern transplantation.

Bones, Teeth, Placentas, Hair
Bones

Since the beginnings of time people have found practical uses for bones, using them as tools, weapons, containers and for art. Animal bones were the most plentiful, and some of them bear names that reflect the uses to which they were put. Thus, *tibia* (the shin bone) means a flute or pipe and *fibula* means a brooch. Bones that are too small actually to be used, such as the bones of the middle ear, have sometimes been named on account of their shape (*malleus* = hammer; *incus* = stirrup).

The human skull can be used as a container. It is not an ideal drinking vessel, but Herodotus (450 BC) had reported on its use for this purpose and the idea appealed to Lord Byron, who wrote 'Lines Inscribed on a Cup Formed from a Skull':

> Start not – nor deem my spirit fled:
> In me behold the only skull,
> From which, unlike a living head,
> Whatever flows is never dull.

I lived, I loved, I quaff'd, like thee:
I died: let earth my bones resign:
Fill up – thou canst not injure me;
The worm hath fouler lips than thine.

Better to hold the sparkling grape,
Than nurse the earthworm's slimy brood;
And circle in the goblet's shape
The drink of Gods, than reptile's food.

Where once my wit, perchance, hath shone,
In aid of others' let me shine:
And when, alas! our brains are gone,
What nobler substitute than wine?

Quaff while thou canst: another race,
When thou and thine like me are sped,
May rescue thee from earth's embrace,
And rhyme and revel with the dead.

Why not? since through life's little day
Our heads such sad effects produce;
Redeem'd from worms and wasting clay,
This chance is theirs, to be of use.

The hard surface of bones has made them useful surfaces for drawings, scratchings, engravings and carvings. It may seem odd that human bones could themselves be considered as art objects. Yet this is the case, in the sense that there are certain collections of human bones, arranged in patterns, and referred to as decorative ossuaries. They are more than a mere collection of bones in a bonehouse. Examples include the 8,000 thighbones and 2,000 skulls in the crypt at St Leonard's Church in Hythe, Kent, England, and the display of bones in the Capuchin Church of the Immaculate Conception, Via Veneto, Rome.

Frederik Ruysch (1638–1731) took things further. He was Professor of Anatomy at Amsterdam, and established a museum of human and comparative anatomy. He was a skilled preserver of bodies, and an expert injector of red wax and mercury into blood vessels to demonstrate their anatomical course. Ruysch prepared some magnificent decorative tableaux, constructed from assorted organs and bones. They were designed to give a picturesque effect rather than to educate or illustrate biological principles. He was the baroque artist of death. In one of his arrangements a human skeleton plays a violin made up of arteries and a thighbone, and another holds lengths of bowel arranged in graceful whorls.

Teeth

Teeth last the longest of all body parts, and at one time, before the development of artificial teeth, human teeth were used to make dentures, as noted in chapter 11. In the old days an occasional tooth transplantation was done, using the unerupted teeth of children, but such a procedure is unnecessary, generally unsuccessful, and unjustifiable. Human or animal teeth have also featured in *ornaments* such as bracelets and necklaces.

A bizarre dental hoard was accumulated by an Italian, Brother Giovanni Battista Orsenigo. He was a monk but also a dentist, and between 1868 and 1903 he extracted and kept in his collection a total of 2,000,744 teeth. This works out as a rate of nearly six extractions a day.

Peter the Great of Russia (1672–1725) was an avid collector of natural history curios, some of which are still to be seen in the Leningrad Museum of the Institute of Ethnography. He fancied himself as a dentist, and in his collection are thirty-five human teeth. He was a powerful and cruel man and when the mood took him he is said to have demanded that anyone, even a casual passer-by, sacrifice a tooth.

Placentas

Although not a regular part of the body, the placenta is a human organ. Weighing about 500 g, it is a noteworthy object, if only for its value as food. Indeed, the word placenta comes from a Greek word meaning cake. The female of every mammal except the human eats the placenta after birth of the infant, thus enjoying a worthwhile meal of protein, iron and other good things.

Human placentas are recognized as powerful, magical objects in the folklore of many peoples, and are often used in medicine and witchcraft. At one time there was a belief that the healthy growth of the child or the well-being of the mother could be ensured by burying the placenta. Hundreds of apparently empty pots were recently discovered, buried in the cellars of houses in south-west Germany in the sixteenth to nineteenth centuries. Scientists found that they all contained amounts of cholesterol that would be expected in placentas at the end of pregnancy. The placentas themselves would have decayed and disappeared.

In the modern world placentas have been used as sources of human cells for culture in the laboratory. Some of the early work on polio viruses was carried out in human placental cells. Placentas are also a souce of certain drugs. In South Africa an ethical row erupted when it was found that placentas were being disposed of to a pharmaceutical company as 'organic waste', without informing the women from whom they were taken.

Hair

Human hair grows very long compared with that of other primates, but it does not last for ever, and wigs (see box) have been worn since the time of the ancient Egyptians. They have never been easy to make and until recently only the wealthy could afford them.

Spare Part Surgery: Organ and Tissue Transplants
Historical Beginnings

Transplants of a person's own tissues from one part of the body to another is not much of a problem. It was carried out by early Hindu surgeons as long ago as the sixth century BC, and introduced into Europe in the sixteenth century by the Italian surgeon Caspare Tagliacozzo. For instance, skin from an arm was used to reconstruct a nose. Nowadays surgeons can sew back on a severed finger or a whole limb. The first successful result with a limb was achieved at the Massachusetts General Hospital in 1962. The arm of a twelve-year-old boy had been severed just below the shoulder: it was packed in ice, taken to hospital with the boy, and grafted back into place. Such an operation is very delicate and demanding work, joining up nerves and small blood vessels, but two years later the patient had regained almost full use of the arm and hand.

It is only in recent times that it has become possible to transplant tissues and organs from one individual to another. The fresh corpse then suddenly became a valuable source of live organs and tissues. These are used for the benefit of sick people and have saved tens of thousands of lives. The corpse, rather than being plundered, is truly serving human-kind.

The first transplants were of skin, cartilage, teeth and corneas. Corneas have the advantage that they do not have the blood vessels that bring in the trigger-happy immune cells (lymphocytes) that cause rejection. The

Wigs and body hair

Many men lose their scalp hair with a characteristic pattern in middle age. Bald men can be attractive if they have other assets, but for the most part baldness signifies advancing age – plus a slight hint of lost sexual potency. Going bald depends on male hormones, yet eunuchs do not go bald.

Louis XIV of France (1638–1715) became bald when he was in his thirties and was very conscious of this. He wore a wig: it was handed to him in bed through closed curtains when he got up in the morning, and he handed it back when he retired to bed in the evening. At about this time wigs became very popular in Europe and before the end of the eighteenth century there were 850 wig-makers in Paris alone.

In the eighteenth century wigs literally reached the peak of their glory, being up to two feet high in women. They were such a trouble to put on that it was customary to sleep in them. We must remember that 200 years ago soap was not cheap or easily available. Wigs (and the home-produced hair underneath them) inevitably collected all manner of human parasites, especially lice. Samuel Pepys describes his wife's visit to the barber, and the wondrous colony of creatures that had made their home in her wig.

Where did the hair come from? The best hair was from other humans, and poor people could earn a useful sum by surrendering long lengths of beautiful hair. Even today there is said to be a religious community in India which maintains its schools and churches by cutting off the long black hair of the girls and selling it to Western countries for the manufacture of wigs. Corpses, of course, were an alternative source.

Beards and moustaches have come and gone since the days of the Babylonians, and have been worn as a mark of distinction as well as of masculinity. Luckily they do not suffer the fate of scalp hair, and stay in place to reassure bald men. Some women find beards and moustaches attractive, and would say that a kiss without a moustache is like an egg without salt. At times they have been ceremonial, perfumed or artificially shaped, and cutting off a man's beard could be a humiliation comparable to cutting off a woman's hair. Arabs have sworn 'by the beard of the Propher'. The English monarchs Edward VI and Elizabeth I tried to impose a tax on beards, but were unsuccessful.

first successful corneal transplant was carried out in 1905 by Eduard Zirm in Olmutz, Moravia, but it was another fifty years before the process was widely performed. Bone and cartilage are also easier to transplant because they only act as a structural framework, eventually being replaced by the recipient's own materials.

Modern Transplantation

Transplant technology is advancing so fast, raising such basic ethical and legal questions, such matters of life and death, that it merits treatment at some length here.

- In 1954 the first successful *kidney* transplant was made, from a healthy twin to the other twin, who was gravely ill with kidney failure.
- The first *liver* was transplanted in 1960. At first, survival rates were poor, and rejection was a problem (see box). This operation is a demanding one, takes up to eight hours, and needs many doctors, nurses and technicians. But difficulties were overcome, and now more than 1,000 livers are transplanted each year in the USA.
- A *heart transplant* was first attempted in 1967, by Christian Barnard, and the recipient lived for eighteen days. The second lived for eighteen months, and over the next seven years more than 100 heart transplants were performed in America alone, at fifteen centres. Yet it was expensive, required highly skilled practitioners, and rejection continued to be a problem. By 1981, good drugs had become available to suppress the immune system and prevent rejection, and now there are more than 150 heart transplant centres in the USA. Five-year survival rates are more than 70 per cent. Worldwide, 3,000–4,000 hearts are transplanted each year, the recipients ranging from newborn babies to seventy-year-olds, and the patient usually returns to school, work or active retirement.
- Also, by 1981, *hearts plus lungs* were being transplanted, which is not so daunting as might be imagined because the major blood vessels connecting heart to lungs are transferred intact.
- Transplantation of *pancreas* (for diabetics), and of sections of *bowel* (for those suffering from intestinal diseases) is now quite common. Segments of lung, pancreas or liver can come from live donors, although taking them is not free from risk.
- Human *embryos* can be used as sources of tissues and cells. This has begun but has generated heated discussion and legislation. People feel strongly about it and the subject is tied up with their feelings about abortion. Faced with the prospect that scientists seem to be able to accomplish almost anything, they respond with a deep repugnance to gross deviations from the natural order (the 'Yuk' factor).

Keeping Organs in Good Shape

After brain death (see chapter 5), different tissues stay alive for different periods. Kidneys, livers and hearts, if they are to be transplanted, have to be removed within thirty minutes of death, and will remain alive and transplantable for up to six hours (longer in the case of kidneys) when

How the body rejects transplanted organs and tissues.

Organs soon die without a blood supply, and joining up the arteries and veins of the transplanted organs to those of the recipient is the first thing that has to be done. This is a mechanical task for the surgeon, and it takes great care and skill. But the immune inspectors (the T-lymphocytes or T-cells) then soon arrive, travelling in the blood, and recognize the transplanted organs as foreign, just as they would recognize an invading foreign parasite or microbe. They go into action, setting off the formation of antibodies and immune killer cells specially directed against the transplanted materials. Hordes of immune cells now pour into the organ, the blood supply is interrupted and the organ is killed, rejected. The immune weapons that enable us to fight off potentially lethal infections have been brought to bear on the foreign organ.

How does recognition take place? Each human being has a set of unique personal labels on all cells in the body. These are the labels (HLA or human leucocyte antigens) that are identified when we do tissue typing. In the body they are constantly being inspected by the T-lymphocytes and if any are identified as foreign ('non-self'), the cells carrying them are destroyed by a powerful army of immune weapons (killer cells, antibodies and immune substances called cytokines).

The unique personal labels are formed under the instruction of the individual's genes. Therefore, if the transplanted organ comes from an identical twin (formed from the same fertilized egg as the other twin) all the genes, and thus all the cell labels, are identical. The immune system does not recognize them as foreign and there is no rejection. Hence identical twin transplants are the most successful. The next most successful are transplants from siblings or close relatives, because they have many genes in common. But the immune system needs to be suppressed, all the same.

Organs from unrelated people may be acceptable if the tissue typing shows that the donor labels are not too dissimilar from those of the recipient. The immune system is suppressed with powerful drugs such as the steroid prednisolone, or azathioprine, or cyclosporin A. Unfortunately, the suppressed immune system is now less able to resist infection, and transplantation has been plagued by a host of infectious complications. This is why immunosuppressive teatment must be delicately balanced, with careful choice of drugs and doses, and the patient must be monitored for infections or other complications.

stored at ordinary refrigerator temperature (5°C). If a kidney is joined up to an apparatus that will pump fluid through its blood vessels it can be preserved for as long as two or three days, but this is rarely done.

Corneas should be taken within six to twelve hours and can be stored for a few days, and skin within twelve hours with storage for up to seven days. With skin, which gives life-saving coverage for the body after bad burns, very thin slices of the patient's own skin (which can regenerate itself at the place from which it is taken) are generally used. Live, laboratory-cultured skin cells are being developed as an alternative (see below).

Blood and plasma have a much longer life, measured in weeks, and, unlike organs, can be successfully stored in the frozen state. This is also true for fertilized eggs, very early embryos and semen. Transfusions of blood (blood cells or plasma) and of bone marrow play a vital part in the treatment of certain diseases.

Medical Problems with Transplantation

The problem of rejection is considered in the box. In addition, as mentioned there, when organs and tissues are transplanted there is always the possibility that something undesirable is going to be inadvertently transplanted at the same time. Infectious agents present in the donor have been the problem. They are not often transmitted by solid organs but the risk is higher with blood or blood products. Infections transmitted in this way have included malaria, syphilis, hepatitis, the AIDS virus (HIV) and various other viruses. Luckily the risk can be greatly reduced by screening the blood of donors and eliminating those carrying the infection. Questionnaires are used as well as tests, and a donor to the American Red Cross blood-bank may have to answer questions such as 'Have you . . . at any time since 1977 had sex with a man who has had sex with another man even once since 1977?' In the case of blood products that do not contain cells (plasma and other materials), the risk is further reduced by treatments that kill off infectious agents, but which would damage live cells.

Where Do Transplanted Organs Come From?

Who are the donors? This is where important ethical issues have to be faced. In Western countries the public are familiar with organ procurement as portrayed in Mary Shelley's *Frankenstein* or the novel *Coma* (see box). These are fantasies, but many people, although more understanding and cooperative than in the past, retain lingering doubts and suspicions about the activities of enthusiastic doctors and scientists.

The first concern is how to decide when a live person turns into a corpse that can be used as a source of organs. In other words, how do you define death? This key question was discussed earlier in the book (chapter 5). At one time the main worry was that people might be buried before they had actually died, but this can be avoided if you delay the burial and give the corpse an opportunity to show signs of life. For organ transplantation, however, time is of the essence, and fear of premature burial has been replaced by fear of premature organ removal. The concept of 'brain death' was introduced to assuage this anxiety.

The mad scientist and the manufacture of monsters

The idea of a surgeon making a new person or a monster by joining up (transplanting) living bits and pieces has long fascinated science fiction and comic writers.

The English novelist Mary Wollstonecraft Shelley (1797–1851) eloped with the poet P. B. Shelley when she was sixteen, and married him three years later. In the summer of 1816 the Shelleys were staying on Lake Geneva with Byron and another friend, and Byron suggested a ghost story competition. As her contribution Mary began to write *Frankenstein, or The Modern Prometheus*. It was published in 1818, when she was twenty-one. She wanted the story to be horrific, 'to make the reader dread to look round, to curdle the blood and quicken the beatings of the heart'.

Frankenstein was the name of the scientist who manufactured an 'android' (human-like, as opposed to robot or bug-eyed) monster out of corpses from graveyards and dissecting rooms. He gave the monster life by passing a powerful electric current (from lighting) through it. Frankenstein's creation is a fairly intelligent, well-spoken 'being', with fears and anxieties of the human kind, but it eventually gets out of control and turns on and destroys its creator, a parable for the idea that scientific inventions can backfire and damage humanity.

Frankenstein became the prototype 'mad scientist' of science fiction and comic books. The surgical construction of monsters is the basis for the H. G. Wells story *The Island of Dr Moreau*, in which animals are vivisected in an attempt to humanize them. The transplantation/body construction nightmare reaches contemporary expression in the novel *Coma* by Robin Cook (1977), or the concept of a 'bioemporium' where brain-dead bodies are kept alive for use as organ donors, experiments or the training of doctors.

Only a small proportion of the world's population have access to organ transplantation. Modern, well-equipped hospitals are for a privileged minority. But even in developed countries there is a chronic shortage of organs. Furthermore, the supply of organs is limited by religious attitudes and customs, as well as by practical considerations. Most important of all, the law has not yet caught up with the transplant surgeons. Medical advances have been rapid, and when we try to find out what is and what is not permissible we enter a legal minefield.

Nevertheless, organs give new life to others, the results are excellent, and transplant surgeons are enthusiastic. Naturally, more transplants are done where a high proportion of the population has health insurance or access to government health services. Practices differ in countries with different religions, so that in discussions of what is acceptable and what is not, religion and ethics as well as economics and medical technology must come into the picture. The use of foetuses and anencephalic infants poses a particular problem (see below).

Cadavers as the Main Source of Organs

About 90 per cent of organs transplanted are taken from cadavers (which means brain-dead donors, or those whose heart has stopped beating), and the rest are from live donors. Kidney transplantation is an established, successful way to treat kidney failure and tens of thousands of patients have thereby been restored to health. Unfortunately the number of kidney transplants is still limited by the number of kidneys available, and hundreds of people die unnecessarily because of this shortage. Problems include lack of publicity, adverse publicity after scares about over-hasty pronouncement of death, failure to get permission from next of kin (about two-thirds give permission when asked), and delays in removing and preserving the kidneys.

Possible donors are those dying with from such causes as head injuries or strokes, who are usually patients in accident and emergency, intensive care, coronary care or neurosurgical units. But family members confronted with the sudden death of a close relative are overwhelmed by emotion, and at this time a decision on organ donation can be traumatic for them. It is estimated that a large proportion of the number of kidneys needed could be obtained by 'more efficient harvesting' of the potential supply. A voluntary organization called BODY (British Organ Donors Society), set up in the UK in 1987, gives help and advice for donors and encourages the carrying of donor cards. An example of the immense benefit that can come from the body of a single donor is provided by the case of a twenty-one-year-old Englishwoman who died in a car crash in America at the end of 1996. Her bones provided about seventy grafts for hips and spines, her corneas gave sight to two blind people, and her heart valves gave new life to two patients with heart disease.

In the USA in 1996 31,000 patients were waiting for kidney transplants and another 1,230 for kidney plus pancreas transplants. In the UK 5,246 were waiting for kidneys on 30 June 1996, a sad figure in view of the excellent results obtained and the life given. The waiting list grows and on one or two occasions transplanted organs have even been re-used.

Because of the shortage of organs, older people (more than sixty years) are often used as donors. The results are not quite as good as with organs from young people, and it has been suggested that their kidneys, because they are old organs, should be given where possible to older people. When Deng Xiaoping, the leader of China, died on 20 February 1997, he decreed that transplant surgeons should be able to use any part of his body. But he was ninety-two and had been ill for a long time, and his organs were perhaps not in the best shape. A Chinese citizen might be honoured beyond his wildest dreams by acquiring such organs, but the real value of Deng's offer was that it might encourage others to do the same thing.

People needing a new kidney are kept alive by dialysis (artificial removal of waste material from the blood) while they are waiting for a transplant. The desperately ill patient needing a heart transplant can be kept alive for up to four months by an artificial blood-pumping machine, ready for a new heart when it becomes available; but only the wealthy or the very important would get an opportunity like this.

With brain-dead donors on life-support machines, the blood is still circulating through the organ at the time of removal, and results are better than when the organ is removed after the heart has stopped beating, for instance in people pronounced dead on arrival at hospital. In the latter case the organ has had time to deteriorate. Further steps can be taken to ensure that organs from brain-dead people are kept in as good condition as possible. Patients dying from bleeding into the brain can be transferred to intensive care units for a brief period and maintained on 'elective ventilation', during which brain-stem death is confirmed and organs then harvested. This sounds like an ethically acceptable practice, but its legality needs clarification, and it has not been widely adopted.

In the UK organs can be taken only if the person has previously given permission, for instance by carrying the organ donor card introduced in 1971 (26 per cent of the population have these but a smaller percentage carry them), or by making a request in the presence of two witnesses. It is part of a 'contracting in' system, and an individual can request donation of his organs after death for transplantation, education (dissection) or research. Nevertheless, surviving relatives can still overrule the deceased's wishes. The Human Tissue Act of 1961 authorizes organ donation but doesn't require it. The exact status of organ transfer is still obscure, from a legal point of view.

Who Owns a Dead Body?

This question was discussed at the beginning of chapter 11. There is still some confusion about who exactly is in lawful possession of the body. In the seventeenth century it was accepted in common law that there were no property rights in a dead body. Officially no one owned it, and legally the dead had no rights. But the surviving spouse, children or next of kin had a right of possession (a sort of trust), for disposal (burial) of the body. In 1989 a widow in the USA was awarded $150,000 for the emotional distress suffered on learning that her husband's brain had been removed during autopsy without authorization. The right of possession of the body means that it must be received intact for burial. In other words, no dissection or removal of organs may take place without authorization.

Because of the shortage of organs for transplantation there have been moves to change the law so that organs can be removed from any cadaver in the absence of an expressed wish against it. This would be what is called a 'contracting-out' or opting-out arrangement. It would yield a substantial increase in organs, resulting in many lives saved, and evidence from a Gallup survey in the mid-1990s shows that only 28 per cent of adults in the UK would oppose it. In Belgium, where an opting-out scheme was introduced in 1987, only 1.5 per cent of the population objected, and the number of donors and transplants doubled. In The Netherlands, however, with a similar well-developed social security system and a large number of well-equipped hospitals, the legal basis for using organs is the same as for doing post mortems: it requires informed consent, as in Anglo-American law. Dutch organ procurement rates (from road accident victims dying after admission to intensive care units), are much lower than in Belgium. Germany has similar difficulties, with 4,000 donors needed each year to match demands but only a tiny fraction of suitable or available cadavers.

In Brazil in 1997, a new law says that anyone's organs can be taken after death unless the individual has previously documented an objection: in other words, an opting-out scheme is in place. Brazil has an organ shortage in spite of the fact that in São Paulo alone there are 5,000 violent deaths a year as potential sources of organs. Unfortunately the new law may make little difference without better facilities for collecting and transporting organs. Under these circumstances a certain amount of illegal organ trafficking seems inevitable

Persistent Vegetative State

The use of patients in a 'persistent vegetative state' as organ donors is controversial. The term itself is unfortunate because it implies a 'vegetable-like' condition. It is a state of 'wakefulness without aware- ness' following severe brain damage, which has been present for at least a month. Such patients can breathe unassisted, but they often seem to be asleep and are incontinent of urine and faeces. They make chewing or other movements, but there is no evidence of communication or purposeful movement, and the cortex of the brain shows only very low levels of activity. The condition resembles that of infants with anencephaly. Is it a 'living death', and does it show no respect for the individual when a meaningless life is artificially maintained in this way? On the other hand, can we be sure that these patients cannot experience anything? Can they be regarded as cadavers?

The trouble is that there have been cases of recovery of awareness after many months, even a year. Although this is rare, and in such cases the persistent vegetative state may have been wrongly diag- nosed in the first place, it means that great care has to be taken. Some say that if the patient's wishes are known and the family asks that life support be terminated, then the patient could become an organ donor. Others disagree and maintain that this would be the first step on a slippery slope, which could end with Alzheimer's disease patients being used as donors. In the UK and USA it is illegal to take active steps to accelerate death, but if the individual is allowed to die by withdrawing treatment, the organs then become unsuitable for transplantation.

In 1997 there were about 1,000 patients in the UK who had been vegetative for over three months, so the problem is not a rare one. Are they no more than a 'second type of corpse' that still breathes, or are they still real human beings with rights? A recent survey showed that nearly all doctors believe it is in order not to treat acute infections and other life-threatening conditions in such patients, and a smaller proportion say that it is sometimes appropriate to withdraw artificial feeding. However, the official BMA guidelines, adopted by the courts in 1993 after the case of Anthony Bland, who suffered severe brain damage during the Hillsborough football stadium disaster, say that decisions to withdraw treatment should be taken only after twelve months.

Organs from Live Related Donors

The fresher the organ the better, and kidneys from live donors give better results than those from dead ones. In the USA in 1994, 35 per cent of transplanted kidneys were from live related donors. Kidneys from relatives are more likely to give a good tissue match, although with the latest immunosuppressive treatments kidneys from unrelated donors are almost as good. Donations from relatives, however, are more commendable, and are allowed by law, although in the UK the law fails to consider donations by spouses or partners ('cohabitees').

Luckily, nature has provided a spare kidney which can be used in this way. When the organ is unpaired (as with the liver or heart) the donor's life would be forfeited, and English law does not recognize that a donor can consent to this. It would be homicide. Some have argued, not unreasonably, that if you are allowed to give your life for your country, or to save another person from drowning, why should you not be able to give it to save a loved one who is dying for want of a transplant?

Another difficulty is when the donor is a child. For instance, several countries allow kidney transplants from one twin to another, as long as the transplant is necessary to save life, and as long as the donor twin understands and consents to it. But it remains a difficult field legally, and in Australia any tissue donation by a live child is normally prohibited.

Organs from Live but Unrelated Donors: The Problem of Commercial Trafficking

When the live but unrelated (that is, not genetically related) donor is a spouse we can only admire the donation as a loving act. In the USA in 1994, 196 of 8,114 kidney transplants were from live unrelated donors. Results are better, the organs being fresher, than with cadaveric donors. On the other hand, it is with this type of donor that the sale of organs comes into the picture, and there is a widespread belief that organ selling and trafficking should be stopped. Commercial dealings in organs are prohibited in the UK by the Human Organ Transplantation Act 1989. However, such trading has been common practice in Asian countries. In India, for example, a healthy adult with a daily wage of 30 pence can earn up to £750 by donating a kidney, enough to raise his entire family from the depths of poverty. His religion seems to support this sacrifice, because Hindu mythology has stories in which parts of the human body are used for the benefit of other humans. In Indian law, the moment of death was defined as the cessation of the heartbeat, and organs could not be taken from donors certified as brain-stem dead if the heart was still beating. This encouraged the use of live donors, and 70 per cent of transplanted kidneys came from paid unrelated donors. In 1994, however, the Indian parliament passed a law outlawing trade in human organs and redefining death so that brain-stem-dead donors could be used. Humans have two kidneys, as noted above, so these are the organs donated; there can be no trade in livers or hearts from live donors. Yet 10,000 patients in India are diagnosed each year as needing liver transplants. Obviously only the very rich can afford to go abroad for their transplants; perhaps a few more will be performed in India because of the new legislation. A change in attitudes will be needed, because relatives generally do not offer cadavers for organ transplants.

Islamic doctrines say that the dead body must be respected, and the prophet Mohammed noted that 'the breaking of a bone of a dead person is equal in sin to doing this while he is alive' But scholars and jurists from Muslim countries are divided about the rights and wrongs of transplantation, which is not dealt with in the Qur'an or the Hadith. Those from Arab countries see it as permissible, whereas those from the Indian subcontinent say it is not. In 1995 an edict gave permission for Britain's two million Muslims to donate their organs for transplanting. So far there have been more than a hundred kidney transplants in Kuwait, using organs of paid donors from India and elsewhere. One organ recipient from the Middle East felt so bad at the sight of his emaciated donor that he put him into a hotel for a month to nourish him before the operation.

According to Buddhism there is nothing intrinsically sacrosanct about the human body, dead or alive: donation of parts of it is an act of compassion and can help in the realization of spiritual progress. In Japan most kidneys come from living relatives. It is illegal to sell organs, and Japanese Buddhism says that they should be given without self-praise and 'with pleasure, forgetting oneself as the giver, the one who receives it, and the gift itself'. Only a quarter come from cadavers (compared with about three-quarters in the USA), and cadaveric donations are being encouraged. But since the idea of brain death has been officially recognized in 1997, thousands of Japanese people could be taken off life-support machines and the supply of fresh organs improved.

The Jewish religion takes the view that organ donation is acceptable. After all, the first recorded transplantation was in the Book of Genesis when God, seeking a partner for Adam, removed a portion of his body and converted it into Eve. In Israel, health ministry regulations originally prohibited people from giving an organ to anyone who was not a first-degree relative. Now, two-thirds of all transplanted kidneys are from live donors and only a third from cadavers. There is a shortage of organs, with 1,000 waiting for kidney transplants but only 160 transplanted each year. Allegations that impoverished Palestinians had sold kidneys for transfer into wealthy recipients were investigated and culprits dealt with severely. New regulations introduced in 1996, however, have allowed other genetic relatives (grandparents, grandchildren) and certain non-relatives (spouses, cousins) to donate their kidneys or a lobe of lung or liver, and it is hoped that the number of donations will double. Nevertheless, many Israelis, whether out of superstition, fear of religious laws or fear of removal of organs before death, remain reluctant to bequeath their organs.

Commercial trafficking in organs causes great controversy. It was outlawed in the UK following the publicity given to the transplantation in London a few years ago of kidneys from paid, unrelated donors from Turkey. Nevertheless, the practice is likely to continue in certain countries and it is perhaps best to aim at control rather than abolition. It could be referred to as 'rewarded giving', to take place only within strict medical and ethical guidelines. It can be argued that a person has a right to sell something that he clearly owns more exclusively than he owns his worldly property. A parent selling a kidney to allow purchase of life-saving health care for a child is surely no different from a parent donating his own kidney to his child.

On the other hand, it is immoral when a person is driven by sheer poverty to make such a sacrifice. In Europe, people wait an average of three years on kidney dialysis before having a transplant. Not surprisingly, there have been reports of Europeans going to India and elsewhere and paying for transplants from live donors – a type of 'organ tourism'. But the results are often poor, because transplant practice tends to be below standard in such places. Organs should be donated altruistically, as a true gift. When money comes into the process, the road is open to exploitation of the poor and vulnerable. Horrifying stories of unlawful trading in organs (mostly kidneys and corneas) removed from the unclaimed dead or from kidnapped people have come from Latin America and Russia. In October 1996 *The Times* reported the case of the mortuary director of a public hospital in Cairo, arrested for removing eyes from corpses and selling them for transplantation. If foetal organs and tissues (see below) were bought and sold, there would be the spectre of women conceiving just to produce a saleable product. Commercial trafficking in foetal organs and tissues has been banned in the USA since 1988.

Organs from Executed Prisoners

In some countries the organs of executed prisoners are taken and used for organ transplants. This has been done in China, and also in Taiwan where since 1990 organs have been taken from twenty-two of a total of fifty-one executed prisoners. In the case of a potential organ donor, execution, which is normally by shooting through the heart, is converted to shooting through the head, avoiding damage to the corneas.

Organs from Anencephalic Infants

About 1 in 2,000 newborn babies are anencephalic. This means that all or most of the forebrain is missing, but the rest of the body is normal. They are nursed, but nearly all of them die after about a week. Their hearts, in particular, can then be used to give life to desperately ill children. However, according to the law a live-born anencephalic infant is *not* brain dead. At times, although not generally in the USA and UK, these infants, when they fail to breathe on their own, have not been artificially ventilated and have been used as a source of organs.

Is the anencephalic infant to be considered a non-person, with no rights and, from a legal point of view, 'dead'? Can we agree with Dr M. R. Harrison, who suggested in the *Lancet* in 1986 that 'the ability to transplant foetal organs may now give us the chance to recognize the contribution of the doomed foetus to mankind'? We know that a baby born in this condition will die with cardiac arrest after a few days, even when artificially ventilated, and it can be argued that using the organs to give life to another may comfort distressed parents after the birth of such a disabled infant. In some countries (for example, Germany and parts of the USA) more active steps are taken. Anencephalic infants are 'harvested' by inducing birth and keeping them resuscitated for an hour or two before, at a convenient time, taking organs. The anencephalic is judged never to have been alive. It may be noted that, looked at this way, many animals – for example lizards, shrimps – are 'dead', and clearly different standards are being applied to humans.

Although anencephalics are doomed to die, one can argue that this is not the point, strictly speaking. Infants with grossly abnormal hearts or lungs are also going to die (unless given transplants), but we would not sacrifice them and use their organs. Another twist to the debate is that anencephaly, from an anatomical point of view, does not necessarily mean complete absence of the forebrain. There is a failure of normal development, but it is not an all or nothing condition, and some anencephalics may have rudimentary development of the forebrain. The difference is quantitative rather than qualitative. The anencephalic child, moreover, can often cry, swallow or withdraw from a needle, and conceivably the midbrain allows it to have certain experiences. It is certainly not equivalent to a guillotined individual, with no brain at all.

It seems that the ethical problem about the anencephalic donor, which has aroused great discussion about brain death and brain absence in transplant donors, is going to be solved by the condition becoming much rares. Now that anencephalic foetuses can be identified early in embryological development and aborted, anencephaly is becoming less common in the USA, some European countries and Australia.

Organs from Foetuses

Recent advances in medicine are making the aborted foetus a rich source of living tissues that can be used to treat disease. Foetal skin has been used to cover severe burns, foetal pancreas has been transplanted into patients with diabetes, and there are promising attempts to treat Parkinson's disease by injecting foetal brain cells into affected areas of the brain of patients. In Britain about 100,000 people suffer from Parkinson's disease, and the injected foetal brain cells appear to settle down nicely, establish connections and help the patient. If this method of treatment is established, the demand for feotal tissues will soar.

Obviously the tissues should be obtained from a dead foetus, incapable of independent existence, and the mother must give her consent. The best results are when the donor foetus is in the first trimester (first three months), preferably at nine weeks.

In the past, hospitals have not always given enough thought to the feelings of parents when pregnancies were terminated. At present there are about half a million miscarriages (expulsion of the foetus at twelve to twenty-eight weeks) each year in Britain. The law, which states that stillbirths and neonatal deaths must be recorded, followed by burial or cremation, has nothing to say about a foetus under twenty-eight weeks. In most hospitals cremation has been the method of disposal, with facilities to take photographs of the dead foetus and have a special religious service. A 1992 report from the Stillbirth and Neonatal Death Society recommends that all bodies from foetuses under twenty-eight weeks are disposed of individually, preferably by cremation, with the opportunity for parents to see the body and attend a service.

The human embryo

At three to four weeks the heart and tail are forming, the tiny buds that are the future limbs have appeared, and there are a tail and gill pouches. The embryo is 4 mm (less than a quarter of an inch) long and does not look greatly different from the embryo of a fish, tortoise or chicken.

At five to six weeks the structure of the major organs (heart, brain, eye, ear) becomes clear. The embryo is now more than half an inch long but still looks broadly similar to the embryo of a calf, rabbit or pig.

By seven or eight weeks the size has increased considerably, to more than one and a quarter of an inch long. The tail no longer sticks out, the internal organs are better developed, ovaries and testes are formed, and the emerging face, hand and feet give the foetus a human appearance.

From now on, as the embryo grows it looks more and more like an infant. During the fourth month hair appears, and the mother is conscious of its movements (so-called 'quickening').

Ethics and Foetal Tissues

At what stage of development does a foetus become a person, with the associated moral status? It is very difficult to say that at any particular time after fertilization of the egg the embryo turns into a human being. Some would maintain that the tiny embryo becomes a person when the brain has reached the stage of development that allows 'brain life', and that this is at about twenty weeks of gestation. But others say this is not logical, and a law accepted not long ago by the US Supreme Court says that the life of each human being begins at conception. The legal rights of the fetus are unclear. Recently a judge in New Jersey, USA, ruled that the unborn child of a female prisoner seeking an abortion must have legal representation.

The question of foetal age and humanity has been influenced by medical advances, because very small foetuses once incapable of independent life can now be saved. In special hospitals 75 per cent of twenty-eight-week foetuses (weighing 1,100 g and measuring about 35 cm long) survive one year. A few (7 per cent) of those as young as twenty-three weeks, weighing a mere 700 g and 30 cm long, with the air sacs in the lungs only just capable of functioning, have been successfully reared. Unfortunately, many of these very premature infants will suffer some sort of physical or mental handicap. On one occasion a brain-dead pregnant woman was kept on a life support system for sixty-four days while the foetus grew, so as to increase the chance of her delivering a live infant.

It is not surprising that many people feel uneasy about using the early foetus as a source of organs or cells and ask difficult questions. Does the foetus feel pain? Does it not make movements in response to noxious stimuli? Even the early foetus has a brain, but how can you find out whether a foetus is brain dead? And when the enthusiast answers that in the early foetus the cerebral cortex (forebrain) is not developed, and that the movements are just reflexes, it can be asked whether a functioning cortex is indeed essential for pain to be felt. And what about the anencephalic infants? Don't they feel pain? According to 1997 guide-lines drawn up by a group of experts in the UK, the foetus does not have the right nerve cells and connections to feel pain before it is twenty-six weeks old. Hence it is felt that foetuses more than twenty-four weeks old should be sedated or anaesthetized during operations, whether these are being done to save them or abort them. Abortion at this stage of pregnancy, however, is rare.

People also worry about the legal arrangements, and want to be assured that organ and tissue donations are managed in a humane and responsible way, and that the law is not just being rewritten in the interests of those who are going to benefit by receiving the organs. As might be expected, people feel more concern about the brain than about kidney, heart or lungs. The brain contains the 'real person'. If you swap brains, or maybe even brain cells (enough of them), why shouldn't behaviour, character, be transferred? Clearly, such considerations lead us into an ethical and moral quagmire. As is often the case (for instance, in the genetic engineering of foods and of domestic ani-mals), science has advanced rapidly, before the ethical braking systems have been thought through. The law is even further behind. It has been said that science has made us gods before we are even worthy of being men.

Ethical Problems with Pregnant Women

There are special legal difficulties here. For instance, it seems proper that a woman should be able to start a pregnancy and, after a planned abortion, donate her foetus to save the life of a very ill family member. After all, a woman who becomes pregnant accidentally can choose to have an abortion and provide foetal tissue. But this is to say that it is reasonable to start a life, knowing that it will be sacrificed. Is this just the beginning of a slippery slope that leads to the unattractive prospect of 'breeding for spare parts'?

Foetal Tissues and the Future

Because of the problems associated with the use of fetal tissue, new regulations in Europe and America have made it hard to obtain. Although President Clinton reversed the Bush administration's 1988 ban on the transplantation of foetal tissue, it is still not widely available. Tissues from elective (planned) abortions are of better quality than those from spontaneous abortions or ectopic pregnancies, and the ethical problems crop up again here. Is the foetus being aborted just to provide foetal tissue?

Although much of the research work is still in the developmental stage, the need for foetal tissue is likely to increase. The most suitable plan is to have a foetal tissue bank, centrally run on a non-profit basis. This ensures more uniform quality and safety testing, and allows a complete separation between those who undergo and carry out abortions, and those who receive the tissues and do the transplantation. Sale of tissues encourages unacceptable abuses and should be prohibited. A foetal tissue bank has now been established in the UK. A very matter-of-fact approach was taken in the former Soviet Union, where the use of foetal tissue was pioneered. A large abortion clinic in Moscow provided enough pancreatic tissue to transplant into 4,000 people suffering from diabetes. Russian hospitals were performing 1–2 million abortions each year in 1993, and women could sign a document agreeing to the use of their foetus in this way.

For all the gravity of the ethical and moral problems, the goalposts are likely to change in response to the rapid advance of medical science. Even now, most people accept the use of foetal tissues and organs as long as it is done in a responsible and humane way. The views of the outright objectors must be heeded, but they are in a minority and should not be able to prevent others from going ahead in a lawful fashion. Commercial exploitation of human embryos must be prohibited, and it seems right that attempts to produce hybrids between human and other animals should be outlawed. Confronted with the complex ethical issues in transplantation, euthanasia (chapter 3) and care of the dying (chapter 15), it is legitimate to ask what the role of the Hippocratic Oath is today (see box).

Hippocrates and medical ethics

The Greek physician Hippocrates (460–375 BC) was a native of the Island of Cos, and is regarded as the father of medicine. He maintained that disease was due to an imbalance of the four humours of the body, which were phlegm, blood, yellow bile and black bile. The adjectives phlegmatic (phlegm, seen as cold and moist), sanguine (blood, hot and moist), choleric (yellow bile, hot and dry) and melancholic (black bile, cold and dry) refer to the effects of the different humours.

Hippocrates set out good ideas about the effects of the environment on health. In describing the ideal doctor, he said that he should be

> of moderate stature and symmetrical limbs . . . chaste and courageous, no lover of money . . . His hair should be cut neatly and symmetrically, and he should neither shave it nor suffer it to grow too luxuriantly . . . When summoned to a patient he should sit down cross-legged, and question him about his condition with becoming gravity and deliberation . . .

Hippocrates also gave us the earliest statement about medical ethics. Although he forbade the use of pessaries to procure abortions, he was full of praiseworthy rules for the benefit of the patient. The 'Hippocratic Oath' includes the commitment to help but not to harm the patient, and to keep all things confidential. The oath in its modern form developed in the mid-twentieth century; though some form of oath of this kind is said to have been sworn in the Middle Ages, it subsequently changed greatly, losing much of its philosophical flavour. Even in modern times it was never universal, nor legally binding. Hippocrates, of course, had not encountered the situations faced by today's physicians, when confidential matters can affect the good of the public or the work of the police.

In the UK today most new physicians are not expected to swear to this oath, although many people think they do so. However, various modified forms are used in some British medical schools and in most in the USA. Perhaps it is good to have some form of ritual and ceremony to remind new doctors of their ethical obligations, and to make some sort of statement about medical ethics to the public. Most students of medicine absorb the principles of the oath without formal instruction. Is it unreasonable to suggest that there should be a universal set of rules governing all doctor–patient relationships? The basics (not to harm, not to exploit or seduce, etc.) are universal, and all would agree that to the physician nothing human should be strange or repulsive.

But physicians differ about matters like euthanasia and legalized abortion. The World Medical Association wants to have an up-to-date wording that could be used internationally; the trouble is that an oath to suit all circumstances is too wordy, and sounds as if it had been written by a committee of lawyers.

In the UK the disciplinary body with a 'code of practice' is the General

Medical Council, which at present is planning to insert extra clauses into the code, urging the doctor to do something about government policies that harm public health or are contrary to patients' interest. This means speaking out about laws on smoking or the advertising of smoking, and complaining about the funding of health services. Other desirable additions to the code of practice would be greater consultation with people about their illness and its treatment, and the need for informed consent.

Meanwhile, as we become accustomed to the use of foetal materials and their often dramatic benefits to the living, we will perhaps come to accept the unacceptable. For example, foetal cells could be maintained in culture after IVF (in vitro fertilization), and their number vastly increased by multiplication to provide a rich rource of healthy human cells, to be used for worthwhile purposes that have hardly yet been dreamed of. The transplantation business is still in its early stages and we have to face the future.

Organs from Animals ('Xenogenic' Transplants)

Many of the problems associated with use of foetal tissues will disappear if scientists master the art of using organs and cells from animals. The animals would be bred for the purpose, maintained free of harmful microbes, and their use for this purpose seems more worthy than breeding them to be killed and eaten.

Transplantation from non-human animals had always been thought of as an unlikely possibility because the organ would be so vigorously rejected, and occasional attempts had not been successful. Chimpanzee kidneys, for instance, had lasted only three to nine months when transplanted into humans, and baboon kidneys just one to three months. In any case, primates are now carefully protected. However, tissues without a proper blood supply, as explained above, are not easily rejected, and heart valves, cartilage, etc. from pigs and kangaroos have been used successfully.

The immunologists and molecular biologists may now have solved the problem. It is the foreign 'label' on foreign cells that is recognized by the immune system, telling the recipient that the material is foreign and is to be rejected. Now, by a miracle of genetic engineering, animals have been produced whose cells have human rather than foreign labels. This is done by transferring into animals the human genes responsible for producing the labels. The human genes are injected into fertilized pig's eggs, under the microscope, and the eggs then transferred into a sow. They develop into embryos, every one of whose cells carries human tissue-typing genes and the corresponding labels. A colony of so-called 'transgenic' pigs with these labels is soon formed.

Pigs were chosen because their organs resemble human ones and are about the right size. They are also tame, easily bred, and although viruses could possibly be a problem, they can be reared free of other harmful microbes and parasites. If this process turns out to be successful the ethical problems of transplantation of human organs would be overcome, because using animals' organs would arguably be no worse than eating them.

Using Dead or Renewable Human Material

Blood, plasma, bone marrow, semen and ova are in a different category from organs. The donor suffers at the most a minor inconvenience and no permanent deficit. The material taken is soon replaced. Perhaps it is because the donor remains healthy and litigious that such tissues have often been the subject of legal proceedings. But the immune system recognizes foreigners whether they are separate cells or solid organs, and there are the same problems with rejection. On the other hand, most of these tissues and cells can be taken at a convenient place and time and stored in the frozen state, which makes things much easier: the transplant surgeon, in contrast, often has to travel to inconvenient places at inconvenient times to remove organs and take them to the recipient.

Hair

Human hair is an important resource for some communities in India. Hindus shave their heads as a religious custom and one temple in Tirupati in Andhra Pradesh collects 4,000 tons of hair a year. Much of this can be sold for export and the manufacture of wigs. Hair is also a profitable source of many amino-acids, including cysteine and tyrosine. A Japanese – Indian partnership decided to build a hair-processing plant at Pondicherry in south India that would use 1,200 tons of human hair a year.

Blood

In 1818 a London *accoucheur* and physiologist, James Blundell, suggested that blood be transfused into haemorrhaging women. Blood-letting was then common so it was a practical possibility. Until then there had been no more than the occasional transfer, either between animals or from animals to humans, and the death of a patient in the seventeenth century and the prosecution of the physician had led to a ban that lasted for 150 years. But the results of these and other nineteenth-century attempts were generally disastrous, and it was not until 1907 that Janssky discovered that there are four types of human blood, which must be matched between 'giver and receiver if serious reactions are to be avoided.

Paid donors are major sources of blood in the USA. This differs from the trade in organs for transplantation (see above) because it can be argued that in this case the recipient is more vulnerable than the donor. The donor suffers no ill health and is financially rewarded, while the recipient runs the risk of being infected by any microbes (HIV, hepatitis viruses) present in the blood. In most Western countries this risk is very small, but elsewhere in the world it is a real one. In 1993 the Europeans imported 650 million dollars' worth of blood plasma (6.3 million litres) each year, one half of it coming from paid donors in the USA. In the UK, voluntary unpaid donors provide a generally adequate source of blood.

Bone Marrow

Transplantation of bone marrow, which contains the dividing precursors of red and white blood cells, is a life-saving proceedure for children with leukaemias and other cancers. Taking the bone marrow from the top of the ileum (hipbone) can be uncomfortable (and generally needs an anaesthetic) but is not dangerous. An alternative source of these dividing ('stem') cells is the umbilical cord, from which blood (from the foetus and the placenta) can be taken simply and painlessly. The first transplantation (transfusion) of cord blood was carried out in Paris in 1988, and by 1997 about 300 such operations have been carried out on children and twenty on adults. It can be stored frozen in cord blood banks, and will be more readily available than bone marrow. Indeed, one US company is trying, unsuccessfully one hopes, to patent the method for handling and storing cord blood.

Skin

Perhaps one day cultivated cells will offer a realistic alternative to whole organ and tissue transplants. For instance, human skin cells can be grown in laboratory culture to provide a 'living skin equivalent' that has been used for skin grafting. The cells multiply in the culture chamber and the cells from six foreskins can be expanded to yield a million million million million cells: enough cells from a single foreskin to cover 23,000 square metres! A commercial product containing expanded skin cells from circumcised babies is already on the market.

Semen and Eggs

Although the focus of this book is on death, it is nevertheless relevant in this chapter to mention the methods by which modern science can start life. Ethics are involved here too, and at times it is hard to separate out what is immoral from what (by our ancient laws) is unlawful. The thorny question of exactly when an embryo becomes a person (or has a soul) is discussed earlier in this chapter.

First, some acronyms need explaining:

- AID (artificial insemination by donor) means the injection into the womb of sperm from the husband, from another donor, or sometimes a mixture of the two. In the UK, 1,500–2,000 infants are born each year by this method.
- IVF (in vitro fertilization) is when sperm from husband or donor is added to eggs (from wife or donor) in the test tube, and after fertilization the tiny embryo is introduced into the womb. Embryos cannot be kept alive in the test tube for much more than about a week. So far in the UK more than 1,000 children have been born this way.
- GIFT (gamete intra-fallopian transfer) is when the eggs and sperm are injected into the Fallopian tubes so that fertilization and growth take place inside the mother.

In intracytoplasmic sperm injection that sperm are injected directly into the egg, in the laboratory, and then placed in the mother. This is a very new technique, used to treat certain types of infertility. Several thousand children have been conceived and born in this way. The problem with this procedure is that major birth defects are increased in these children, presumably because the injected sperm may be abnormal, not the sort that would otherwise have been able to fertilize the egg.

A special set of problems arises when a surrogate mother carries the embryo on behalf of the wife, whether after sexual intercourse with the husband, after AID or after IVF. Under these circumstances it may be that neither the husband nor the wife has contributed any genetic material to the child.

Donations of semen and ova have been of great value in the treatment of infertile couples. In the UK the only payments the donor can receive are for expenses incurred in giving the donated items. The donation should be a gift that is freely and voluntarily given.

Frozen human semen has been available for twenty years. One of its uses is to treat infertility due to cancer of the testis, the semen being taken from the patient as soon as the cancer is diagnosed. When the patient has been treated, or after any other condition is treated (by chemotherapy or radiotherapy), resulting in reduced sperm production, the stored semen can be used. In September 1997 the UK Human Fertilization and Embryology Authority gave permission for tissue to be removed from the testicle of a two-year-old boy who was about to be treated with chemotherapy. The chemotherapy could result in him being sterile when he grew up, and the stored testicular tissue would enable him to father children. In the USA semen is often taken from dead men, at the request of their partners and families, and stored for future use.

Semen, however, can be the source of hot debate. In the UK in 1997, a woman persuaded doctors to remove sperm from her husband as he lay in a coma with meningitis. He died, and she wanted to use his stored semen to produce a child. It would be a posthumous pregnancy. They had agreed beforehand that this would be a good idea. But the law forbade this, following which there was such a public outcry that eventually permission was granted as long as the insemination was done in another country.

Ova present more of a problem. They are not obtained quite so easily as sperm, and there is a shortage. In addition, present storage methods are unsatisfactory and only about one in 100 eggs survive freezing. The medical response has been to seek other sources of ova or to use very early embryos (see below), and here we enter yet another ethical and legal minefield. An alternative solution is to pay people for giving ova (or semen). In the USA this is accepted and established, and the donor is handsomely rewarded. Some would argue that ova will always be in short supply in countries where there is no financial reward.

Use of Embryos or Cadavers as Sources of Ova

It is all very well to use ova washed from the uterus of living donors, but what about ova from dead women or from embryos? Tissue from the ovary of live donors has been successfully used in animals, and will soon be possible in humans. A tiny piece of ovary, taken from another woman at any stage of the menstrual cycle, can be used restore fertility in someone whose ovaries have stopped making eggs or have been removed.

It will soon be possible to retrieve eggs from foetal or cadaveric ovaries. In neither case will the donor have given consent, but the main objection is that the child resulting from such procedures will have to come to terms with its origin, which in the case of the foetus is a person who never existed as a mother. How would a child react to the news that its mother was an aborted foetus? In addition, there is the common hostility to anything under the heading of genetic engineering or embryology. Must we go, people ask, to whatever limits science is capable of, and is not the creation of life from an aborted foetus a step too far? In response to these dilemmas the use of foetal ovarian tissue for fertility treatment has been made illegal in Britain, and the use of ovarian tissue from cadavers is under review.

Use of Whole Embryos for Fertility Treatment

The word embryo here means a tiny ball of four to eight cells produced after IVF. They can be frozen in case of future need, but unfortunately fewer than one in ten of them survive after thawing. The first pregnancy that came from transfer of a frozen and thawed embryo was reported in Australia in 1983, and the first live birth took place the same year in The Netherlands. Unfrozen embryos freshly formed in the test tube (by in vitro fertilization) give a higher rate of success, that is to say higher pregnancy rates. Problems arise in deciding how long to keep the embryos once they have been frozen, and there are similar problems in the case of frozen ova or sperm. Once frozen they last for very long periods (decades), but for how long *should* they be kept?

At least 30,000 embryos were being preserved by freezing in the USA in 1995, and 52,000 in the UK in 1996. In the case of embryos the thorny question of destroying life arises if they are 'discarded' at a later date. The maximum time for storing embryos in the UK was increased in 1996 from five to ten years, as long as the couples consented – the longest in all the countries that have specific laws on this point. In July 1996, when the five-year period was up, and it was time to allow 3,300 frozen embryos to perish, a large proportion of the couples concerned either had moved and could not be traced, or failed to respond to registered letters. Their 2,100 embryos were destroyed. Those who did respond either allowed them to be destroyed, donated them to other couples, kept them for their own future use or offered them for research. Some maintain that it would be better to extend storage to the end of the donor's natural reproductive life, until they are, say, fifty years old, but there would still be the problem of then throwing the embryos away.

Across Europe there are wide differences in the regulations covering research, storage and donation of embryos. A prohibitive approach is typified by Germany, where many practices allowed elsewhere in Europe are punishable by imprisonment, whereas in Italy there are almost no statutory regulations at all. The strength of the Catholic church in Italian society and politics perhaps makes it difficult even to discuss legislation on these matters.

The use of embryos for fertility treatment will inevitably raise legal questions. For instance, a recent judgment in the Australian state of Tasmania has recognized that children born (in a surrogate mother) from a frozen embryo after their genetic parents have died have a right to inherit. This seems logically correct.

Medicinal Uses of Corpses

Folklore contains many tales about the miraculous effects caused by seeing, touching, kissing, even eating parts of the body. King Charles II of England, in his last illness, drank a potion containing forty-two drops of extract of human skull. The Yanonami Indians of the Amazon think it barbarous that Westerners do not drink the ashes of their relatives.

The use of mummies in medicine has already been referred to in chapter 9. Physicians have been precribing mummy for patients, in pulverized form or as an extract, since AD 1100. For a while, when demand exceeded supply, there were one or two mummy factories that converted modern corpses into mummy powder.

Trouble with Growth Hormone

Certain hormones have been extracted from corpses, purified and used to treat diseases. Very occasionally children suffer from a deficiency of growth hormone, which is produced by the pituitary gland, an organ the size of a pea behind the nose at the base of the brain. These children are destined to become dwarfs, but can be enabled to develop normally if they are regularly injected with growth hormone. In the 1960s and 1970s the growth hormone was extracted from the pituitary glands of cadavers, and hundreds of glands were used to make each batch. The children to whom the hormone was given developed normally, but unfortunately the pituitary gland of a very small number of cadavers had contained the infectious agent causing Creutzfeld–Jakob Disease (CJD, similar to BSE), and this contaminated the whole batch. The result was that a few unlucky children (fifteen to twenty worldwide) developed CJD, which is a fatal neurological disease, when they grew up. At the resulting court cases it was alleged that the pituitary glands, from tens of thousands of cadavers, had generally been taken without permission, and without due care. The growth hormone is now produced pure in the test tube by genetic engineering.

The Amazing World of Human Relics

Human bodies or parts of bodies have been considered to have a special significance when they belonged to holy people such as religious leaders, saints and martyrs, or to famous kings, tyrants, pop singers, etc. They are often kept, preserved and put on display, so that they can be venerated by worshippers and admirers. In 1977 there were attempts to steal the embalmed body of Elvis Presley from its marble mausoleum. Although such relics are not used in any practical sense, they are undoubtedly made use of.

Kings and Queens

The embalming and public display of the corpses of kings and queens is referred to in the last section of this chapter and also in chapter 9. In modern times the same treatment has been accorded to people like Mao Tse-Tung, Lenin and Eva Peron. When Lenin died in 1924, his brain was first removed and cut into tens of thousands of sections for study by the Soviet Brain Institute. Presumably these sections failed to yield results of great neurological interest. The body was then embalmed, but it was not well done and he had to be re-embalmed in 1926. Finally he was put on permanent display in a special mausoleum in Red Square, Moscow, where he remained, virtually deified. Citizens could file past and pay their respects. A team of embalmers were employed to make twice-weekly check-ups on the corpse's cosmetic needs, attending to the face and keeping it in immaculate condition. Eventually, after the dissolution of the Soviet Union and to symbolize Russia's break with its communist past, it was decided in 1993 to remove the body and bury it in an ordinary grave, but so far (1998) this has not been done.

The Skull – a Powerful Relic

In art and literature the skull is commonly used as the image of a dead individual or as a 'memento mori', a more general reminder of human mortality. Skulls are clean and durable. Whether from enemies, friends or famous people, they can be kept more or less indefinitely as emblems or trophies. When he died in Vienna in 1791, Mozart was given a third-class burial, sharing a grave with fifteen other bodies. But the sexton of the Viennese church, an admirer of the composer, had marked the body with a piece of wire, and when the grave was re-opened in 1801 he took Mozart's skull and kept it as a sacred relic. After being studied by the anatomist and anthropologist Josep Hyrtl, it was finally acquired by the Mozarteum in Salzburg.

Collections of skulls and other bones have sometimes been stored in artistic profusion in holy places. The Capuchin crypt, off the Piazza Barberini, Rome, houses the ossified remains of 4,000 monks in a series of small chapels. Skulls, pelvises, vertebrae, thighbones and clavicles are separately arranged in sweeping patterns in what amounts to a macabre art form. The idea, apart from providing a communal resting place for these holy men, was to remind everyone of life's brevity and the need to pay attention to the soul. Many of the bones in the catacombs of Paris (see chapter 6) are arranged in similar patterns.

Another memorable site is the crypt of St Leonard's Church in Hythe, Kent, England. This contains the bones of about 4,000 people, mostly from graves in local churchyards, and dating from AD 900 to 1500. There are 2,000 skulls, a thousand of them side by side on shelves, staring at the visitor in their eerie anonymity. The idea of showing the bones here was probably to attract the pilgrims passing through the port of Hythe in their thousands on their way to the holy relics in Canterbury.

Religious Relics

The story of religious relics, as told in a fascinating book by James Bentley, *Restless Bones* (1985), is a monument not only to human credulity, but also to the magical power of the human corpse and its constituent parts. The preoccupation with relics reached its peak in the middle ages, but relics have featured throughout Christian history and have incited killing, corruption and deception as well as worship.

One of the earliest relics was in Cologne, which had the richest collection in Europe. A British queen (Ursula), on her way back from a visit to the Pope in Rome was beheaded by the Huns in Cologne in AD

238; 11,000 attendant virgins were supposedly massacred at the same time. Ursula's remains, minus the head, were displayed in a jewelled shrine in St Ursula's Church, Cologne. Assorted bones of the attendant virgins – actually only eleven of them – cover the walls of the Gold Chamber Chapel of the church.

The early Christians officially recognized the cult of relics, and the Church Councils of Gangra (AD 345) and Nicaea (AD 787) said that those who despised or spoke disparagingly of relics were to be excommunicated. The Council of Carthage (AD 801, 813) recommended destruction of altars that did not have relics. In medieval days touching a corpse, body fragment or other relic was thought to confer great powers and could cure diseases. Relics, paraded in religious processions, were believed to protect the whole community. They were the object of pilgrimages; they buttressed the faith of peasants and nobles alike; they were even carried into battle. Relics were bought, sold, fought over, lost, found, stolen. Early Christians took the view that distributing the bodily remains of martyrs round the world would serve to spread their power and influence at the same time. Martyrs were made so easily that those killing a potential martyr made a point of reducing the bones to ashes before the body could be cut up eagerly for relics. Unfortunately for them, even the ashes became relics. When Thomas à Becket was killed by the knights in Canterbury Cathedral on 29 December 1170, the monks and townspeople collected fragments of brain from the floor, and also his spilt blood by dipping pieces of cloth into it. The church made good use of his corpse. Medieval pilgrims would be expected to visit no fewer than four shrines. The first was where he was killed, a sword slicing off the top of his head, and it contained a fragment of brain and the point of a sword. The second shrine had a casket containing the rest of the brain, with an adjacent well whose water was tinged with his blood (available for purchase). The third contained the top of his skull, and the fourth his bones. His remains were moved to a new shrine in 1220, and a fingerbone and other fragments put on display. The tomb was opened and the bones examined by experts after the blitz in the Second World War, and they were placed in another tomb, but the exact fate of all of them is not clear.

Bones are the most enduring but not the only relics. Blood might appear to pose problems for the collector. When St Cyprian, Bishop of Carthage, was about to be beheaded in the third-century AD, his Christian followers threw pieces of cloth on to the ground so that they could be retrieved as blood-spattered relics. Syrian monks in the fourth century AD are said to have poured oil through a funnel into a saint's tomb and tapped it off with a spigot inserted underneath. Essence of saint would have been much sought after by pilgrims. But what about larger volumes of fluid? Naples Cathedral contains two glass vials of black material, said to be the blood of St Januaris who was martyred by beheading 1,700 years ago. The blood sometimes liquefies, signifying that the martyr is alive in heaven and it is to be a year of prosperity for Naples. His head stands between the vials and the rest of the body in a special crypt in the cathedral.

Aachen Cathedral has a shrine housing what are said to be the swaddling clothes of Jesus, the loincloth he wore on the cross, plus a thorn from the crown of thorns. Jesus figures prominently in relics although according to the Bible he was carried straight up to heaven leaving no trace of his body behind. Most are therefore 'secondary relics' in the form of bits of the cross, the nails or the crown of thorns, though sometimes claims were made for fingernails, milk teeth, hair, blood or holy tears. At least eight places had claimed to possess the holy foreskin. One of these foreskins, together with the supposed holy umbilical cord, was once kept in a crucifix filled with oil in the private chapel at the Lateran, Rome. But items like this became an embarrassment to the church, and in 1900 the Holy Office outlawed relics and threatened excommunication to anyone writing or speaking of the holy foreskin.

Jesus's mother, Mary, was another favourite source of relics, including locks of hair, girdle, menstrual towels and milk. There were pieces of John the Baptist, such as part of the jaw or the top of the head, in at least seven places in Europe. A large proportion of relics, obviously, are fakes, but this did not diminish their influence. Those who had relics enjoyed papal indulgences, that is to say reductions in the time to be spent in purgatory (see chapter 16).

Relics were valuable; they fetched a price. Indeed, they counted among the richest loot in war, and the crusaders who sacked Constantinople in 1203 loaded carts with them to sell when they got home. In 1425 the Provost and Fellows of Eton College handed over to King Edward III a list of college assets; these included a bone of St Anthony, an arm of St George, a tooth of St Nicholas, parts of St Stephen, and so on, as well as pieces of the holy cross.

John Calvin (1509–64) attacked the cult of relics and ridiculed its abuses. He pointed out that, as well as the 'anthill of bones', there were enough pieces of the True Cross to 'build a good ship', enough thorns from Christ's Crown of Thorns to make a hedge. Martin Luther (1483–1546) added his voice to the criticism, in spite of the fact that his patron, the Elector Frederick, had a collection of 5,005 fragments of saints and secondary relics. Yet relics had their serious-minded apologists. The French scholar Charles Rohault de Fleury calculated in 1870 that the True Cross would have had a volume of about 178 million cubic mm, whereas the total volume of all the existing relics was only 3.9 million cubic mm.

The status of existing relics is uncertain. Some of them could be authenticated by the Vatican using methods such as carbon-dating, but this has not often been done.

An example of the gradual dismemberment of a corpse for relics is provided by St Francis Xavier, a Catholic missionary who died in China in 1552. His disciples covered the body with quicklime and shipped it to Goa. Bits of the body were removed at intervals. A visiting Portuguese noblewoman succumbed to her adoration (or her yearning for a relic) and bit off the little toe of the right foot. She returned it when, miraculously, the toe-stump was seen to bleed, and it was placed in a separate reliquary. Various internal organs were removed in 1636 and distributed to churches around the world, and later the right arm was taken. In 1694 the body was reported to be in a state of remarkable preservation. The remains, in a marble and silver casket, are put on display every ten years, the last occasion being in 1984, when they attracted tens of thousands of tourists and pilgrims.

13

The Abuse of Corpses

Corpses are not always put to rest with dignity and respect, not always treated kindly or used for the benefit of others. Abuse of the corpse or of the dying person has been a recurring feature in history.

Punishing the Corpse

If one believes that the corpse is really still alive, in the spiritual sense, it is a not a big step to decide to administer to it any appropriate punishment. In England in the seventeenth century there are even instances of corpses being arrested for debt. The corpse might be beaten. For instance, the Shinto doctrine in Japan generally forbids injury to a dead body, which is dangerously polluted and must be purified by appropriate rites. Yet in Japan until the 1940s the corpse of a suicide would sometimes be beaten as a punishment. In England until 1821 the legally prescribed treatment for suicides was to be buried at the crossroads on the highway, with a stake through the heart (see chapter 3). The ultimate punishment, an insult to the corpse and thus to the dead person, was to send it for dissection. Few post-mortem fates were worse than being physically violated, literally taken to pieces, in this way. The Murder Act of 1756 in England arranged this as an extra punishment for the corpse of a murderer. The subject is discussed at greater length in chapter 11.

The same feeling about punishing corpses led to greater abuses. Crimes that were not adequately punished by mere killing merited a more outrageous treatment of the corpse or of the dying person. In a crueller age, barbaric severities were possible. The extremes were reached in the treatment of people such as John Owen. Owen, who had been excommunicated by the Pope, was unwise enough to declare that it was lawful to kill the king. The sentence for this outrage, passed in 1615, decreed that the following punishment be given, and in this order:

1   He be drawn to the place of execution. (Rather than be allowed to walk, he was to be dragged through the streets on a hurdle.)
2   His privy members to be cut off.
3   His bowels to be burned. (This meant that after a short period of hanging, so that he was still alive, the bowels were to be removed by a butcher and burned.)
4   His head to be cut off.
5   His body to be dismembered. (In other words the arms and legs were to be cut off, a procedure often referred to as 'quartering'.)

In the familiar phrase, the victim was hung, drawn (disembowelled) and quartered.

Dismembering the body implied an extra ecclesiastical penalty. It was thought that resurrection on the Day of Judgment would then be impossible. Edward VI (1537–53) allowed peers of the realm to escape these severe penalties: even when convicted of crimes like murder or treason they could die by beheading.

Another gruesome episode was that of Sir Oliver Plunkett, Catholic Archbishop and Primate of All Ireland, who was put to death in 1681 for his role in an alleged Popish plot. He was first hanged at Tyburn, then cut down before being dead and disembowelled. His entrails were burned while he watched. Finally he was beheaded and quartered. Subsequently different parts of the corpse were retrieved and retained as holy relics (see chapter 12) in various containers in various parts of the world. Each arm ended up in a different convent or church, and a piece of the skull in Boston, USA.

Oliver Cromwell's corpse was given all the pomp and ceremony of a state funeral when he died in 1658, but suffered a 'retrospective' punishment after the restoration of the monarchy in 1660. His body was exhumed from Westminster Abbey and dragged on a sledge through the streets to Tyburn, where it was hanged. The body was buried beneath the gallows, but the head was cut off and displayed on the end of a pole attached to the roof of Westminster Hall, where it remained for ten years.

The corpse of Cicero (AD 106–43), the Roman orator and statesman, was punished after he had been murdered. Cicero's verbal brilliance had led him to attack Mark Antony in a series of speeches, which was unwise because Mark Antony wielded great power in Rome after the assassination of Julius Caesar. Antony declared him an outlaw, and had him captured and then killed by bounty hunters when he tried to escape. Cicero's head and right hand were brought back to Rome and nailed up in the same public rostrum on which he had stood to make the offending speeches. It was a gruesome reminder that it was the head that had thought and spoken, and the hand that had written, against Mark Antony. For good measure Antony's wife Fulvia thrust a bodkin through Cicero's tongue.

A recent example of the abuse of a corpse occurred in England after a woman in her twenties was hit by a train and killed near Billericay, Essex, on 6 February 1997. The body, covered with a tarpaulin, remained on the tracks, and train drivers were ordered to run over it rather than disrupt peak-hour services. About twenty trains passed over it before it was removed.

Cannibalism Revisited

Eating human corpses surely counts as abuse, and the general subject of cannibalism has been covered in chapter 8.

An interesting cannibalistic suggestion was made by the writer Jonathan Swift, who in 1729 set out an unconventional solution to the poverty and population problems faced by Ireland at that time. It was called 'A Modest Proposal for Preventing the Children of Poor People from being a Burden to their Parents or the Country and for Making them Beneficial to the Public'. In this period, Ireland was being used as a source of raw materials and cheap food, yet was forbidden to trade on its own with other countries. Basing his proposal on the same exploitative argument, Swift claimed that the problems of poverty and overpopulation could be solved by one and the same means, and he spelt out his scheme in startling detail.

Of the 120,000 children born in Ireland each year he suggested that 100,000 were sold to butchers for 10 shillings each. A year-old child of 28 lb, he pointed out, would provide nourishing and wholesome food and could be stewed, roasted, baked, or boiled. The skin could be used to make gloves for ladies and summer boots for fine gentlemen. It sounds horrific – and was meant to. Swift wrote it as a biting satire, in which he expressed his concern for the misery of the oppressed Catholics and his contempt for the unconcern of the English for their plight.

The story of Fritz Haarmann from Hanover, Germany, provides a chilling echo of Swift's 'Modest Proposal'. In 1919, when Germany was suffering severe food shortages in the aftermath of the First World War, Haarmann murdered at least twenty-eight people aged between thirteen and twenty, dismembered them, and sold the meat as sausages. He was apprehended in 1924, tried, and sentenced to death by decapitation.

A possibly more acceptable fate for a cannibalized human is on display in the museum of the Royal Naval Hospital at Haslar, Hampshire, England. The exhibit is a kayak. An Eskimo who had eaten his companions in order to survive in sub-zero temperatures had constructed the kayak from their bones and skin.

Understanding Human Sacrifice

Sacrifice is a religious rite found in the earliest forms of worship in all parts of the world. The word comes from the Latin term *sacrum facere*, to make a sacred ceremony. Sacrifice is a profoundly significant human act, performed by different communities for tens of thousands of years, much studied and discussed by anthropologists, archaeologists and psychiatrists. Many books, many careers, have been devoted to the subject. Animals such as goats, bulls, or rams were common victims, and one of the most potent objects that could be offered was blood. Human blood was especially potent; what more valuable offering could be made to a deity than a human life? Human sacrifice was practised in parts of India before British rule, in Peru, and in pre-Buddhist Tibet, but generally involved only a small number of victims. The Aztecs of Mexico were the greatest practitioners (see below).

Why Sacrifice?

A sacrifice is made to please, to thank, to question, to ingratiate or otherwise communicate with a god. One could argue that its widespread use is yet another example of illogical thought by humans. You carry out a ritual, and the thing you were worried about turns out all right. Clearly then, the ritual works. But there is a logical fallacy, and it applies as much to, say, the treatment of disease as it does to sacrifice. The fallacy can be exposed as follows. We say: if it is true that the far side of the moon is made of green cheese, the sun will rise tomorrow. We find that the sun does indeed rise tomorrow, therefore we conclude that the far side of the moon is made of green cheese!

Examples of motives and occasions for sacrifice are:

- in thanksgiving for a successful harvest, or success in battle;
- at a time of danger, sickness or crop failure, to plead for redress;
- when a building is being constructed, to consecrate the building and repel evil influences: at one extreme a sacrificed human being may be buried in the foundations; at the other, blood from a sacrificed animal may be poured in, as in various West African cults;
- to ensure fertility (of the soil, or of domestic animals); in ancient days nothing was considered more efficacious in this respect than a ritual slaying of a human being;
- to appease the anger of the god;
- as a gift to the deity, hoping of course for a return favour or gift;
- as a technique for transferring guilt to the sacrificed object, so that the guilty are purified and a bond established with the deity: in this way Christ's death is believed by Christians to have been a sacrifice on behalf of humans, whereby in a single act he wiped out their sins.

Child sacrifice nearly 200 years ago

In his 1827 account of the Tonga people of the South Pacific, William Mariner describes the sacrifice of a child. A man had killed another in the heat of the moment in a consecrated place, and for this sacrilegious act the chiefs decided that a child (of a male chief) had to be sacrificed.

The child was accordingly sought for but its mother, thinking her child might be demanded, had concealed it. Being at length found by one of the men who were in search of it . . . the band of gnatoo was put round its neck . . . two men then tightened the cord by pulling at each end, and the guiltless and unsuspecting victim was quickly relieved of its painful struggles. The body was then placed upon a sort of hand-barrow . . . carried in a procession . . . to various houses . . . one priest sat beside it, and prayed aloud to the god that he would be pleased to accept of this sacrifice as an atonement for the heinous sacrilege committed, and that punishment might accordingly be withheld from the people. When this had been done before all the consecrated houses, the body was given up to its relations, to be buried in the usual manner.

Sacrifices could be offered on a table, for the deity, or they could be consumed. The ancient Mexicans propitiated their sun god Huitzilo-pochtli by symbolically consuming him. A statue of him was made out of beetroot paste mixed with human blood, cut up, shared out and eaten. The Eucharist, for early Christians, can be looked at in this way, because Jesus is thought to be present in the bread and wine that are offered and then consumed. The bread and wine are converted ('transubstantiated') into the body and blood of Christ. In a sense, it is cannibalism. Often the sacrificial object is burned. It goes straight up into the sky, an excellent way of delivering it to the god. Therefore fire is an important feature of many types of sacrifice.

Sacrifice is a feature of agricultural rather than hunting communities, probably because of its significance for the fertility of the earth. Many things are offered, but the release of blood, the sacred life force, has greatest power. The knife or sword is therefore the usual implement, and it is not always necessary to kill the victim. Less often the sacrifice is by strangulation or drowning. In ancient Peru women were strangled, the Celts sacrificed women by immersion, and among the Maya in Mexico young maidens were drowned in sacred wells.

Human sacrifice often occurs in earlier, primitive societies and is later replaced by 'more civilized' but less powerful forms. It was a feature in Dravidian villages in India, but during British rule animals were used as victims. After the coming of Buddhism to Tibet, pieces of dough were offered as a sacrifice instead of human beings, because blood sacrifice was prohibited by Buddhism.

Whatever the nature of the sacrifice it called for special people to do the job, who could be relied on to carry out the necessary rituals. It was usually done on an altar (probably from the Latin *altus*, a high place). The Aztec temples were certainly high places, and the temple of Uitzilopochtli had 114 steps leading up to the sacrificial site. The English poet Tennyson (1809–92) uses the word in a more general sense when he refers to: 'The great world's altar stairs/That slope through darkness up to God'.

Human sacrifice was practised on the greatest scale in the Aztec civilization in Mexico before the Spanish conquest. The Aztecs believed that the sun needed nourishment in the form of human blood, and in the fourteenth century as many as 20,000 victims were killed each year in their rituals (see box).

Who Was Sacrificed?

In ancient times in the Middle East servants and slaves were often buried with their royal master, because he (or she) needed an appropriate retinue in the next life. The victims may or may not have been killed before burial. This habit continued in China until the seventeenth century. Children were often chosen. The Assyrians, Canaanites and at times the Israelites sacrificed children, sometimes by burning them, and in Vedic

India the followers of the goddess Kali sacrificed male children. In Peru, high in the Sarasara mountains, the frozen remains of at least four children, boys and girls, have recently been excavated (see chapter 10). They died a violent death at the time of the Incas, probably by blows to the head, and were then buried as part of a religious ritual called Capa Coche.

Although the leader or king was often the master of ceremonies, he could also be the one to be sacrificed. Killing an old, weakening king might be a good way to ensure that the soil and fertility did not suffer the same enfeeblement. Some of the bog corpses described in chapter 10, judging by their appearance and their uncalloused hands, were probably royal or 'aristocratic' people. The king of the Jukun, in Nigeria, was in theory allowed to rule only for a period of seven years. He was put to death after reaching this allotted span (although if he was popular, and if the harvests were satisfactory, he could stay in office for a longer period). The ritual decreed death by strangulation, using a noose and the end of a long piece of cloth. This was so that his blood was not spilt, and so that the executioners did not have to look into his eyes as he died.

Using Corpses for Fertilizer, Fat, Leather

Dead human bodies contain valuable materials, and their use for such purposes as medical teaching and transplantation, as discussed in chapters 11 and 12, is considered acceptable. Corpses are also a source of elements such as carbon, nitrogen and phosphorus. These elements, which are released when the corpse is cremated or as it decomposes, are recycled by nature. They are needed for plant growth, and act as fertilizers. The corpse can thereby be said to have produced something useful. If dead bodies were 'composted' in special containers (see chapter 8) then methane, a useful fuel, could also be collected. (In London at present even the carbon, nitrogen and phosphorus from *sewage* are wasted. The final rich product from the great sewage works north and south of the river Thames is taken on board ships and dumped in the North Sea.)

However, exploitation of corpses for commercial gain is another matter. According to a letter to *The Times* of 28 June 1870, human dust and ashes (presumably from mummies) were being shipped from a ridge of rocks near Alexandria and imported into England where the mixture fetched £6 10s a ton. The writer hoped that this loathsome trade in human guano would be prohibited.

The Aztecs

The Aztecs, at the time of the Spanish conquest in 1519, were at the peak of their civilization. Their art, architecture, knowledge of astronomy, system of government and religious observances had reached high levels. The fact that they had not yet invented the wheel was surprising. Their enormous empire of more than a million people depended on the fertility of the land. The main city, Tenochtitlan, was founded on a island in the middle of a swamp, and surrounded by a network of canals and lakes. They ate maize, beans and sage, plus fish and other aquatic creatures, and also the two domesticated animals the turkey and the dog. The latter was a hairless variety, fattened for eating, but not quite so tasty as the turkey.

The Aztecs believed that to keep the sun moving in the heavens, so that the seasons would come and crops would continue to grow, it was essential to feed it with its natural food, which was blood. It had to be human blood, and without it the very life of the world would stop. Human sacrifice was a necessity for the welfare of mankind.

The victim was led up to the top of the temple and, with thousands of people watching from below, was stretched out on his back on a slightly convex stone. His arms and legs were held by four priests, while a fifth ripped open his chest with a flint knife and tore out his beating heart. This was not done with a single sweep of the knife but called for a minute or more of dissection to expose the heart and free it from its connections in the chest. The heart was held up to the sky and the rest of the body cast down the steps below. Heads were removed and impaled on racks beside the temple. It was an exceedingly messy procedure, with its blood-drenched priests, and the Spaniards were shocked by it. When the Emperor Montezuma II took Cortes and his companions to the top of one of the temples to enjoy the fine view they could not help being put off by the blood smeared over everything and the nauseating odour of rotting flesh. Yet the Spanish conquerers in their turn readily massacred, burnt, mutilated and tortured their victims. The Aztecs, on the other hand, killed for the benefit of the people. They were horrified by the tortures brought by their conquerors from the land of the Inquisition. They had no doubt their shedding of blood was the only way to deal with the uncertainities of a continually threatened world. Sacrifices took place each month, often hundreds at a time. At the dedication of the enlarged Great Temple 20,000 were sacrificed together.

But where were the victims to come from? Thousands were needed. The Aztecs got them by making war on their neighbours, the whole point being to take prisoners who could be used for sacrifice. Soldiers were armed, but they tried not to kill, and were accompanied by specialists with ropes and nets to secure the captives.

The Aztecs were civilized people, with greatnesses and weaknesses. Their ordered society, their sensitivity to beauty, their strength of religious feeling, existed side by side with their obsession with the mystery of blood and death and its accompanying cruelty. Their culture was destroyed for ever by a mere handful of European adventurers who were cunning, courageous and equipped with horses, muskets, cannon, crossbows and steel swords. All these were unknown to the Aztecs, and in battle the Spaniards were invincible.

In *Things Observed 1847–1848* the French writer Victor Hugo maintained that British farmers were grinding up bones from French, English and Prussian soldiers slain on the battlefields of Austerlitz, Leipzig and Waterloo in 1815, and using the powder to fertilize their fields. More than 50,000 died at Waterloo and nearly all were buried on the spot. Thirty years later the bones, plus the bones of hundreds of horses, were being shipped to England.

It is a giant step into barbarism when the corpses are obtained by systematic murder, and used to manufacture products that are readily obtained from domestic animals. From Nazi extermination camps there are accounts of corpses being used to extract fat for making soap, or skin being taken to make lampshades. There are one or two other equally appalling episodes in human history, including the use of human skin to make leather.

Necrophilia: Macabre Eroticism

Very rarely the corpse is used as an object of desire. Intercourse with the dead is mentioned in the writings of the Marquis de Sade; it occasionally figures in obscene jokes, and the sexual abuse of cadavers has been referred to in certain books about sexual perversions. Its rarity is indicated by the fact that it is not mentioned in most textbooks of psychiatry.

14

Identifying Bodies and
Parts of Bodies

When a corpse has decayed, or when only part of the body is available, it may be difficult to identify the individual. Identification is asked for by relatives or descendants, in the case of a famous person, or for medico-legal purposes.

Each year in the UK tens of thousands of people go missing. Some turn up later, some have been murdered, others are living somewhere unknown to their friends and relatives. All cities have to deal with unclaimed, unidentified, unwanted corpses. Each year about a thousand people die in London who are unidentified, with no known friends or relatives. They include not only the mentally ill, the drug addicts and the derelicts, but also people living on their own, anonymously. They generally die at home and are not discovered until someone notices a peculiar smell. In addition, about fifty bodies a year are recovered from the River Thames: mostly suicides, jumpers from bridges. Usually they are in a good enough state of preservation to be identified without difficulty, if anyone is available to do this. If not identified they are given a number such as DB (dead body) 27, and buried (without a headstone) or cremated and the ashes scattered. Identification is important after catastrophes and disasters because it allows relatives to settle estates, look after dependants and collect insurance money.

Murderers generally hide the body by burying it. Finding a buried body, knowing where to dig, is something that has to be worked out by forensic pathologists and the police. The buried body, as it decomposes, releases methane gas, and some of this leaks up to the surface and can be tested for with special equipment. Dogs can be useful, and in Britain the police have seven sniffer dogs trained to detect the gases of decomposition coming from bodies under water. The decomposing body also releases heat, and aerial sensors, developed for military purposes, can detect localized warm spots by infra-red photography.

Whose Body?

A freshly dead body with the face more or less intact is easily identified by someone who knows the person or has a photograph of them. This is not always so easy for those who perish in fires and accidents, or for those whose remains are recovered a long time after death.

In one extraordinary murder investigation the upper portion of the victim's body had been embedded in cement in the foundations of a building. A cast was made of the head (skull) and from this a face was constructed (see chapter 11). The face was shown on television, and when someone came up with a photo of such a woman the victim was identified. By carefully pouring silicone into the part of the cement containing the hand the investigators even got some fingerprints. Then a fragment of a cigarette package found with the body enabled the police, with the help of the cigarette manufacturer, to date the immersion in cement to the year 1974–5. In the end a suspect was charged with the murder.

The unique parts of the body, anatomically speaking, are faces, ears, teeth and fingerprints. Each of us, having scanned hundreds of thousands of human faces during the normal course of our lives, can confidently identify one face in millions. We pay less attention to ears and teeth, and most of us wouldn't know where to start if we were asked to analyse fingerprints. The unique feature at the molecular, chemical level is our DNA, and this can be used to distinguish one human being from any other, with the exception of an identical twin.

The shape of the ear is unique when examined by an expert. The suspicion that Hitler was using doubles (lookalikes) at public functions during the Second World War was confirmed when ear patterns in photographs of the Führer were seen to be different on different occasions.

Fingerprints are the pattern made by the tiny ridges projecting from the skin surface. They function mechanically, helping to prevent fingers slipping on surfaces. The ridges are slightly greasy because of oil from local glands, and they therefore make a pattern on objects that are touched. Between the ridges, in the valleys, are the openings of sweat glands. Sweat glands are formed in the four-month-old embryo and stay the same throughout life. The sweat also is deposited on the touched object, and even if there is no grease the sweat pattern can be seen when special powders or stains are applied. Fingerprints on a piece of paper can be identified years after they were left. Fingerprints are the classic tool for identifying criminals. The odds against two people having identical fingerprints are astronomical, about ten million million to one. Even identical twins have slightly different ones because of minor alterations in the pattern during development in the womb. The pioneering discoveries about fingerprints were made by Sir William Herschel, Sir Francis Galton and Sir Edward Henry in the late nineteenth century. They showed that fingerprints stayed the same throughout life and were unique for the individual, and they worked out systems for identification. Fingerprints were adopted by Scotland Yard as a means of detection in 1901. Nowadays the processing of prints is automated. Instead of laborious manual comparisons having to be made with several million other prints that are on file, the computer gives the answer in minutes. Fingerprints often remain detectable on a putrefying corpse, and when the outer layer of the skin has been lost by prolonged immersion in water the pattern can sometimes be seen on prints taken from the tissues under the skin. Palm prints are mainly for fortune tellers, but palm and sole prints are occasionally useful to the police. Every criminal, of course, knows that he must not leave fingerprints behind at the scene of the crime, but it is less widely known that prints can be recovered from the inside of rubber gloves!

Other highly characteristic features include lip crease patterns and the pattern of veins on the back of the hands, but these do not have the well-studied reliability of fingerprints. Occasionally a characteristic scar can provide the answer (see box).

Teeth are the hardest part of the body and are often still there after the rest of the corpse has decomposed or been burned. During the Second World War many otherwise unrecognizable victims of bombing raids were identified by their dentists, who had records of their special pattern of fillings, missing teeth, and so on. The bite pattern, as well as the dental pattern, is characteristic of the individual and is sometimes important in medico-legal cases. For instance, a thief was once identified by the bite marks he had made on an apple, and bite marks on other people have been used as evidence in court. Hitler, his wife Eva Braun, and Goebbels killed themselves with cyanide at the end of the Second World War, and their bodies were buried and subsequently burned, but Hitler's jaw with its golden bridges was kept for identification.

Perhaps surprisingly, it may take an expert to distinguish between human and animal bones. The bones in a bear's paw, for instance, are very similar to those in a human hand. When the skull, pelvis and femur are available the sex of a person can be decided with 90 per cent accuracy. In the female the pelvis is quite different, the forehead is higher

The case of Dr Crippen, caught by radio

The trial of Dr Hawley Harvey Crippen for the murder of his wife Cora began at the Old Bailey, London on 18 October 1910. Crippen was in love with his typist, Ethel Le Neve, and was short of money. He had no affection for his wife, and would inherit from her estate if she died.

His wife disappeared, and he said she had gone to America where she had died and been cremated. But the police were suspicious, and five or six months later, in a hole in the cellar in their house, human remains were found. They were headless, limbless and boneless, and had been mutilated in a way that suggested the murderer had some acquaintance with human anatomy. A lethal quantity of the drug hyosine was reported in the remains. Identification seemed impossible until, on a piece of flesh from the abdominal wall, an old scar was seen, in exactly the region of an operation Mrs Crippen had had eighteen years earlier. At this stage Crippen fled and boarded a boat for the USA. But the police obtained news of his whereabouts, and radioed this information to America. Crippen was arrested on the boat just before landing. He was disguised, and his lover was dressed as a young man.

The trial attracted immense interest of a morbid and vulgar kind. Crippen was convicted, sentenced to death and hanged. His lover was acquitted. This was the first time wireless telegraphy was used to catch a criminal.

and steeper, the skull a bit thicker, the nasal bones less prominent and the ridges where the muscles attach are less pronounced. The age, at least up to twenty or twenty-five years, can be determined by the pattern of growth and fusion of the epiphyses (growing ends of bones).

Special tests can distinguish a human bloodstain from an animal one. Blood is characteristic of the individual, and it can be examined for the presence of blood group substances, which are present also in sweat, saliva, semen and gastric juice.

DNA Fingerprinting

This is by far the most reliable method of identification, and because of technical advances it has taken over from all the earlier laboratory methods such as serology (reaction of specific antibodies with tissues). Tests focus on certain parts of the DNA (the so-called 'repetitive' base sequences) or on mitochondrial DNA (see chapter 11). The DNA profile is unique for the individual. Only identical twins have the same pattern. DNA is present in the nuclei of all cells and the specimen can be any part of the body with nucleated cells, whether blood, semen or material from old bones. Any part of any organ would give a reliable result.

DNA takes the form of very long molecules, and the total human DNA, when stretched out in a line, would be about a metre long. There is too much DNA in a cell to examine all of it. Only a tenth of it is actually used biologically, and the rest is redundant and repetitive. But the redundant bits just next to the used bits are characteristic of the individual, and this is what is tested. What is done is to extract the DNA from any part of the body, break it up into fragments, and then separate the fragments on a special jelly. On the jelly each fragment can be seen as a line, and it is the pattern of lines that is unique for the individual. It is similar to the pattern of parallel bars on the 'bar-code' used to label goods in the supermarket. If DNA tests on blood, semen etc. from the suspect match those on material from the victim, it is accepted as evidence.

Using the polymerase chain reaction (PCR) in which a single DNA molecule can be amplified many million times, thus providing enough material for a test, a single hair is sufficient. The bulb of a hair has many cells attached to it, which were torn out from the bottom of the hair follicle. This means that a murderer could be incriminated because he had left behind on the victim's body a single, solitary hair. In January 1997 a couple in their seventies grabbed a few hairs from a retreating burglar, who became the first in the UK to be identified and convicted by this method.

Recently, the techniques have improved so much that enough DNA to identify an individual can be recovered from a swab taken from the palm of the hand. Furthermore, there is enough DNA on objects touched regularly by hands (handles of leather briefcases, pens, car keys, telephone handsets, etc.) to make an identification as good as a fingerprint. Our skin cells are constantly being shed from the skin surface in the form of dead skin scales, about a hundred million of them a day. The fine white dust your finger picks up when you draw it over the surface of a shelf or a piece of furniture consists mostly of these skin scales. A palm print has enough of them for the DNA in the dead cells to be analysed. The method is so sensitive that merely shaking hands with someone can transfer enough of your DNA to their palm for it to be identifiable.

Use of DNA Tests to Identify Human Remains

DNA tests are helping to solve a variety of problems about human remains. For instance, nine members of the Russian imperial family were shot or bayoneted to death by the Bolsheviks in 1918 and buried in an unmarked grave at Yekaterinburg. Recently, what were thought to be their bones were excavated. Mitochondrial DNA was analysed and the patterns, when compared with those of living descendants, confirmed their identity. The remains have been reburied in an appropriate place. The same technique was used to show that the bones of a man who drowned in Brazil in 1979 were those of the Auschwitz doctor, Josef Mengele. Also, using material from live people and with a more practical aim in view, it was used to identify the families of children who were orphaned by the murder squads during Argentina's dictatorship. The children could then be united with their families.

DNA fingerprinting may one day give an answer to certain historical questions. For example, are the small skeletons in an urn in Henry VII's chapel in Westminster Abbey indeed those of the princes (King Edward V and Richard, Duke of York) who were supposed to have been murdered in the Tower of London in 1494 on the instructions of their uncle, later Richard III? Are the bones recovered in 1992 from a Bolivian graveyard really those of Butch Cassidy and the Sundance Kid, who either shot each other or were shot down by soldiers in Bolivia more than eighty years ago?

The US armed forces, which at one time used fingerprints or metal tags ('dog tags') to identify all US soldiers, sailors and airmen, are now to switch to DNA profiling. This will give foolproof identification of bodies, however disfigured, that are left on battlefields or in hastily dug graves. Governments of the future might wish to take things further and tag all citizens, and thus increase control and surveillance of the individual (especially those who make things difficult for the authorities). A British Conservative Home Secretary recently proposed that DNA profiles of all those who have ever been convicted of a criminal offence should be retained in a government library; about one in ten of the population would be eligible.

Murder Investigations and the Corpse

When a dead body is found under suspicious circumstances, an expert called a forensic pathologist is called in by the police. Was it murder, accident or suicide? These people carry with them a 'murder bag', which contains waterproof apron, gloves, thermometer, syringes, swabs, dissecting instruments, jars for specimens, lens, torch and camera. They have to be available at any place, any time. Listed below are some of the things they must do, questions they must try to answer. Much of the relevant background information is to be found in chapter 5.

Examining the Scene of the Death

Are there blood stains in the vicinity? Was the body dragged to the site? Fatal wounds do not necessarily kill fast. A man with a stab wound through the heart has been known to run a quarter of a mile before collapsing. The wound in the heart gets partly sealed off by contraction of the heart muscle. Damage to the main arteries supplying the heart is more likely to be instantly fatal.

How Long since Death?

This is an important question and numerous methods are used to answer it. Many factors influence the calculation, and charts are used to make an estimate. The only foolproof indication of the time of death is when a bullet stops a clock.

Rigor mortis. This is described in chapter 5. A rough and ready guide is the following:

body warm and flaccid = dead less than 3 hours
body warm and stiff = dead 3–8 hours
body cold and stiff = dead 8–36 hours
body cold and flaccid = dead more than 36 hours

Sometimes, when death takes place under conditions of intense physical activity or emotional stress, muscles can go into spasm almost immediately. For example, a drowning or falling person (or a soldier dying on a battlefield) grasps at nearby objects. This cadaveric spasm is rare, but it means that the individual was still alive at the time of drowning or falling.

Body temperature. This is taken in the rectum, the nostril or, possibly best of all, the ear. Interpretation is not always easy. Clothing, location of the body, size of the body (small corpses cool faster), presence of water, atmospheric humidity and temperature all have to be taken into account. The outside temperature is another factor (there may be no cooling of the body in the tropics). The temperature of the person at the time of death has an obvious effect. A fever or muscle activity raise body temperature a few degrees, while it is lower in hypothermia.

Stomach emptying. This depends on knowing the time of the last meal, and it is unreliable. It takes two to three hours for the stomach to empty after an average meal, but fatty items, strong spirits, nervous shock and stress lead to delayed emptying, while more fluid in the meal causes earlier emptying.

Eye changes. The appearance of the retina and chemical changes in eye fluids (vitreous humour) give indications as to the time since death.

Changes in the blood and in the bone marrow can help to determine the time since death.

The Post-mortem Examination

Once the on-the-spot examination has been completed and any necessary specimens taken, the corpse is put in a 'body bag' and taken to the mortuary for post-mortem examination (autopsy). But first a blood specimen is taken to be tested for hepatitis viruses and HIV (the AIDS virus). The results will tell whether the examination should be carried out with special precautions.

Then the body must be identified. This is done by relatives or acquaintances, when available. They do it by looking at the face. Other identifying features are described above. Tattoos can be useful. Coloured ones tend to get blacker over the years because the black pigment (Indian ink) is more resistant to removal by the body's white blood cells. Scars or old burns can help. At an early stage scars are brownish red, but they fade and get narrower over four to six months, becoming whitish after a year.

Sometimes specimens help in an unexpected way. A man was found dead by the roadside with injuries to the chest. It looked like a road accident until bovine hairs were found on the corpse, and it turned out that the victim had been crushed by a bull in an adjacent field, then tossed over into the road.

Preliminary X-rays are necessary in the case of gunshot wounds, to help recover the missile and with shotgun injuries to determine where the pieces of shot are.

A routine pathological examination is then carried out to check especially for injury, heart disease, haemorrhage, cancer, etc. If burglars break into an old man's house and tie him up, and he dies with a massive brain haemorrhage an hour or two after being released, the question will be 'Did he die of a natural disease of brain arteries, and how was it related to the assault and tying up?' Similar questions arise in reaching a view on the employer's liability in the case of death associated with work. Deaths due to heart attacks can also come into this category. We know that people who lead sedentary lives are six times as likely to suffer cardiac death during or within an hour of heavy physical exertion such as shovelling snow, jogging or sexual intercourse. Old men, however, may die of what is called 'senile myocardial degeneration' with only moderate disease in the heart.

Cause of Death

Gun wounds. Guns are frequently the culprits in homicide, suicide and accident alike, and sophisticated methods are available for piecing together the story from a study of the victim's body. What happens depends on the velocity of the missile. At velocities of up to about the speed of sound (340 metres or 1,100 feet per second), which includes crossbows, air rifles, and most revolvers, the missile thrusts aside body tissues along a narrow track and causes local damage. At velocities greater than the speed of sound (for instance military-type weapons at 980 metres per second), the missile sends a shock wave of compression ahead of it as it passes through the tissues. This very brief period of pressure causes a wide zone of damage along the missile track. The bullet generally passes right through the body and out the other side. When the bullet hits bone it may shatter it and send a fragment out through the skin. When infantry-men stood in tight rectangles, as at the Battle of Waterloo, it was possible to be injured by a flying fragment of bone from a wounded comrade.

The tiger shark, the arm and the modern-day Jonah

In the late 1950s a live tiger shark was caught off the Australian coast near Sydney. It was brought ashore and placed in an aquarium, where it vomited up a human arm. Tiger sharks have a powerful bite (a force of 3 tons per square cm), but the arm had clearly been cut off before being eaten by the shark. Although it had been in the shark's stomach many days, distinctive tattoo marks were visible, and these were used to identify the person, who turned out to be the victim of a gangland killing. The murderers had dismembered the dead body and thrown it into the sea, where normally it would have been disposed of within a day or so. It was their misfortune first that the very shark that ate it was caught, and second that it vomited up the incompletely digested arm.

Another denizen of the seas off the coast of Australia is the sperm whale (the story is told in the Killer Whale Museum, Eden, New South Wales, Australia). This whale grows up to 50–60 feet long and has very large teeth, but only in the lower jaw. In February 1891 a harpooned sperm whale smashed one of the small whaling boats to pieces with its tail and a crew member, James Bartley, was found to be missing. At the end of the day the dying whale was hauled to the main ship, the *Star of the East*, and processing began. They thought they saw movement in the stomach, and on slicing it open, out came the missing man. He was unconscious, his skin was bleached a deathly white, and he was almost blind. He had been inside the whale for fifteen hours, and although he lost his hair and was delirious for two weeks, he recovered and lived for another eighteen years.

The actual nature of the wound depends also on a lot of other things, such as whether the weapon is rifled or smooth-bored, the presence or absence of a 'choke', and the nature of the propellant. At short ranges (less than about 18 inches) soot, smoke and hot gases enter the wound. As might be expected, the exit wound tends to have a more ragged appearance, with flaps of skin directed outwards. The pathologist is usually asked to estimate the range, which he does from the appearance of the wound. This is easiest in the case of shotgun wounds, where the shot are more spread out as the range increases.

Cutting, stabbing, and chopping. Clearly, different weapons (sharp knife, axe, icepick, etc.) will cause different types of wounds. A cut throat usually means suicide, and the classic wound (for a right-hander) starts high on the left side of the neck and severs blood vessels and sometimes the windpipe as it comes across, often tailing off a bit on the right hand side. There may have been be a few tentative preliminary cuts. Stab wounds can mean murder or suicide. Obviously, if there are multiple stab wounds in different parts of the body it is unlikely to be suicide. Also, in the case of suicides wounds are generally made on the chest where the heart is, rather than to the head or abdomen, and a suicide usually pulls the clothes aside before doing it.

Burns. Death is often due to the inhalation of smoke and other fumes, such as cyanide fom the combustion of plastic materials, rather than primarily burning. Burns can occur after death as well as before it. If the air passages in the lungs contain sooty particles, it means the person was alive in the fire and breathed them in; he was not placed there when already dead.

There have been very occasional reports of what was called 'spontaneous human combustion', as described by Charles Dickens in his novel *Bleak House*. The human body is burnt to ashes without much damage to surrounding objects. Even the victim's shoes may be spared. Suggested explanations of this almost apocryphal phenomenon have included large amounts of alcohol in the victim's body, or the slow combustion of body fat with the clothing acting as wick, like a candle. It is seen when the body is near an open fire or chimney, so that there is a good upwards draught.

Electrocution. The amount of damage depends on the actual amount of electricity flowing through the body, in amps per second. This will depend on the pressure, that is to say the voltage, and on the resistance to the flow of current in the tissues. The skin is the most resistant part of the body, and more resistant when dry than when wet. This is why bathrooms are such dangerous places, with good opportunities for electrocution (electrical gadgets, water in the bath) – as well as for drug overdoses (the bathroom cupboard), suicide by throat or wrist slashing (razor blades), and injury (slipping on a wet surface, hitting the head, and then drowning or dying of the head injury).

A current of 240 volts is more likely to kill than one of 110 volts. When 50–80 milliamps pass across the heart for more than a second or two it is usually fatal. If the current passes through the chest, spasm of chest muscles and diaphragm can inhibit breathing. The head and neck are less commonly involved, but the current can act on the brain stem and paralyse the heart and respiration.

Strangling. Strangling by hand is not too difficult to recognize. It is almost impossible to strangle yourself, but it may be hard to tell whether a ligature was put in place by a suicide or by a murderer. The question of suicide arises with certain auto-erotic practices. A little cerebral anoxia (lack of oxygen) can result in erotic hallucinations or enhanced orgasm. The anoxia is produced by putting the head into a plastic bag, or more often by placing a ligature round the neck which is tightened by hand. As the erotic reward is experienced, consciousness fades and the ligature slackens. Obviously things can go wrong, and although the circumstances will generally make it clear whether it is accidental suicide, this can be difficult if a sexual partner was present. A tightly drawn noose round the neck can cause death by reflex inhibition of the heartbeat.

Hanging. If the victim was strangled by wires, ropes, or cords, did he hang himself or was he throttled to death? Hanging is nearly always suicidal, and in this case the rope or cord-mark on the neck rises as it approaches the point where the knot is, and there is a space between the cord and the skin where the rope leaves the body surface.

Drowning. Bodies found in water are often in a bad state. Was the person already dead on entering the water? Did the dead person have an ordinary heart attack in the water? The pathologist will be expected to give an answer.

During true drowning, water is shipped into the lungs, and as the person struggles for breath the water and air form a foam which further interferes with breathing. If it is fresh water it passes from the lungs into the blood, and dilutes it. If it is sea water things tend to go in the opposite direction: water leaves the blood and enters the already waterlogged lungs. Diatoms in the blood are referred to in chapter 5; if they are present it suggests the victim was not dead on entering the water, but was trying to breathe. Once the struggle is over the corpse sinks, with the head, arms and legs, the heaviest parts, downwards. After putrefaction and gas formation it rises again and floats – and is not a pretty sight.

Not all people who die in the water have drowned in the classic way. In some of them the sudden immersion and the rush of cold water into mouth and nose stimulates a reflex that stops the heartbeat (cardiac arrest). They die within a minute or so.

Jumping from heights. This mode of death is described in the section on suicide in chapter 3. The person may have jumped, fallen or been pushed. The place and the circumstances are obviously important. When suicides leap from buildings, the outcome is more reliable when the launch is from at least the fifth floor.

Hypothermia. Shivering ceases when the body temperature falls below 32°C (normal temperature is 37.5°), and if it gets below 28° death nearly always follows. The very young, the very old and the intoxicated are especially susceptible. So-called 'death by exposure' usually involves hypothermia, and hypothermia is what can kill drowning people or those clinging to the wreckage for long periods. The pathologist will look for characteristic changes in the skin, although these can be due to the dead body being stored in a refrigerator.

Alcohol and other drugs; poisons. When it looks as if toxic substances have played a part in the death, the pathologist will have to take samples of blood, urine, vomit and stomach contents, and also a piece of liver (where drugs are concentrated) for analysis. The possibilities are almost endless, with more than thirty groups of poisonous agents to be considered, although the common ones are fewer in number. He may be helped by relatives, friends, workmates or witnesses, and if he is lucky there could be a suicide note, even a labelled drug container in the vicinity. Poisoning is the commonest method of suicide in developed countries. The body as it lies may offer help: needle marks, the smell of certain drugs, burns round the mouth due to corrosive poisons like paraquat can provide clues.

Alcohol (*ethyl alcohol*) is the most commonly used drug in the world, and a powerful trigger of violence and homicide. It diffuses through the whole body with the exception of fatty tissue, because it doesn't dissolve in fat. Females have more subcutaneus fat, so when they drink the same amount of alcohol (in relation to body weight) as males, a greater proportion of it is present in blood and other tissues. They are therefore more vulnerable to intoxication.

Legal limits for blood alcohol levels in drivers differ in different countries (relatively high in Ireland, low in Sweden): in Britain the limit is 80 mg per 100 ml (35 mg in the breath test and 107 mg in urine). The sequence of events as blood alcohol levels rise is as follows:

30–50 mg per 100 ml of blood	driving skills impaired
50–100 mg	less inhibited, talking, laughing
100–200 mg	drunkenness, nausea, unsteadiness
200–300 mg	vomiting, stupor, maybe coma
300–350 mg	danger of death, breathing paralysed

But individuals differ. You can achieve more than 300 mg per ml by taking in 300–500 mg/100 ml alcohol, which means between a half and one litre of whisky, gin or other spirit in less than one hour. Heavy drinkers tolerate higher levels than the rest of us. A chronic alcoholic once survived with a blood level of more than 1,500 mg per 100 ml.

Thre are many other drugs that can be tested for, including heroin, cocaine and Ecstasy, any of which may be relevant for the police investigation. The rate of metabolism (speed of disappearance in the body) obviously matters.

Antidepressant drugs, so widely prescribed, are often used for suicide and cause about 10 percent of drug deaths in England and Wales. Until the 1980s *barbiturates* accounted for 20 per cent of all drug-related deaths in England and Wales. Now they are no longer prescribed as sleeping pills. Barbiturates can be detected in the corpse for long periods, even after years in the exhumed corpse.

Paracetamol is the favourite suicide drug in the UK today, killing 100–150 each year (aspirin causes about fifty deaths a year). The amount that can be bought over the counter without a prescription was reduced in September 1997. There have also been suggestions that the antidote to paracetamol, a substance called methionine, should be incorporated into the pill.

Arsenic, antimony, cyanide and strychnine, the murderer's standbys in the old days, are rarely used today because they are so easily detected. The classical one was *arsenic*. After acute poisoning, say with 100 mg arsenic trioxide, vomiting and collapse occurs after half an hour to an hour, followed by death. Chronic poisoning can be detected by demonstrating arsenic in nail or hair samples. *Antimony* is similar to arsenic and is present in 'tartar emetic'.

Cyanide was quick-acting and reliable, although unpleasant for the victim. Hydrocyanic acid (Prussic acid) is present in certain insecticides and rodent killers, and in a few types of polish. After inhaling the gaseous form (hydrogen cyanide), death occurs within minutes. It stops oxygen getting into cells, and the surplus oxygen in the blood gives the body a brick-red colour. Cyanide has a special 'bitter almonds' smell, but three-quarters of the population are born unable to smell it. Cyanide is easily concealed ('suicide pills') in rings or in hollow teeth.

Strychnine is a cruel poison, causing an agonizing death with muscle spasms, convulsions, and eventual exhaustion and death.

Insulin has many times been used for murder, and often for suicide, especially by doctors and nurses because they have access to it. A big dose lowers the blood sugar level, causing brain damage and death within a few hours. It is not easy to test for, but it can be done.

Carbon monoxide is often the culprit in suicides, sometimes in homicide and accident, and can be detected in the body for up to six months after death. It still kills up to 1,000 people a year in England and Wales, usually coming from vehicle exhausts or leaky heating systems. In many of the Nazi extermination camps the victims were herded into large containers and the exhaust from diesel vehicles pumped in. It was cheap and quick-acting; the bodies, after removal of gold-bearing teeth, were buried or burned.

Whose Cells?

We have already considered the problem of who owns a body. Similar problems can arise at the cellular level. When cells are cultivated in bottles they cannot divide more than fifty or sixty times, as mentioned in chapter 4. But they can change in character and become immortal, that is, by dividing indefinitely. Cancer cells do this. One day in 1951 Henrietta Lacks was in the operating theatre of an American hospital undergoing surgery for cancer of the uterine cervix. Some of the tissue from her cervix was handed to an enthusiastic pioneer of cell culture, who managed to cultivate the cells in bottles. The cells (called Hela cells, after the donor) became immortal and before long had been distributed to almost every laboratory in the world. Their total volume must eventually have exceeded that of Henrietta Lacks herself, who thereby achieved a type of immortality.

Sometimes the question 'Whose cells?' arises as a legal rather than a truly biological question. In 1976 the cells from the enlarged spleen of a patient with leukaemia were taken for testing in an American hospital. They were found to contain a new virus (human T-cell lymphotropic virus type 2) which was of great interest. As they multiplied in culture, the cells also produced one or two biologically important molecules that could be purified and sold to other laboratories. The cell line (called 'Mo' after the patient, John Moore) was the subject of a patent, US patent 4,438,032. But John Moore had never signed the consent form and, when he was better and had gone home, he took out a lawsuit against the original hospital laboratory doctor, claiming a share in the profits because they were being made with his cells. Eventually the California Supreme Court ruled that the laboratory doctor had erred in failing to reveal the financial value of the cells, but added that John Moore no longer owned them. In the end a settlement was negotiated between the doctor and the patient.

Not all patients do so well. In 1995 the US National Institute of Health was granted a patent on a cell line derived from a patient in a remote part of Papua New Guinea. The cells came from a blood sample taken as part of the Human Genome Diversity Project. They contained a virus, and had potential use for a vaccine and for diagnostic tests. But in this case the person supplying the cells is unlikely to benefit; nor, indeed, will his people, because they certainly will not be able to afford either the diagnostic kit or the vaccine.

Legal problems about the ownership of cells arise also with frozen and stored ova, fertilized ova, semen, and sometimes with blood. Semen and ova, surely, belong to the donor; but what if stored semen from the husband is used by the widow after his death? An instance is given in chapter 12. What are the legal status and inheritance rights of any resulting children? These are matters for the lawyers, but it seems clear that advances in biomedicine (gene therapy, pre-natal genetic testing, treatment of defective newborns) are going to daunt even the bravest legislators.

Who Owns DNA Sequences?

Having the DNA sequence of a gene means that you have the key to manufacturing the substance made by the gene, whether that substance is insulin, erythropoetin (used to treat certain anaemias) or a virus protein used for vaccination. In the world of commerce a gene sequence can be a valuable possession, and people have already started patenting gene sequences. The possibilities are rapidly increasing as the DNA sequence of human beings is charted (see chapter 10), and there have been particular misgivings about the patenting of human gene sequences. If someone patents a human gene (DNA) it would give them a legal monopoly over that piece of DNA. Is that morally acceptable? Might it be possible for them to prevent the patented material from being reproduced – in other words, to prevent the person who was the source of the gene sequence from having children?

Some of the objections are emotional reactions against scientists and DNA research. The scientist is viewed as a 'gene snatcher', and many are afraid of a slippery slope, not knowing where it will end. Other objections are more reasonable. In the first place, patenting is a matter of money and ownership rather than human well-being; and in any case a gene is not really an invention to be patented, but a discovered product of nature, a part of the human body. Furthermore, is it not possible that patenting (owning) a gene will restrict the free access of information to other scientists and act as a barrier to scientific advance? On the other hand, it is argued that the rush to patent a commercially valuable gene is a great stimulus to investment and research, which is good for science. Perhaps a compromise will be reached in which certain DNA sequences, the ones where the gene product is of practical use in diagnosing. treating or preventing disease, can be patented. A balance will have to be struck and, as always, the lawyers will force us to think harder and more clearly.

In May 1997 HUGO (the HUman Genome Organization) came out against patenting human DNA. Results on human DNA, they say, should be freely available to scientists except in the case of certain genes which are involved in disease and are targets for drug development, and when research along these lines calls for high-risk investment.

Removing Parts of the Body

Most corpses are buried or burned in an intact state, undamaged apart from a possible post-mortem examination or the onset of early putrefactive changes. There is an almost universal feeling that one's body, if it is to be buried, should be allowed to rest in peace without being interfered with. Pope Boniface VIII issued an edict in 1299, saying that anyone dismembering a corpse would be excommunicated. Nevertheless, organs or other portions of the body have often been removed from corpses, for various reasons.

To Prevent Putrefaction or as a Preliminary to Embalming

There were good reasons for removing the bowels (evisceration), outlined in chapter 5. It was standard practice for nobles and kings killed on foreign battlefields, and meant that the rest of the body, or the bones after the flesh had been boiled off, could be brought home and decently buried. When the corpse was to be embalmed the viscera were often placed in special viscera chests.

A part of the body may be buried separately because the owner is still alive. One of the English commanders at the Battle of Waterloo had one of his legs shot off. The leg was recovered and given a local burial and its own tombstone, while the rest of the body returned to England and lived to a ripe old age. In the nineteenth century a hated Mexican dictator had his leg was amputated, and he arranged for it to be given a proper cathedral burial. At a later date he was deposed and the leg was dug up by his enemies, only to be reburied with full ceremonies when he came back to power again. Muslims generally bury any part of the body that is removed by the surgeon. An amputated leg, for instance, is taken by relatives and buried in the place where, one day, it will be joined by the owner of the leg. In Chinese imperial courts the eunuchs, after they had been castrated, kept their genitalia (it was not a delicate operation with careful removal of just the testicle) so that they could be buried with them. They could then go to paradise as complete men. The last eunuch died in a Buddhist temple in 1996, aged ninety-four.

The practice of separate placental burial (see chapter 12) also comes under this heading.

For Relics

Relics are not only collected from saints and martyrs; they are also much sought-after from emperors, kings, famous authors and artists, etc. (see chapter 12) When Napoleon died, neglected, on St Helena, local people eagerly sought souvenirs of the great man. These included his clothes and his hair clippings; and someone was bold enough to remove his genitals, a particularly potent relic from such a glorious emperor. An object looking like a seahorse that was said to be the shrivelled renmants of his penis was auctioned at Christies in 1972. His heart was put in a silver vase, his stomach in a silver pepper pot (he had stomach cancer), and a portion of the intestines found its way to the Royal College of Surgeons, London. The rest of the corpse was finally returned in triumph to Paris in 1848, twenty-seven years after he had died, and buried at Les Invalides. Tutankhamun, who is referred to in chapter 9, also lost his penis to relic hunters, probably shortly after the mummy was found in 1925; the young king's mummified penis, together with a golden penis sheath, had disappeared by the time he was reburied in 1926.

Some interesting portions of the body have been collected as relics. When the Empress Josephine visited the tomb of Charlemagne (742–814) in Aix-la-Chapelle she was presented with his shoulderblade. It was set in a piece of the True Cross, enclosed in crystal, and attached to a gold chain that had hung round Charlemagne's neck. Another separately preserved shoulderblade was that of St Andrew in Edinburgh. The bones of Galileo's index finger reside in a glass case in the Natural History Museum in Florence. The finger is slightly curled and points upwards as if the great man is making an uncomplimentary sign to his ecclesiastical opponents. An unusual relic is a tooth of Isaac Newton (1642–1727), sold in London in 1816 to be set in a nobleman's ring; doubtless it was hoped that this relic would enhance the wearer's cerebral processes. The ancient Romans practised *os resectum*, in which a part of the body such as a finger was cut off from the corpse and buried, while the rest of the body was cremated and the ashes put into cinerary urns. The buried part was not so much a relic as a remnant of the dead person that would be saved from immediate destruction in the fire: a sort of compromise between cremation and burial.

As War Booty

The scalping of the dead on the battlefield was practised by the Scythians, Celts and Teutons. Although associated with Native American tribes, it seems to be something they had learnt from their European enemies. Some states offered bounties for Indian scalps, and settlers scalped Indians as evidence that they had killed them. Massachusetts gave £40 for each scalp of a male Indian over twelve years old, and £20 for that of a woman or child. Unusual war relics include the noses of thousands of Koreans cut off by the Japanese during their 1597 invasion of Korea. In 1993 20,000 of these noses were decently buried.

Parts of the body may also be removed to punish or abuse the corpse Beheading, drawing and quartering of the corpse was described in chapter 13.

To Carry out the Wishes of the Dead or to Honour Them

The heart has always had a special status as the seat of the emotions, of courage, and of piety. Since the Stone Age it has at times been buried separately from the rest of the body, although it was often included in the viscera and removed with them. Richard I of England was killed in 1199 and most of him was buried at the feet of his father at Fontevraut Abbey, France. The viscera were buried locally at the Abbey St Saviour at Charroux in Pottou. His heart, however, was sent to Rouen Cathedral as a reward to the citizens for their loyalty to him. William the Conqueror (1027–87) died of wounds suffered during the siege of Nantes. His bowels were buried in Chalus, France, his heart in Rouen Cathedral, and the rest of his body in the Abbey of Les Dames, Caen. In the twelfth century during the crusades, many of those who died in the Holy Land wanted their hearts to be buried there. This was the beginning of the practice of heart burial by Christians, and it was adopted by European royalty and by others. In seventeenth-century England at least 190 heart burials are on record. The Scottish missionary and explorer David Livingstone died in Alala, Africa, in 1873. His heart was buried there, under a tree, together with his viscera. The rest of his body was filled with salt, smeared with brandy, and left to dry for fourteen days. Then, enclosed in a cylinder made of bark, covered with sailcloth and lashed to a pole, it was carried about a thousand miles to the African coast at Zanzibar. From here it was shipped to England and buried, nearly a year after he had died. Shelley's heart, which would not burn (see chapter 7) is in Boscomb Manor House, Hampshire. Thomas Hardy's ashes rest in Westminster Abbey, next to those of Charles Dickens, but his heart is buried with his first wife in the churchyard of St Michaels at Stinsford, Dorset.

Heart burial has not lost its appeal. The body of Tsar Boris III of Bulgaria, who died in 1943, was lost, presumably destroyed by the communist secret police. But the heart was recently recovered, in its lead capsule, from a Bulgarian graveyard and reburied.

The head carries more of the person's individuality, as described in chapter 15. In 1996 a widow in Egypt dug up her dead husband's skull because she was lonely and missed him. She was arrested by the police for her pains. A complicated clerical burial was that of St Thomas à Becket in Canterbury. His brains were in two separate tombs, with two more tombs for the skull and for the rest of the bones (see chapter 12). The saga of Oliver Cromwell's head is described in chapter 13.

For Organ Transplants

The removal of organs for use as spare parts is described in some detail in chapter 12.

For Study

The brain has been the most popular organ for study. Are the brains of the great intellectuals, artists and scientists bigger, or do they differ in other ways from normal brains? It turns out that mere brain weight has little to do with it, as shown by the following brain weights in ounces:

Trotsky 56; W.M. Thackeray 58; Marilyn Monroe 51; Robert F. Kennedy 51; Walt Whitman 45; Anatole France 35. The average adult male brain weighs 49 oz (1,370 g) and the female 44 oz.

What about the convolutions (wrinkles) on the surface of the brain? They are formed because the front of the brain (forebrain), basically a dense network of special nerve cells in the shape of a hollow tube, has expanded so much during our evolution. It now covers much of the rest of the brain, and the blown-out tube has also collapsed (forming wrinkles on its surface) so that it could fit inside the rigid skull. More wrinkles mean a larger area of forebrain, which means more nerve cells. Our brains have extensive convolutions, but geniuses have not been shown to be different from the rest of us. Porpoises, moreover, have more convolutions (and a heavier brain) than we do.

But brain weight and surface convolutions are crude things to be looking at. Might there be differences if more detailed, subtle comparisons were made? It seems unlikely. Mere brain structure sets us apart from other primates but it does not explain intellectual differences between humans. In the search for something special in the brain of a genius, Einstein's brain was removed before he was cremated in 1955, and stored in Princeton Hospital, New Jersey. The brain was subsequently moved around to different hospitals and studied by expert neuropathologists, but nothing unusual was noted. There has still been no final published report and the brain, preserved in formalin, sits in a jar somewhere awaiting further study.

PART IV

Death and Afterlife

15

Death and the Corpse:
The Emotional Impact

The greater part (of mankind) must be content to be as though they had not been born, to be found in the Register of God, not in the record of man.

Sir Thomas Browne (1605–82)

The Prospect of Death

No one would deny that death (one's own death, at any rate) is difficult to face up to. La Rochefoucauld said that man could no more look steadily at death than at the sun. It is easy to think of one's own death with a horrible sinking feeling, with apprehension and dread. But death can also be contemplated from a more philosophical point of view. The philosophers cannot help being interested; as Socrates said, quoted in Plato's *Phaedo*, 'Then it is a fact, O Simmias, that true philosophers make death and dying their profession.' We can philosophize that death belongs to life, is immanent in life (see chapter 1) and that none of us is going too get out of it alive; but this doesn't make it easier. Death is the conqueror and leveller, and awaits us all, regardless of glory, wealth or beauty. Inwardly, we should be prepared for it. Nevertheless, we cannot accept the total obliteration of the self, if this is what death means. Death, after all, is no abstract idea but the great, apparently insuperable threat to our continued existence. The emotions aroused by death include sorrow, anger, defiance, resentment and resignation; but the commonest one is fear. From the point of view of the survival of the species, this is not unexpected. By their nature, living things have to do their utmost to survive, reproduce and avoid death. Being afraid of dying can be thought of as an advantage in life's struggle.

Many people would admit, after careful consideration, that they fear the dying process as much as death itself. Almost everyone would hope for a swift, peaceful exit from life. Who would not sooner die suddenly of a heart attack than die slowly of cancer or starvation? The fear of pain is not the only ordeal in the dying process. There is also the knowledge that before long you will be extinguished like a candle; and there are anxieties about leaving loved ones, anxieties about things left undone.

On the other hand, it can be argued that the thought of being born would be an even greater terror if we knew that it was about to happen. The purely physical pains and stresses of birth are tremendous. Luckily, we are not in full possession of our faculties and are not committing things to memory at that traumatic moment. Francis Bacon (1561–1626), the English philosopher and statesman, remarked that: 'It is as natural to die as to be born; and to a little infant, perhaps, the one is as painful as the other.'

In medieval Europe the church used many devices for reminding people about death. The idea was to inspire fear, the fear of death, in sinners and the unrepentant. Frightful visual images called *memento mori* (Latin = reminder of death) were displayed to reinforce the words of the preachers and the prayer book. Memento mori were paintings and sculptures showing skulls, skeletons, or skin and bone figures of death. The corpses were sometimes portrayed being eaten by worms, or with worms crawling out of the eye sockets. European artists in the eighteenth and nineteenth centuries delivered striking reminders of death by painting a body or face with one side alive, but the other side dead, disfigured and decaying. Death was democratic, and it awaited both rich and poor. The image was often enhanced by symbols of death such as scythes, hour-glasses, candles and skeletons. Surely the fear of death and the horrors of hell would make people behave better. But however reasonable it might seem to suppose that people can be terrorized into good conduct by the prospect of hell, the evidence is poor. Christians are not demonstrably more virtuous, kinder to others, than those who practise religions with no hell, or those with no religion at all.

Yet the possibility of an afterlife, with or without heaven and hell, has an obvious effect on attitudes to death, and this topic is dealt with in chapter 16. Several studies have shown that those with a religious outlook tend to have a more forward-looking, optimistic attitude to death. Those living alone fear death the most, while those living with a spouse tend to evade the issue. Normal adults, however, are not plagued by the fear of death. If they think about it they may admit to being worried, and fear may lurk in the unconscious, but it is not an everyday anxiety. It is mainly children (over the age of five), old people and the mentally disturbed who have a high level of concern.

The poet Walt Whitman (1819–92) seems to have been optimistic about death:

> To die is different from
> What anyone supposes
> And luckier.

Those who share this optimism may have what seem to them to be good reasons for this, apart from religious beliefs. They could, for example, look forward to being frozen into immortality and brought back to life at a later stage, as described in chapter 10. For them death is an unnatural, unnecessary event, an imposition on the human race. For others, the optimism may be pure bravado. They will call themselves 'sixty-five years young' instead of 'sixty-five years old.' Others make jokes about it. Winston Churchill died aged ninety-one, and on his seventy-fifth birthday is reputed to have said: 'I am ready to meet my Maker. Whether my Maker is prepared for the ordeal of meeting me is another matter.'

As an example of a bold, practical view, William Hazlitt (1778–1830) wrote in 'On the Fear of Death' from *Table Talk*:

> Perhaps the best cure for the fear of death is to reflect that life has a beginning as well as an end. There was a time when we were not: this gives us no concern – why then should it trouble us that a time will come when we shall cease to be? . . . To die is only to be as we were before we were born . . . And the worst that we dread is, after a short, fretful, feverish being, after vain hopes, and idle fears, to sink to final repose again, and forget the troubled dream of life!

One might think that dying is easier if life has been busy and much has been achieved; but it is not so, because there always seems to be more that could have been done. Sir Isaac Newton (1642–1727) invented the binomial theorem and also discovered calculus, the properties of light and the law of universal gravitation. His impact on science was immense; in the words of Alexander Pope (1688–1744),

> Nature and Nature's laws lay hid in night:
> God said, Let Newton be! and all was light.

Newton's book *Principia* is one of the greatest feats of the human intellect; he was revered and honoured by his contemporaries and was

President of the Royal Society for twenty-five years. Yet he remained a shy and modest man, suggesting in his dying words that there was much more to be done.

> I don't know what I may seem to the world. But to myself I seem to have been only like a boy playing on the seashore and diverting myself in now and then finding a smoother pebble or prettier shell than ordinary, whilst the great ocean of truth lay all undiscovered before me.

It can be argued that the sovereign remedy for the fear of death is to look it constantly in the face; to get used to the idea of dying and accept it. This is the strategy advocated in many of the great religious beliefs of the world, as outlined in chapter 16. You can, they say, avoid death in a manner of speaking because your spirit then leaves the body and enters a World of Light, or the Universal Consciousness, the sort of state described in near death experiences (see chapter 4).

This is perhaps easier to do when a serious illness makes us come to terms with death, for a while. Mozart, in a letter to his father on 4 April 1787, when he was thirty-one, wrote:

> Since death (to be precise) is the true end purpose of life I have made it my business over the past few years to get to know this true, this best, friend of man so well that the thought of it not only holds no terrors for me but even brings me great comfort and peace of mind.

Seeing the Body

Those who have seen a dead body know that it is a sobering experience. Yet the power of the corpse to arouse emotion falls away in a definite sequence as it decays and disintegrates. Naturally the fresh corpse is the most powerful reminder of the dead person. Emotions are readily kindled by this sight, and in earlier centuries viewing the corpse had an important place in family and religious life. After it has been tidied up, perhaps embalmed, its power is if anything enhanced. We are in the presence of the dead person.

Mummification adds a degree of distance, but the personal appeal is still there. In Roman Egypt (fourth century AD), coffins had a lid that slid back so that you could communicate face to face with the mummified body. Death masks or life-like effigies are next in the sequence, acting as artificial reminders. Sculptures or paintings have a similar impact, while a photograph, especially when the person looks directly at the camera, is a more powerful force because it captures an actual moment of life. It is the real person who is looking at us. Photographs of dead faces are different: more moving because we know they are dead; the eyes are shut and they look peaceful.

The face and the head have a greater impact than other parts of the body. Through the face the individual expresses his uniqueness. As it happens, the head is also a part of the body whose removal automatically brings life to an end, as contrived with admirable efficiency by Dr Guillotine. The ears, the fingerprints and the sole patterns are also unique, but only on expert examination and they do not arouse strong feelings. It is because of the impact of the face and head that the heads of executed criminals were at one time publicly displayed as a lesson to the rest of the community. Intact heads made more of an impression than skulls, and heads could be soaked in brine to stop birds pecking away the flesh and thus converting the head into a more anonymous object. Whatever the circumstances, the severed head makes a unique impression, as exemplified in the cases of John the Baptist or Oliver Cromwell (see chapter 13).

Other parts of the body are less easily recognized and accordingly have less emotional impact. People are often able to identify their partners when only the regions below the knees are visible, but for humans the main focus is always on the face. The heart, hands and hair have an emotional pull – and indeed, as we have seen, the heart was sometimes given a separate burial (see chapter 15) – but these parts are not so readily identified as belonging to the deceased. The remaining pieces of the body such as muscles, bowels, lungs and kidneys are anonymous. They have a powerful impact if we know they are from a special person, but much less so than the face or the fresh corpse.

As the body skeletalizes, the head retains its supremacy. The skull is not so personal, but it is an image of death and also a formidable reminder of the dead person. This is reflected in the skull's pride of place in ossuaries, and in the practices of head hunters and head shrinkers (see chapter 9). Of the rest of the skeleton the hands probably come next, but there is little to choose between the other bones: they are significant if we know they belong to the dead person, but less so otherwise. A dying English king is reported to have asked his followers to carry his bones before them as they marched, because the enemy would not be able to endure the sight of him, dead or alive. Yet bones, because they last longer, have an important place as mementos.

Up to this point, the changes in the corpse can be described as quantitative, in so far as less and less of the original structure remains. Nevertheless the skeleton or an identifiable fragment of the body is still visible. The extraordinary preoccupation with relics (see chapter 12) attests to the psychological power of visible and tangible human remains. It is after the next stage in breakdown, after incineration, or complete dissolution of the body in the soil, that there is a qualitative change in our attitudes. There are now no bones, no identifiable remains. Ashes have no individuality, except when we know whose they are. The process of dissolution is complete, and the person has become no more than a memory. But the ashes remind us of that person, we still feel emotional, and the connection between the living and the dead is still there.

Scattering the ashes is perhaps less moving than a regular coffined burial, but it is nevertheless an emotional ceremony. If there were no ashes and the body disappeared totally in the furnace, relatives and friends would be deprived of a natural outlet for their feelings. It might then be worth providing fake ashes. As the ashes are scattered, the thread is broken. There is now nothing to see. The corpse has lost location as well as substance, and the body's constituents re-enter nature's elemental cycles.

Corpses and Death Avoidance

In European art from the fourteenth to the sixteenth century, death was represented by powerful, macabre images. Half-decomposing corpses, called 'transi', were frequent themes in tomb sculpture, and rotting bodies occur in the paintings of this period, as seen for instance in the Masaccio fresco in the church of Santa Maria in Florence. Fifteenth-century paintings of the 'danse macabre' depicted men dancing to the grave in a confrontation with their own death. The message was that since we all share the same destiny, we should prepare for it. The clean skeletons of later centuries seem attractive objects when compared to late medieval corpses with their putrefying flesh. Skulls and thighbones (crossbones) were acceptable images in the seventeenth to nineteenth centuries, and as 'memento mori' they figured in rings, dishes, spoons, snuff-boxes, even furniture. Death was a skull or skeleton, with the scythe or hourglass as badge of office.

The Victorians were obsessed with death. Infant mortality was high, and many died before old age as a result of disease, injury or infection. Both in the family and in the community there were frequent encounters with death and with the dying. Later in the century, family photograph albums might include photos of dead children. In contrast to this, the topic of sex in Victorian times was subject to taboos and prohibitions. It was not to be mentioned. As scholars have pointed out, we have now reversed this state of affairs in Western society. As far as sex goes, almost anything is permissible, whereas death has been swept under the carpet. People die in hospitals or institutions, and we are spared regular encounters with death. In the UK 80 per cent of deaths now take place in hospitals, and half of the remainder in retirement or nursing homes. Most of us have never seen someone die, nor seen a corpse.

Death needs to be brought out from under the carpet and nearer to the centre of life. It should be something recognized, something we can come to terms with; not something unmentionable, because that is what breeds an unnatural fear of death. Perhaps the Mexican Day of the Dead, a day of the year when everyone celebrates death, is taking things further than necessary, but a fuller acceptance of death would be healthy for modern society. Most of us are not adequately prepared for coping with dying and bereavement, and even less well equipped to face our own death. Yet wisdom concerning death is as likely to be needed as expertise in first aid, which is the subject of special courses.

We should move in the direction of Buddhist and in particular Tibetan attitudes to death. According to *The Tibetan Book of the Dead*, meditation on death leads to familiarity with it and acceptance of it, so that it does not come as a shock and a terrible threat. We should overcome our natural, animal fear. Death need not be the ultimate personal disaster, and understanding it is the key to spiritual well-being.

The Comparison of Death with Sleep

This is a common comparison, and many poets liken Death to its brother, Sleep. The word cemetery is derived from a Greek word meaning a resting or sleeping place. We might think of sleep as the normal state. It is experienced at the beginning of life (in the womb), and again in sleep-ridden old age, with wakefulness in the interlude. If you now take the idea further, and regard non-existence as sleep, then, as implied in William Hazlitt's words, life itself can be thought of as an island of dreams in an ocean of sleep. Shelley's poem *Adonais*, inspired by the death of Keats, contains the lines:

> Peace, peace! he is not dead he doth not sleep –
> He hath awakened from the dream of life –
> Tis we who, lost in stormy visions, keep
> With phantoms an unprofitable strife,
> And in mad trance strike with our spirit's knife
> Invulnerable nothings.

There is, incidentally, a medically important connection between sleep and life. You can deprive someone of the special type of sleep in which there are dreams, rapid eye movements and characteristic electrical patterns in the brain. This is done by waking them up every time that sort of sleep begins. After a while they may suffer hallucinations and serious psychiatric consequences. In other words, you need that type of sleep for mental health. People who do not sleep normally due to medical conditions that disrupt their sleep complain that their illness is more profound, and their quality of life is worse, than in those with the same condition who sleep normally. Shakespeare sang the praises of sleep in *Macbeth* (Act II, Scene 2):

> Sleep that knits up the ravelled sleave of care,
> The death of each day's life, sore labour's bath,
> Balm of hurt minds, great nature's second course,
> Chief nourisher in life's feast.

Could it be that we need sleep because we need to dream? Sigmund Freud (1856–1939), who 'invented' the subconscious, had no doubts about this, and insisted that: 'The interpretation of dreams is the royal road to a knowledge of the unconscious activities of the mind.' Dreaming, in other words, is an essential part of psychic metabolism. The need for dreaming and sleep was echoed by Goethe, the German poet, dramatist and philosopher (1749–1832), who said that: 'Man cannot persist long in a conscious state, he must throw himself back into the unconscious, for his roots lie there.'

What about the subject matter of dreams? Men and women have talked about their dreams for centuries, finding them worrying, terrifying or amusing, but often regarding them with superstitious apprehension. Dreams are still taken seriously, and questions asked about their meaning. The young woman who dreamt that her husband was making love to someone else was furious. When reminded that it was only a dream, she pointed out that 'If he does that sort of thing in my dream, what will he do in his own?' Dreams have long fascinated

Animals asleep, yawning, hibernating

Different animal species seem to adapt their sleep requirements according to their life in the wild. It has been suggested that those that are seldom attacked (bats, cats) tend to sleep a lot (fifteen to twenty hours a day), whereas those in constant danger of attack (sheep, pigs, cows) sleep less (three to four hours a day) – keeping a lookout, presumably. Humans, with eight hours a day, are in between.

But there is another explanation of these differences. Herbivorous animals, especially ruminating (cud-chewing) animals, need to spend a lot of their time eating, and they cannot afford to sleep for too long. In contrast to this, carnivorous animals, the attackers, are used to infrequent meals, and when they are not hunting they often sleep. We are familiar with slumbering lions, tigers, cats, dogs, just as we are used to the sight of constantly feeding cows, sheep, horses. The slumbering carnivore seems relaxed, positively lazy. This brings us to yawning.

We yawn when we are tired or bored, and a good yawn, like a good laugh, can be infectious. We also yawn when we wake up. That deep drawing in of the breath, plus the stretching of limbs that goes with it, boosts the movement of blood from the veins to the heart, improves the circulation and gets the body ready for action. This is useful for deep sleepers, and it is our carnivorous cats and dogs rather than our herbivorous cows and sheep, that yawn. The fact that we yawn could mean that we have a basically carnivorous nature. Yawns have been given all sorts of meanings. G. K. Chesterton said that a yawn was a 'silent shout', and others have gone further, suggesting it signifies aggression, a symbolic attempt to eat a victim.

Whatever theories we invent about sleep have to take into account the fact that animals need it too. When deprived of sleep they too become irritable, and may engage in vicious fights with one another. The typical sleep associated with dreaming is observed in cats, rats, goats, lambs, rabbits, birds and chimpanzees. Animals have dreams, as anyone will know who has seen a sleeping cat twitching its tail and whiskers, or a dog pedalling its feet, twitching its nose, even barking.

The most extreme form of sleep, although it is not the same as sleep, is seen in hibernation, practised by hamsters, hedgehogs, bats, woodchucks and squirrels. Hibernation is a profound biological adjustment that enables the animal to survive a season of cold and food shortage. There is a general shut-down of body activities, and the animal temporarily becomes cold-blooded. It is a useful adaptation to winter (*hibernus* being the Latin word for winter, comparable with the Sanskrit world *hima* = snow, which goes with *alaya* = residence, to give us Himalayas), and there are built-in safety devices. If the body froze it would die, so if the temperature approaches freezing point the animal wakes up and begins to generate body heat.

anthropologists and psychoanalysts. Freud's book on *The Interpretation of Dreams* is a classic, and Jung's views (as expressed, for example, in *Modern Man in Search of a Soul*), have had a great influence. Mythology is full of dreams.

The conventional view is that we need sleep partly to recuperate from daily wear and tear, and partly to conserve energy and look after basic bodily processes. Cell division and growth are at a peak during sleep. What can we learn about sleep from animals? This is set out in the box.

An Aside on Narcolepsy

Most of us are familiar with sleeplessness and it is interesting to consider its opposite. Narcolepsy is a rare sleep disorder that tends to run in families, and which seems to have something to do with the immune system. The affected person drops off to sleep perhaps four or five times a day (though it can be as often as thirty times), each episode of sleep lasting from a few minutes to an hour or two. Of course, anyone may fall asleep in a warm room, when bored or after a heavy meal. What makes narcoleptics so bizarre is that they may drop off while eating or standing up, even during sexual intercourse. Their urge to sleep is irresistible, and they often give up going out to see friends or plays or movies. They sleep less at night, so that the total amount of sleep (and the time spent in dreaming) over a twenty-four-hour period is much the same as in normal people. They are successfully treated with stimulants. Unfortunately, many of them also have attacks in which they fall down (paralysis).

Belief in an Afterlife and Anxiety about Death

Does belief in an afterlife mean that you are less worried about death? Perhaps, but many non-believers face death with equanimity. It is the natural conclusion, everyone has to die, and the fact that it is *I* who is involved is no problem. That is not to say that they welcome it, especially if it is premature (i.e. comes before the natural span; see chapter 4). And, as noted above, often the fear is of the dying process rather than of death itself. Most people hang on to life, however unpleasant it may be, and would accept Shakespeare's words in *Measure for Measure*, Act III, Scene 1:

> The weariest and most loathed worldly life
> That age, ache, penury, and imprisonment
> Can lay on nature is a paradise
> To what we fear of death.

Tennyson, in 'The Two Voices', was of the same mind:

> Whatever crazy sorrow saith,
> No life that breathes with human breath
> Has ever truly longed for death.

The prospect of immortality, however, seems almost as bad as death itself. The feeling that living for ever is unappetizing is backed up by the knowledge that death is required by nature (see chapter 1). Tithonus, in Greek mythology, was loved for his beauty by Eos (the goddess of dawn), who managed to arrange with Zeus that he would live for ever. She forgot to arrange also that he enjoyed eternal youth, and when he grew hideously old, Eos turned him into a grasshopper.

Others feel so worried at the prospect that they are prey to disturbing thoughts. For instance, in William Wordsworth's 'Strange Fits of Passion' the poet, on his way to his lover (who looks as 'fresh as a rose in June'), arrives at her cottage door and is overcome by an unreasonable attack of anxiety:

> What fond and wayward thoughts will slide
> Into a lover's head!
> 'O mercy!' to myself I cried,
> 'If Lucy should be dead!'

Hundreds of poems, dozens of books, focus on death. They show a great spectrum of attitudes. Perhaps the wisest is that death is feared because it is unknown territory. For instance, the English poet John Dryden (1631–1700) wrote:

> Death, in itself, is nothing: but we fear,
> To be we know not what, we know not where.

Frederick Cummings (1889–1919), an English naturalist and an expert on lice, expressed a biological, down-to-earth acceptance in *Journal of a Disappointed Man*. He wrote, using the pseudonym of W. N. P. Barbellion:

December, 1912.
Paleontology has its comfortable words . . . It has relieved me of the harrassing desire to live . . . I have discovered that I am a fly, that we are all flies, that nothing matters. It's a great load off my life, . . . For nothing can alter the fact that I have lived; I have been I, if for ever so short a time.

He goes on to meditate on the recycling of his component molecules, which gives him a sort of immortality:

> And when I am dead, the matter that composes my body is indestructible and eternal, so that come what may to my 'Soul', my dust will always be going on, each separate atom of me playing its separate part – I shall still have some sort of finger in the Pie. When I am dead, you can boil me, burn me, drown me, scatter me – but you cannot destroy me; my little atoms would merely deride such heavy vengeance. Death can do no more than kill you.

Cummings' *Journal* charts the progression of his illness (multiple sclerosis) with great honesty. We feel for him, and the *Journal*, with its introduction by H.G. Wells, is one of the great revelations in English autobiography.

Many poets expressed the view that death was the end, and there was nothing more. The English poet A.C. Swinburne (1837–1909), in 'The Garden of Proserpine', seemed in no doubt about it:

> We thank with brief thanksgiving
> Whatever gods may be
> That no life lives for ever;
> That dead men rise up never:
> That even the weariest river
> Winds somewhere safe to sea.

The same message comes from the Bible, Ecclesiastes 9:

> For him that is joined to all the living there is hope: for a living dog is better than a dead lion. For the living know that they shall die: but the dead know not any thing, neither have they any more a reward; for the memory of them is forgotten. Also their love, and their hatred, and their envy, is now perished; neither have they any more a portion for ever in any thing that is under the sun.

The English poet and playwright John Gay (1685–1732) wrote his own irreligious epitaph:

> Life is a jest; and all things show it.
> I thought so once; but now I know it.

Death as it Seems to Children

Children are said to become aware of death from about the age of five, although it may at first be a partial awareness. It is easy for the young child to think that the dead person is asleep; alternatively, they have merely 'gone away' – easy to believe if the child is told the dead person still exists, really, and will be seen again one day. In *The Interpretation of Dreams*, Freud refers to a child of eight years old coming home from a visit to the Natural History Museum, London, and saying to his mother: 'I'm so fond of you, Mummy: when you die, I'll have you stuffed and I'll keep you in this room, so that I can see you all the time.' Small children can have a remarkably mild reaction to death, possibly because they are not fully aware of its meaning. Both my parents died when I was six years old and I recall the weeping adults fussing over us, when all we wanted was to get on with our game. Mummy and Daddy had gone away, so what was all the fuss about? Greater awareness of death comes usually by the age of nine, and death is often thought of as a person or a frightening figure. Perhaps children of this age understand killing more than death.

If the child itself is dying, the main fear is of the separation that this involves. Such children have a better idea of what dying means than their healthy counterparts, because they have experienced illness, separation from parents, and perhaps the death of other children in hospital. In general, school-age children recognize death as a fact and as final, but they often see it as something that happens to old people, not to them. In their fantasy lives of violence and death, as portrayed on television, death is not real. Death comes more quickly than in reality, and the dead are routinely brought back to life in games of cops and robbers. Adolescents have a deeper understanding of death but not always an understanding that their own personal deaths are inevitable.

The inability to deal with death is surprising from one point of view. The Victorians were more likely actually to see death and dead people, but we are saturated with images of death and killing in film, television and newspapers. The average person, adult or child, must be exposed through the media to hundreds of deaths and killings each year. It is the familiarity with such images that reduces their impact. The deaths have become mere symbols, abstracted from the real world of dying and the dead.

Old People and Death: Hospices and the Care of the Dying

Older people no longer have illusions of their indestructibility; they have had parents and friends who have died, and they have gone to funerals. It has been said that women are more prepared for old age, and possibly death, because of the menopause, a more dramatic reminder of ageing and death than the mere loss of hair, wrinkles and so on experienced by men. Whatever the case, there is certainly a tendency for old people to accept death more easily. This is perhaps because age makes them feel differently about death rather than just because they have 'had their innings' and are ready to retire from life. But they reflect on their life and their successes and failures. If they can look back with satisfaction and few regrets to a meaningful life with some achievements they can often accept death more readily. If, on the other hand, they see too many lost opportunities and personal misfortunes, death seems more fearsome.

The hospice movement (see box) has been a great success and has made the dying process for many more dignified and comfortable. People with terminal illnesses are often unwilling to face up to death and may request euthanasia not so much because of pain as out of the fear of being abandoned, being a burden and losing control. Old people are beset with negative images from the media. In the UK at the moment youth is celebrated while old age is denigrated. The emphasis is on immortality, and the elderly can easily feel unappreciated and neglected. Older people have trouble getting jobs. Surrounded by these views, they are perhaps confused when asked to accept death as natural. Perhaps the time of the 'oldies' will come, as the proportion of old people in the country increases and their importance as consumers and voters is more widely recognized.

Many patients, when told that their illness is terminal, refuse to accept it. Their first thought is to seek second opinions and alternative treatments. Many of these treatments are harmless, aimed at increasing general health, and they do give the feeling that something positive is being done. Other treatments may be more aggressive, generally using unorthodox or alternative methods. Cancer has attracted much attention, with emphasis on a holistic approach. The body is purified and detoxified by colonic irrigations, nutritional needs are attended to by giving vitamins, enzymes and minerals. Immunotherapy is popular, the cancer being attributed to a weak immune response which can be restored by treatment. Psychological approaches include getting the

Hospices and the care of the dying
In the UK there have been hospices since the nineteenth century, but the modern hospice movement began when Cicely Saunders founded St Christopher's Hospice in England in 1967. She began as a nurse, later becoming a doctor, and she emphasized the control of physical pain as a key in the management of terminally ill people, something that had been overlooked in the past. She aimed at care of the patient and family as a unit.

Volunteers were employed, and there was a follow-up programme for family members after a patient's death. Until then, end-of-life care had not seemed important in medicine. Indeed, in the medical literature there are still many more articles on cell death than on human death.

As a result of her pioneering example, round-the-clock pain prevention is now widely accepted. There are seventy or eighty hospices operating in the UK today, many of them supported by cancer charities. A hospice can hand pick staff who are specialists in terminal care. They are comfortable with dying patients and give them affection and acceptance.

Currently there are moves to have more day-care centres where people can be given specialized attention as well as join in social activities, if possible being kept in the community rather than moved too early into beds in hospices. One of the great fears about death and dying is the feeling that you are not wanted, and bringing dying people back home reassures them about this. They have not been deserted, and are among familiar surroundings. If the family wants this and can handle it, it is often the best idea.

The first American hospice was founded in Connecticut in 1973, initially voluntary and staffed mostly by social workers and clergy. It soon attracted physicians and the movement grew to become part of regular medical care. There are now more than 2,500 hospice providers in the USA. To be accepted, the patient should have an estimated six months or less to live, and agree to have palliative rather than curative treatments. Hospice programmes can be based in hospitals, some of which have hospice units, or in nursing homes, but 80 per cent receive hospice care at home. Hospice treatment is regarded as the gold standard for the care of the terminally ill, and it is funded by Medicare. The goal, as in hospices everywhere, is to provide continuous pain relief and a dignified, comfortable death, plus proper care for the patient and the family as a whole.

New attitudes to the dying are gaining strength. In 1994 the philanthropist George Soros founded the US Project on Death in America, and so far he has donated 15 million to programmes on death and dying.

patient to develop a positive attitude, and spiritual methods may focus on harmony between the patient and nature. But some of the alternative treatments can actually be harmful, as was the use of Krebiozen in the 1960s and Laetrile in the 1970s. The doomed patient will turn to anything, and a new and more or less untried treatment gives hope. The fact that conventional medicine rejects it seems irrelevant. Alternative treatments can certainly boost morale, although in the case of cancer there is no medically acceptable evidence that any of them is effective; and last-minute grasping at straws may simply put off coming to terms with death.

It is difficult to accept the fact that one will soon die. It is something that happens to others, like serious injury or death of a spouse, but not, except in a distant theoretical way, to oneself. Mortally ill people probably suffer more, in addition to any physical suffering, through not coming to terms with their death. The confrontation can be put to one side for a while by make-believe about recovery, but it must come. Epicurus had wise things to say about death (see box).

Epicurus and death

The Greek philosopher Epicurus (341–270 BC) wanted to remove death from the list of things we worry about. He wanted people to enjoy the pleasures of life, especially mental pleasures, and to bear life's ills cheerfully. Human happiness is the highest good, and what cannot be prevented must be endured. With fewer expectations there will be fewer regrets. The gods enjoy a perfect happiness, and this would be interrupted if they had anything to do with the control of human affairs. Therefore the gods have no influence! A similar endearing logic led him to point out that 'Death, the most terrifying of ills, is nothing to us, since so long as we exist death is not with us; but when death comes, then we do not exist. It does not then concern either the living or the dead, since for the former it is not, and the latter are no more.' Epicurus advocated a simple, balanced life, but in later times the Romans added extra features, so that 'epicureanism' came to refer to those who gave themselves up to the enjoyment of sensual pleasures.

We return to the same conclusion. It is best to accept death gracefully, although many are reluctant to do so and fight it to the end. There can be a distressing struggle as they are torn from life. Acceptance of death with calmness, resignation, and submission is advocated by some of the great religious teachers from India. The same attitude is expressed unforgettably in the Christian Bible, in Luke's Gospel 2: 29: 'Lord, now lettest thou thy servant depart in peace.'

Mourning and Grieving

In our relatively safe Western world, death, especially when it is premature, is an unexpected visitor. It was a familiar one in past centuries, first because premature deaths were common, and secondly because it generally happened at home. Today nearly everyone ends up in a hospital or other institution, and most people have never seen a person die. Road traffic accidents and heart attacks are the usual 'public' deaths. The emotional impact of seeing death, even when it is the death of a stranger, is profound, and is discussed below.

Death of a close friend or relative, perhaps dearly loved, brings a separate series of powerful reactions. Personal grief is now added to the shock of death itself, and it is worse when it is actually witnessed. There are certain recognized stages of grief, and some or all of them have to be gone through before the loss is fully absorbed so that life can begin again. Although it may seem artificial to separate off these stages, they give an idea of the complex responses to the death of a loved one.

Stage 1: Numbness, shock. It seems as if the world is carrying on but you are not part of it. Initial disbelief and denial are common: This isn't happening to me. It is a mistake, a dream from which I'll soon wake up and find that everything is all right. This stage may last a few days. To face up to the reality of the death it can be important to see the body. If this isn't possible because death was due to an accident or because the body was lost, it can give rise to anxiety and lengthen the mourning process. After a long illness, the initial feeling is often of relief that the person's sufferings are over, and this can turn to guilt at the thought that their death has brought relief rather than sorrow.

Stage 2: Sorrow. Sorrow takes over, overlapping with the shock, as the irreversibility of the event comes home. At this stage there is a need to express grief, to weep, to share the feeling with others if possible. At the same time there can be a further episode of guilt. You could have done more for the dead person, been nicer to them. If the death was unexpected, perhaps you haven't been kind enough lately. An illness is always stressful for the carer, and perhaps you weren't always patient enough. Could you have done more for the ill person? You promise to God to make up for it and be good if only you can wake up and find it is all a dream.

These feelings are dealt with by talking to others about the death and the dead person. It may help to see the body again or to visit the grave. The sometimes lengthy process of acceptance begins, while acute sorrow turns into a more chronic ache and guilt which can last a long time.

Stage 3: Anger, rage. This includes anger at God, or with fate, for letting this happen. It may spill over into anger with friends and family for not having suffered as much as you, or not having understood. It is an expression of the natural human wish to blame someone or something for an event which is not understood. For natives in parts of Papua New Guinea forty years ago most deaths were obviously due to sorcery, and the next step was to find who was responsible. Reprisal murders used to be common. This stage is often accompanied by a good deal of aggression and by irrational resentment, and can last for days or months.

Stage 4: Acceptance of the death. The anger has abated, and this is the time to realize what has happened and to experience the true pain of the death. The world has changed and you are going to have to come to terms with it. Often there is a degree of apathy, a reluctance to do anything or to take decisions. This can be a problem at such a time, when there are so many practical things waiting to be done. It may be accompanied by fear that you won't be able to cope. When acceptance leads to depression it is a further handicap for the next of kin, and the depression may require medical treatment.

Stage 5: Recovery and resolution. The worst is over and you begin to put your energy into new roles, new relationships. You face the world again, start the new life.

These are the typical mental and emotional reactions to the death of a loved one. Each person reacts in a slightly different way, and various stages may be missed out, or merge to become indistinguishable.

The effects of bereavement are sometimes long-lasting. As might be expected, deaths of children are particularly traumatic, and sudden deaths cause greater grief than deaths following long illnesses.

Emily Dickinson (1830–86), in 'The Bustle in a House', expresses a matter-of-fact approach to a loved one's death:

> The Bustle in a House
> The Morning after Death
> Is solemnest of industries
> Enacted upon Earth –
>
> The Sweeping up the Heart
> And putting Love away
> We shall not want to use again
> Until Eternity.

It is interesting that a similar series of steps (denial, anger, depression, apathy) tends to be gone through when the individual approaches his own death. The mortally ill person suffers, not only because he fears death, feels cheated, with loss of control over his fate, but also because he senses that people are unconsciously withdrawing from him, already counting him as dead. He has been written off, left to endure his lonely fate. He may also feel that everyone will get on quite well without him, and that he won't be mourned for long. Nowadays his mental suffering is often blurred by prompt pain relief and sedation.

Mourning can be particularly painful when babies die. Each year in the UK about 8,000 babies are stillborn or die in the first month of life. They are mourned, and the mourning may last longer than it does in the case of widows and widowers. There is the feeling that the baby never had the chance to be recognized as a real individual. In the old days the mother was sedated, the corpse whisked off, and all baby items like pram, cot and clothes removed. Nowadays mothers are often encouraged to hold the dead baby, even wash and dress it. Mementos such as a lock of hair, a handprint or footprint, help to soften the sense of loss.

Faced with an unexpected death, people often ask 'Why?' Some of the great religions try to answer this question, but to the non-believer the question has no real meaning. You can generally say what caused the death, but asking why that particular person should die at that time is like asking why an individual wins the lottery, or is born on a certain day, or exists at all. Indeed, the question presupposes that an unknown entity must have had a hand in it, with reasons and motives, rather than that it is merely the chain of cause and effect plus chance.

Mourning Customs

From earliest history, societies have had their own beliefs, customs and rituals to be followed after a death. The funeral with cremation or burial is the routine in Christian communities. It is a public display in which the death is acknowledged as real and final. It encourages support for the bereaved and tribute to the dead, and unites families. The occasion forms an important part of the grieving process, helping the survivors to accept and come to terms with the death, while cherishing the memory. Visits to the cemetery and anniversary grievings serve to remind the living of the dead.

Different religions have different characteristic rituals: for instance the *shiva*, a traditional seven-day period of full mourning in some Jewish communities, followed by one month of partial mourning; or the

Catholic wake, with graveside ritual and special masses. Mourners at a Muslim funeral are often more demonstrative than at Christian funerals, moaning, crying and beating the breast. Professional mourners are sometimes present to encourage the expression of grief. The death is accepted as the will of Allah, and the bereaved take comfort from surrendering to it.

Religious rituals give meaning to the death and provide help for vulnerable survivors. Because the ritual takes place within a familiar framework it makes the death more acceptable and at the same time puts limits on grieving. Wakes were common among the poor in nineteenth-century Wales, Scotland and Ireland, and to a large extent were social occasions. The social function increased as it became customary to have a bit of a feast after the funeral.

In Victorian days death was part of the social fabric and the accompanying rituals served as an occasion for the statement of one's place in society. The rise of the undertaker is described in chapter 5: he supplied, for a price, the 'respectable funeral'. Details like the number and accoutrements of the horses, the quality of the coffin and the size of the funeral procession depended on the money available. Clearly defined rules governed the wearing of black, both what should be worn and for how long it should be worn. The stage at which the mourners could resume normal social activities was also laid down. In England black armbands were common until fifty years ago. At one time mourning jewellery was given away, its quality adjusted according to the social rank of the recipient. Samuel Pepys arranged in his will for the distribution of 128 rings in three price bands, some costing more than a hundred pounds. Tiny strands of the departed's hair were sometimes mounted in rings.

Physical Illness Brought on by Grief

It has long been known that certain physical illnesses can be brought on by grief. Close survivors of the dead person (widows, widowers, sisters, brothers) show fairly well-defined symptoms and signs, including headaches, giddiness, weakness, nausea, breathlessness, weight loss and hair loss. These are not serious, but close survivors can also develop conditions such as ulcerative colitis, arthritis and skin disease. These are true physical afflictions but also have a strong psychological component.

In 1943 there was a great fire in the Coconut restaurant, Boston, USA. Many people died in the fire, and a few of the bereaved experienced massive rectal bleeding at the funeral, caused by acute attacks of ulcerative colitis. There was also an increased mortality in the survivors, the so-called 'broken heart' phenomenon. In a study of 4,486 widowers over the age of fifty-four, it was found that their death rate in the first six months after the bereavement was increased by nearly 40 per cent. Mental illnesses or depression can also be triggered off by the death of a loved one. Those who have lost a spouse within the past six months are much more likely than others to consult a psychiatrist. The link between disease and stress is now accepted by physicians, and the grief reaction is part of the response to stress.

What to Do when Someone Dies

In the midst of death and sorrow it can be difficult to keep one's head and deal with practical things. Luckily there is help at hand, many small but useful books telling about coffins, funerals, cremation, organ donation, probate, and so on. Many organizations offer help, counselling and support to the bereaved. In the UK, addresses are given in *The New Natural Death Handbook* and the *Which?* consumer guide *What to do when someone dies* (see References for details). The Internet, too, is playing its part. Obituaries can be placed on it permanently and without charge at <http://www.cemetery.org/>, and there is a comprehensive index of death, dying and grief resources at <http://www.cyberspy.com/lwebster/death.html>. The Natural Death Centre is at <http://www.newciv.org/GIB/>.

16

The Afterlife and the Future of Corpse Disposal

The poetry of history lies in the quasi-miraculous fact that, once, on this earth, on this familiar spot of ground, walked other men and women, as actual as we are today, thinking their own thoughts, swayed by their own passions, but now all gone, one generation vanishing after another, gone as utterly as we ourselves will shortly be gone, like ghosts at cock-crow.

from the autobiography of G.M. Trevelyan (1876–1962)

The Afterlife and the Soul

Throughout the ages people have known about the putrefaction and dissolution of corpses. They have seen death and have appreciated its finality. The poets in particular have been able to ponder on life's brevity and the certainty of death and bodily decay. But the human mind, conscious of itself and of its individuality, is unwilling to contemplate its permanent extinction. Its response to the possibility that all mental activity, including the soul or spirit, is snuffed out for good when the brain dies, is to invent a better fate. Humans have therefore constructed, with ingenuity and imagination, various systems of belief in an afterlife. These beliefs, interwoven with valuable ethical precepts and laced with a good deal of myth and magic, constitute the great religions of the world. A god or gods, whose actions control the world and give it meaning, are part of most but not all religions.

When you see a corpse you have to conclude that this is the end, as far as the body is concerned. So what happens to the self, to the soul or the spirit? There is a need to propose something that can be understood by ordinary people. It is in response to this that destinies like heaven and hell, or transmigration of the soul, or (more difficult) its union with a God-like cosmic entity, have been developed.

Humans, unlike other species, bury their dead; burial pre-dates cremation. When they bury so-called grave goods (food, tools, weapons) with the body it certainly looks as though they believe in an afterlife. Paleolithic man comes into this category, although it is impossible to get a clear picture of the beliefs of prehistoric people. Certainly from the beginnings of recorded history and in all the earliest religions some sort of soul has been postulated. People had dreams, they remembered things for years, they could have their own thoughts; so there must be more to human existence than just the physical body. To start with, this elusive immaterial thing was thought of as a rarefied substance, like a breath, and one Greek word for it was *pneuma*, the breath or spirit. At later times the experts tried to be practical and locate the soul in some special organ during its residence in the body (see box).

In preparation for an afterlife, corpses were usually left intact, even preserved, but people could still believe in the soul even when the corpse was completely destroyed by cremation. At first sight the ancient Sumerians seem to have been an exception. They believed that the sole function of human beings was to build temples and supply food for the gods, who had created everything. After dying they could no longer serve the gods and there was therefore no reason for their continued existence. There was no question of being rewarded for good conduct or punished for bad. Yet in spite of knowing they had no post-mortem role, they believed in an afterlife. After turning into a type of ghost (*etimmu*) they would inhabit a dark, unpleasant land-of-no-return.

In England the conventional eighteenth-century citizen felt comfortable going to bed with the following words, penned by an anonymous writer:

> Now I lay me down to sleep;
> I pray the Lord my soul to keep.
> If I should die before I wake,
> I pray the Lord my soul to take.

A century later, more searching questions began to be asked. The French philosopher and historian Joseph Renan (1823–90) must have spoken on behalf of a handful of hopeful sceptics when he said: 'O Lord, if there is a Lord, save my soul, if there is a soul.' The word 'soul' virtually disappeared from the psychologist's terminology in the nineteenth century, and it has become a theological term. But the church no longer dominates our life as it once did, and most people are no longer interested in the old philosophical arguments and religious disputes. For them the word 'soul' implies no more than something that goes with the physical body. If the body is a machine, the soul is the rest of it. It is the spirit, the mind, it is consciousness or conscience, and, as the ancients said, it departs (or dies) when the body dies. It is not easily defined. Medieval thinkers spent their lives discussing the nuts and bolts of the idea. For us, poetic imagery comes to the rescue with evocative words such as these by Shelley, in 'Prometheus Unbound':

> My soul is an enchanted boat,
> Which like a sleeping swan doth float
> Upon the silver waves of thy sweet singing.

Where is the soul?

There was a down-to-earth aspect to all the myths and fantasies invented about the self, the soul and the afterlife. Learned people asked themselves exactly where the soul was during its residence in the body. It was a practical question, comparable to the one about the exact description and location of hell. One of the first suggestions was that the soul was in the liver, an organ which at that time seemed to have nothing else to do. Later some of the ancient Greeks, including Aristotle, began to favour the heart, while others (Galen, Pythagorus and Plato) opted for the brain.

The brain remained the most popular abode, and the Greek anatomist Herophilus (300 BC) went further and focused on a small region (the fourth ventricle) above the stem of the brain. The picture was confused because different philosophers had different ideas about the precise nature of the soul. But it was agreed that the human soul was not just in control of physiology but must also be a 'reasoning soul', the basis for human conscience and rational thought. When it departed from the body, you were dead.

The soul was referred to in textbooks of anatomy up to the seventeenth century, and the French philosopher and mathematician René Descartes (1596–1650) thought it resided in the pineal gland, a tiny pea-like object on top of the mid-brain. Here, receiving information from the eyes and the limbs, it could interact with the body and make it work. In primitive vertebrates the pineal had been in the form of two eyes on the top of the head, looking upwards for danger through holes in the skull, but in humans it seemed to have no particular job to do. Descartes wanted to think of the soul in practical, physiological terms.

Descartes had a point, because now we know that many of the higher functions of the brain depend on input from a region of the brain-stem just below the pineal gland. When it is out of action there is 'brain-stem death'; or, in Descartes' terms, 'the soul has absented itself'.

Evidently humans, so far, have been unable to conceive of death as permanent extinction of the self. There has been a universal yearning for a post-mortem existence in some shape or form. But is this a real need, an inevitable feature of human biology? People are now beginning to find the prospect of personal extinction an acceptable one. There is the possibility that the great religions of the world, so important during earlier stages of human history, will eventually die away, having served their purpose. Already some of us are prepared to live without the promise (or threat) of a life hereafter, and without the god that generally goes with it. This makes sense, biologically and logically, and is quite appropriate for humans in the twenty-first century. To put it crudely, in the words of an anonymous poet:

> When you're dead
> Get it into your head
> You're finished, extinguished.
> There's no spirit or soul
> Just a black hole.
> They see your photo, your letters
> But you're not there
> Melted into thin air
> Gone forever
> Whatever
> They tell you.

God could, of course, remain, in some form or other. But continuing to accept life after death would call for difficult compromises with the world of common sense and science. When Leo Tolstoy (1828–1910) was dying he was urged to return to the fold of the Russian Orthodox Church. He replied: 'Even in the valley of the shadow of death two and two do not make six.' Yet whatever our feelings about God and an afterlife, it does not alter the fact that we do not want to die. Jean-Jacques Rousseau (1712–78) wrote: 'He who pretends to look upon death without fear lies. All men are afraid of dying, this is the great law of sentient beings, without which the entire human species would soon be destroyed.'

Religion and the Afterlife

It is impossible, indeed it would be downright presumptuous, to attempt to do justice to in such a short space to the beliefs of the great religious teachers and philosophers. Nor am I qualified to do so. Each of the main religions has many variants, many offshoots. The following very brief account purports to be no more than an imperfect and incomplete introduction to the subject.

From prehistoric times humans have created gods. They were needed to explain the mysteries of life, the season, fire, thunder, birth, death. Voltaire said that even if God did not exist it would have been necessary to invent him. Death was not the end, and the gods were important in the afterlife. In Greece, by the sixth century BC, there were thousands of different gods, so many that 'Feast of the Unknown Gods' was observed to make sure none was missed out.

It is easier with only one God, and most of the great religions of the world acknowledged their own special God or Supreme Being. And each religion had its own version of the afterlife. Nowadays, many people find it hard to believe that the 'person' carries on after death as a soul or spirit. They regard the perennial search for death's 'meaning' as futile. Yet their inability to accept that *they* will one day be permanently extinguished makes them unhappy as complete unbelievers. In addition, there is the natural hope that the dreadful wrongs of this world will be put right in another world, that the wrongdoers will be punished and the virtuous rewarded in something resembling a Last Judgment. These natural human beliefs are parodied in a poem by Rupert Brooke, entitled 'Fishes' Heaven':

> Fish (fly replete in depths of June,
> Dawdling away their wat'ry noon)
> Ponder deep wisdom, dark or clear,
> Each secret fishy hope or fear.
>
> Fish say, they have their Stream and Pond;
> But is there anything Beyond?
> This life cannot be All, they swear,
> For how unpleasant, if it were!
>
> .   .   .
>
> But somewhere, beyond Space and Time,
> Is wetter water, slimier slime!
> And there (they trust) there swimmeth One,
> Who swam ere rivers were begun,
> Immense, of fishy form and mind,
> Squamous, omnipotent, and kind;

> And under that Almighty Fin,
> The littlest fish may enter in.
>
> .   .   .
>
> And in that Heaven of all their wish,
> There shall be no more land, say fish.

It is perhaps because burial was a common practice that the afterworld was thought of as being underground, a dark unpleasant place. It could be a grim repository of dead souls, and in Greek classical mythology Hades was the abode of departed spirits, gloomy but not necessarily a place of punishment or torture. Later and in Christian mythology it was hell, a place of terrible torments where the damned suffered for their misdeeds (see box).

The gods, however, were generally happy up in the bright sky. In religions where there were retribution and rewards after death, this suggested suitable fates for the good and the bad. The good could be sent up with the angels or located on some beautiful island ('Isle of the Blest'), and the sinners banished to the underworld. In the Christian hell there was also the physical torment of fire.

This meant pain rather than warmth, the latter by itself being perhaps welcome in, say, an Eskimo's afterlife. Sinners were burnt alive a few centuries ago, and the eternal flames experienced in Hell made it a terrifying place.

The ancient Egyptians, as outlined in chapter 9, believed there was to be a magical resuscitation of the dead, with a judgment of conduct on earth. Mummification was their way of making sure the body was still there and ready for this summons. Once the idea of an afterlife had been established, then Egyptian mortuary practices followed logically – especially in the case of kings. A passage from an ancient Egyptian text says: 'O flesh of the King, do not decay, do not rot, do not smell unpleasant.' Tombs were constructed during life and mortuary priests appointed, in proper preparation for death and a safe passage to the next world. The Egyptians' greatest fear was to be completely forgotten. It is because of their belief in the afterlife, their mummification and construction of tombs equipped with everyday things, that we know so much about daily life in ancient Egypt.

Zoroastrianism, the ancient religion of Iran, as modified by Zoroaster in about 800 BC, survives in India today as the religion of the Parsis. Its followers believe in rewards and penalties: There is a bridge to be crossed after death, which is broad for the virtuous but so narrow for the wicked that they fall from it into hell.

The Christian afterlife: medieval details

It may be noted that in the English language words describing good or enjoyable things are far less numerous that the words describing suffering and pain. In Roget's Thesaurus a total of 498 nouns, verbs and phrases are listed under the heading of pain and painfulness, whereas only 296 are listed under pleasure and pleasurableness. Evidently unpleasant things evoke a richer vocabulary.

Certainly in visual art (e.g. the paintings of Hieronymus Bosch) the agonies of hell are depicted in infinitely greater detail than the delights of heaven, which in many cases consists of no more than a few angels, saints and cherubs basking or making music in the clouds.

Hell and the devil were described with meticulous care by Jesuits and others in the sixteenth century. The devil's features, including his penile anatomy, were itemized. He was said to carry out sexual acts forbidden to believers, such as sodomy, and was equipped with a forked penis so that he could commit fornication and sodomy at the same time. His demons, lieutenants and other employees were numbered at 7,405,926. With his masses, his church and disciples, his power and his knowledge, he was in many ways a mirror image of the deity. Certain theologians maintained that hell had a topography, a flora and fauna, a climate, and was 200 Italian miles across. In their quest for practical details they decided that a cubic mile was sufficient to hold 100 billion souls, if tightly packed.

In our day Christian beliefs have become so much more reasonable that it is perfectly possible to be a good scientist and a good Christian.

The original Christian doctrine is that humans have forfeited immortality ever since Adam and Eve abused the freedom granted them in the Garden of Eden. Adam and Eve may seem a couple of happy and innocent pioneers to our modern eyes, but they were undergoing a crucial test, without knowing it. They failed (sinned). They had fallen from grace, and Original Sin was then (most unfairly) transmitted down through the generations to all of us. As the Bible says, 'the wages of sin is death', so death is our fate. For the next 2,000 years theologians were kept busy suggesting ways of avoiding or explaining the burden of original sin.

Take the Resurrection and the Last Judgment. God is expected one day to bring the dead back to life, before deciding who is for heaven and who is for hell. Christ himself rose from the grave, and belief in a future world is fundamental to Christianity. The Christian church resisted cremation for hundreds of years because of the problem God would then have had raising people from the dead. But logically, why couldn't everyone be judged individually as soon as they died? Would they all have to wait, possibly for thousands of years, before there was a mass resurrection and gigantic public trial at which sentences for heaven or for hell were passed? Surely this would be unfair for the virtuous, keeping them standing by for so long before their just reward.

Purgatory was an extra post-mortem complication for Catholics. Even if someone dies in a state of grace, with the worst ('mortal') sins cleared by absolution, they have to suffer for a while if they have not paid for any other pardonable ('venial') sins. Purgatory (Latin = a place of cleansing) is where they suffer (by being unable to see God in spite of their love and longing for him) and are purged of these sins. On All Souls' Day (2 November) Catholics seek by prayer and offerings to alleviate the sufferings of souls in purgatory. All Catholics (except saints) had to spent a while in purgatory, but the period could be shortened, or abolished (remitted) altogether by means of indulgences or pardons. The indulgences were a profitable source of income for the church in the Middle Ages. You could buy pious pictures, rosary beads or objects blessed by the Pope, or pay for the performance of masses for the dead. You could also get indulgences by reciting certain prayers or visiting certain shrines. The idea of purgatory was condemned by the Church of England.

Christianity has diversified into many different sects, and in some form or other is the official religion of large sections of the world. Attempts to suppress it in the old USSR were unsuccessful, and Lenin once remarked that 'This damned religion is like a nail, the harder you hit it the further it goes into the wood!' Today, according to most surveys, less than half of the people in the UK believe in an afterlife, and even fewer believe that what happens there is determined by our conduct on earth. The figure differs in different Christian countries, but times are certainly changing.

The Jewish religion has less clear ideas about the afterlife and immortality. Heaven may be a reward for living in accordance with God's laws, but the Old Testament tells little about it. After death, the soul spent about a year in an intermediate state, sometimes called 'the bosom of Abraham', during which it visited the body and familiar places. The afterlife does not figure prominently in modern Jewish thought.

Islam, today, is the largest non-Christian religion in the world, with hundreds of millions of followers. Muslims' sacred book is the Qur'an. Long ago, when parts of the Middle East were fertile and supported dense populations, this part of the world spawned great civilizations and was a centre of trade and intellectual ferment. But the Arabian desert itself has always been a hostile place, where the weak are wiped out by the sun, the sand and the dryness. Tribal gods were numerous, and, long before the rise of Christianity, there was one higher God called Allah. Mohammed, the prophet of Allah, was born in Mecca in AD 570. He became a preacher with a band of disciples, but was at first persecuted and driven out of Mecca. He taught that people should abstain from stealing, fornication, slander and killing newborn infants, and should obey him. In AD 622 he founded Islam (Arabic for 'surrender') and the Muslims, meaning 'those who submit' (to the will of God) were quickly successful. Islam spread thought the world as converts were made, if possible by peaceful means but if necessary by the sword. There are interesting similarities between this religion and Christianity.

Muslims have many beliefs and many duties; for our purposes here, they hold that in the next life there will be a final judgment, followed by appropriate rewards and punishments. Death is likened to sleep, and the soul rises into the throat before leaving the body. The virtuous enter the Garden of Delights, which was described originally in terms of masculine tastes. The unbelievers proceed to hell where they have boiling water poured over their heads. Allah has arranged for damaged skin to be replaced by new skin, so that the punishment continues to be felt. The martyrs of Islam who have been killed in a holy war or driven from their homes have their evil deeds instantly forgiven, and automatically enter the Garden of Delights. Over the centuries the original doctrines have been adapted and reinterpreted, and there are differences in different branches, such as the Shi'ites and Sunni.

Hindus believe in an afterlife, the soul being reborn (reincarnated) over and over again in different bodies, as described on p. 349. Escape is achieved by reaching a certain state of understanding, when the illusion of worldly things has been conquered.

For the Hindu there are four astramas or main stages of life:

1 Childhood.
2 The life of the working adult.
3 Middle age, when the person begins to free himself from worldly obligations, finds a spiritual leader and a discipline.
4 The stage of *sannyasa*, which is the true achievement of life. Now, although he may outwardly be an aged, half-naked beggar, he is inwardly as free as the wind, king of the world, a sage. Few Hindus, however, reach this final stage, or would want to!

The Buddhist, too, believes in a cycle of births and rebirths, but the method of escape is different. He accepts the 'Four Noble Truths':

1   All existence involves suffering.
2   All suffering is caused by indulging desires.
3   With the suppression of desires all suffering will cease.
4   Suppression of desires can be achieved by the right beliefs, right resolve (renouncing carnal pleasures, not hurting living creatures), right speech, etc.

This path of conduct will allow escape into non-existence or nirvana. Nirvana is a Sanskrit word meaning 'extinction', and it implies the end of personal existence, with its desires and passions, and the attainment of perfect tranquillity. The spirit is at last freed from the endless cycle of craving, existence and death. For the Buddhist, an understanding of death and our impermanent nature is essential for spiritual well-being. The Buddha himself said: 'Of all footprints, that of the elephant is supreme. Similarly, of all mindfulness meditations, that on death is supreme.' Ideally, the Buddhist should meditate on death every day.

Zen Buddhism was founded by a Chinese monk in about AD 1187. It is a Japanese offshoot of Buddhism, and, as practised today, a mixture of Buddhism, Taoism and mysticism. Its view of the afterlife is a popular one, and it rejects reincarnation as well as the Buddhist idea of *karma*, the sum of a person's actions that determines his fate in the next existence. Unfortunately for the uninitiated, the teachings of Zen Buddhism, like those of many religions, are full of both obvious statements and paradoxes. It is sometimes described as the path with no railings. Zen teaches that instead of seeking things you should try to get a feeling for the unknowable. This is not too clearly defined, and there are many other words that describe it: the Abyss, Empty Space, The Void, The Clear Light, Nothingness, The Universal Consciousness, even The Kingdom of Heaven. This thing, The Clear Light, call it what you will, is everywhere, but people are too engrossed in striving with demons and gods of their own making to experience it. Ancient philosophers and sages knew about it, and near death experiences (see chapter 4 and below) give a foretaste of this clear, kindly light.

The ancient views on renunciation and a 'drop the body' approach to life after death have persisted especially in Tibetan Buddhism. Because of its physical isolation in a mountainous region, modern life has left Tibet almost untouched. Tibetans were writing, studying, meditating and building monasteries a thousand years ago. The Tibetan literature on death and dying is contained in thousands of texts, many of them now destroyed, unfortunately, by the Chinese. The first to appear translated into English, in 1927, was *The Tibetan Book of the Dead.* This consists mainly of ritualistic texts to be read to help a dying or dead person. It was accepted that people could return from the dead before being cremated or buried, and there are hundreds of accounts of these happenings in Tibetan literature. In Tibet, perhaps more than anywhere else, people have been able through meditation to experience the stages of death long before the time of actual dying. This is the ultimate achievement in death awareness.

The interesting feature of those religions in which worldly things are renounced is that they say, in essence, that if you detach yourself from life during life, death will then be no problem. In a way, you will already have died. Presumably Jung had this in mind when he said: 'As a physician I am convinced that it is hygienic to discover in death a goal towards which we can strive; and the shrinking away from it is something unhealthy and abnormal which robs the second half of life of its purpose.' My own belief is that the religious experiences of the seers, the Zen Buddhists, the contemplatives, are probably manufactured at a physiological level in the brain. The experiences are susceptible to drugs, to damage, and according to this view have no objective existence, no existence outside the brain of the person experiencing them. Many would disagree with this, and the controversies arise because of the immense gap between the neurophysiologist's picture of the brain, with its millions of nerve cells arranged in unbelievable complexity, and the individual's experience of consciousness – the gap between the dense, three-dimensional network of microscopic nerve endings firing off their chemical messages, and the feeling 'I think, therefore I am'. In the words of Henry James (1843–1916), 'Experience is never limited, and it is never complete; it is an immense sensibility, a kind of huge spider-web of the finest silken threads suspended in the chamber of consciousness, and catching every air-borne particle in its tissue.'

Perhaps one day this gap will be closed, but for the moment we have to accept that science cannot explain consciousness. Although that is the case, it is nevertheless possible to believe on the one hand that the individual ceases to exist when the brain dies, yet at the same time to imagine that some part, which is not the self and is difficult to define, survives. And if, in spite of the countless evidences of our own mortality, we can experience this timelessness, this non-existence, during meditation, then the gap will have closed as far as is presently possible.

Reincarnation

This means the rebirth of the soul in one or more successive forms. The soul or spirit transmigrates, and takes up residence in a new human, or an animal, or even a vegetable. The soul is thought to leave the body through the mouth or nostrils and enters, for instance, a butterfly or a future human being. If there is a soul that can leave the body, the idea of settling down in a new place is not unreasonable. In today's science fiction one popular theme is that the soul (mind) of an alien enters and takes over a human body. There is a fairly convincing story by the Czech novelist Franz Kafka (1883–1924) in which a man wakes up in bed one morning to find that he has been transformed in to a gigantic insect (*The Metamorphosis*).

Reincarnation features in many old religions, including Manichaeism, Gnosticism, and the ancient Greek Orphic mysteries. It is also seen in modern beliefs such as theosophy. These religions, like most, say that what you do in this life will have an influence on the next life. Theosophy has a mixed origin, believing in a supreme but incomprehensible deity, and reincarnation is one of its great principles. Man must go through a series of lives before attaining spiritual perfection and reaching the theosophist's equivalent of nirvana. The Theosophical Society was founded in 1875 by Helena Petrovna Blavatsky.

A Hindu believes that birth and rebirth is a cycle that goes on and on until the realization that the individual soul (*atman*) and the absolute soul (*brahman*) are one and the same thing. Only then is the individual freed from the eternal cycle or *samsara* of birth and rebirth. The Sikh adds the extra point that souls are eventually absorbed into God. Buddhists, too, believe that one aspect of the soul, perhaps best referred to as the character, can be reincarnated into a new person. For them, the way to leave the wheel of birth and rebirth is through discipline and meditation, and reaching a state where all earthly desires have been lost.

Near Death Experiences

The things that happen during NDE have been taken as evidence for an after life. NDEs occur when a person nearly dies, whether as a result of illness, during surgery, or after a heart attack or an accident. NDEs have basic features in common. The details differ, but in different countries and in people of different racial and ethnic groups there are similar features. The shared core is extraordinary. Probably more than three-quarters of people who nearly die have NDEs. NDE-type phenomena have been alluded to by philosophers and mystics for centuries. Five stages can be distinguished:

1   A stage of peace and a sense of well-being.
2   A feeling of separation from the body, often looking down on it from above. This occurs in about half the cases studied, and it is often accompanied by emotional detachment and sharpened awareness.
3   A sense of entering the darkness, for instance going down a dark tunnel.
4   Seeing the light, which is usually brilliant white, golden, or sometimes blue.
5   Entering the light and experiencing unsurpassed beauty, love and peace.

At stage 5 there is often a sense of some kind of barrier, and a feeling that if they had gone beyond the barrier it would have meant death. Some people report physical details (landscapes, people, flowers, trees, music) of a world of amazing beauty and are convinced this was the afterlife, and may be resentful at being brought back from it.

Serious illnesses such as pneumonia account for a large proportion of NDEs. For some reason stages 4 and 5 are less commonly reported after attempted suicide.

The NDE is a true experience rather than a dream or a hallucination. The person sees and feels it all with great clarity, not as if they were in a semi-conscious state. It has a profound effect. After recovery, these people often have a reduced fear of death, and are more loving and compassionate. All say it has made a big difference to their life, and about 75 per cent are convinced that there is an afterlife. A few develop powers of extra-sensory perception (telepathy, clairvoyance), presumably triggered off by the NDE. (NDEs are not due to the drugs given to dying patients because they occur in drug-free people.)

A minority of NDEs are unpleasant ('negative'), the subject reporting fear, panic, mental anguish, helplessness, intense loneliness. They are in dark, gloomy surroundings, perhaps on the edge of an abyss, sometimes being dragged down by an evil force. Some of them knew for sure that they were on the edge of hell, with the sickening smell of decay.

One of the first studies of NDEs was undertaken in the nineteenth century by a Swiss geologist and mountaineer, Professor Albert Heim. He had had a few nearly fatal falls, and he collected accounts from other people who had been in mountain-climbing accidents. In 1982 Michael Sabom, a cardiologist at Emory University School of Medicine, published interviews with 116 NDE survivors. He believed that for every phenomenon there was an explanation, and concluded from his study that there was no satisfactory medical explanation for NDEs. Cardiologists, because of their experiences with heart attacks and resuscitation, are well placed to study NDEs, and in 1978 another cardiologist, Maurice Rawlings, described bad NDEs, in which the patient was terrified, convinced he had been on a visit to hell.

NDEs were set on a sounder scientific basis when George Gallup, with the resources of the Gallup Poll Organization, using rigorous methods for sampling and analysis, found that NDEs were not all that uncommon. About 1 per cent were 'hell-like' and many were neutral, in other words not especially good, bad, or memorable; and some survivors from near-death episodes reported no NDE. Others clearly superimposed their own personal beliefs and images on to the NDE, reporting visions of dead relatives, Christ, or guardian angels. The phenomena are not dissimilar to those experienced by Christian mystics, yogis and other holy men, as described in *The Cloud of Unknowing*, a book written by a great English mystic who lived at the time of Chaucer. These experiences included wonder, awe, a sensation of light, communication with a god or a cosmic consciousness, becoming part of a greater whole, and they were often brought on by meditation and the control of breathing.

Is there a simple medical, physiological explanation for such experiences, or must we accept that they are true visions of an afterlife?

Explaining NDEs from a Physiological Point of View

One could suggest that NDEs are a result of cerebral anoxia, when the higher brain centres are starved of oxygen. As the circulation to the brain begins to fail, a number of extra things happen and cause the NDE. It is a matter of chemicals and nerve cells. This seems reasonable and many physicians might think that the question needs no further discussion. Anoxia is known to cause weird experiences even when life is not threatened.

An alternative explanation is that NDEs are due to mild brain 'seizures', in which overactivity of the temporal lobe (the part under the temple) of the cerebral cortex causes the characteristic experiences. It is well known that abnormal activity of the temporal lobe cause hallucinations and 'out-of-body' experiences.

Another possibility is that as death looms near, our ancient response to injury and stress comes into action. When something big and stressful happens, the body responds in a characteristic way. After a sudden devastating injury to an arm, leg or chest, on a battle field or in an automobile accident, the victim may feel no more than a calm acceptance of the event. There is no pain, just numbness, and if the injury is not too severe he may continue to walk, fight or run from the scene, whichever is appropriate. The pain comes on hours later. This is nature's way of handling trauma. At an earlier stage in our evolution a victim who stayed put, howled and nursed the damaged parts would be less likely to survive, whether the injury was due to a predator or an accident. The numbness and the calm allow one to make a life-saving response to the injury. It makes sense in view of our origins. The mechanism for this is as follows. Within seconds there is a stream of messages down the 'sympathetic' nerves to skin and muscles, plus an outpouring of adrenalin from the adrenal gland on top of the kidney. These cause a so-called 'flight or fight' reaction, in which the heart beats faster, blood is diverted from skin and other organs to muscles, and the sensation of pain is blunted: just the responses that are needed for to enable either flight or fight. We all know that wounds are barely noticed during such an emergency. It is nature's way of looking after us. At the same time another type of response takes place in the brain, with local production of pain-relieving, calming substances called endorphins. These are endogenous (produced from within), morphine-like chemicals, and they are formed in certain parts of the brain in many types of stress, including long-distance running and maybe also during acupuncture. They alter sensory awareness and make you feel able to handle the situation. Could the detached awareness of the NDE be based on these responses?

Other Explanations: The NDE as a Religious or Spiritual Experience

The mystics and yogis tell us that you do not have to come close to death to experience a NDE. We know that by meditation you can cut off sensory messages to the brain, and this could cause a loss of the sense of reality and other effects. But NDE subjects report a *greater* than normal sense of reality. Could it be self-hypnosis? Or is it the experience of an archetypal image, the sort shared by all human beings? None of these explanations seems entirely adequate, and one has the feeling that we know too little about the subject to reach a conclusion. At present, then, NDEs are not fully understood or explained except in religious and spiritual terms. In these terms the NDE is a mystical vision of the nature of the universe. But perhaps we should reserve judgment and await further evidence before leaping into exclusively paranormal or physiological exlanations.

In the USA and in the UK there are associations and institutes for Near Death Studies, and research into this difficult area will perhaps one day help us to understand. Meanwhile we can read the cheerful melancholy of Omar Khayyam of Naishapur (the Persian poet, astronomer and mathematician), who died in AD 1123. His *Rubaiyat* was translated by Edward Fitzgerald in 1859.

> Oh threats of Hell and Hopes of Paradise!
> One thing at least is certain – This life flies;
> One thing is certain and the rest is lies;
> The Flower that once has blown forever dies.

> Strange is it not? that of the Myriads who
> Before us pass'd the door of Darkness through,
> Not one returns to tell us of the Road,
> Which to discover we must travel too.

> The Revelations of Devout and Learn'd
> Who rose before us, and as Prophets burn'd,
> Are all but Stories, which, awoke from Sleep,
> They told their comrades, and to Sleep return'd.

> I sent my Soul through the Invisible.
> Some letter of the After-life to spell:
> And by and by my Soul return'd to me,
> And answer'd 'I Myself am Heav'n and Hell.'

Ghosts, Spectres, Hauntings

A ghost, as referred to here, means the spirit of a dead person, although it can also mean any unseen presence, such as the Holy Ghost or the spirit of God. Ghosts can be good or bad. A ghost that haunts is generally one that molests living people. I mention ghosts here because if they are real, and a dead person can sometimes be visualized, perhaps heard talking, then this would have to count as evidence for an afterlife – as long as it was not a hallucination, something merely imagined by the beholder.

Almost all peoples, from the earliest times, have believed in ghosts, spectres or apparitions. They are the very substance of religion, and they are said to communicate with the living, punish them or reward them. In our day, stories of haunted houses in particular can be very convincing. There are thousands of haunted sites in England. Ghosts, by the way, are nearly all witnessed during daylight hours. Ghosts seen to be disproportionately female (a woman in grey, etc), and can be smelt (burning sulphur, incense) and heard, as well as seen. Numerous accounts refer to footsteps, screams; Tedworth House in Hampshire, England, was famous in 1661 for its phantom drummer.

Are ghosts mere illusions, images formed by deranged minds, or are they ever real? A hundred years ago nearly everyone believed in the supernatural and the unexplainable. Spiritualism originated in the USA in about 1848 and soon spread across the world. It was based on the supposition that it is possible to communicate with the unseen world of the dead. This takes place at a gathering of people called a *séance* (French = a sitting), usually round a table. In the presence of a special person, a medium, who is usually in a trance, there are supernatural happenings. Objects move without anyone touching them, or the medium conveys information from dead people. Occasionally the medium appeared to build up a visible human form from a material called ectoplasm which is extruded from the mouth.

Spiritualism had millions of followers in the first half of the twentieth century, and the Society for Psychical Research was founded in London in 1882 for the scientific investigation of spiritualistic phenomena. These phenomena included telepathy, clairvoyance, poltergeists and faith healing, as well as regular ghosts. Many famous scientists were members, and similar societies arose in France and the USA. The Society produced numerous volumes describing its painstaking investigations of psychical phenomena. Many turned out to be fakes or frauds, but sometimes things happened that were hard to explain by the laws of physics. For instance, Daniel Danglas Homes (1833–86) was a well-known Scottish spiritualist, brought up in the USA, who was already holding séances at the age of seventeen and was excommunicated from the Roman Catholic church for his spiritualist activities. Stringent investigation of Homes by the Society failed to account for his feats by known physical laws.

Ghost hunters must be sceptical, and they have to use cameras, tape recorders, thermometers and other more sophisticated methods of detection. In the late twentieth century we may be less ready to believe in ghosts, but we continue to hear about or have personal experience of thought transference and other unexplained phenomena. Most people probably believe in supernatural incidents of some sort or other. Unfortunately, these things cannot be reproduced on demand in the laboratory and are therefore difficult to study. My personal verdict is that even if ordinary ghosts are figments of the imagination, the jury is still out as far as the other 'paranormal' phenomena are concerned.

The Transit to the Afterlife

There is a recurring idea that the dead have to make a journey of some kind to get to the afterworld, often in a boat, and that the journey may have dangers. Some primitive peoples regarded a rainbow as a bridge by which the dead could ascend to the sky, and others believed that they went up in the smoke of the funeral pyre. The ancient Egyptians thought that the dead Pharaoh travelled up to heaven in his solar boat to join the sun god Re.

Spiritualism brings out the big guns

In 1876 a celebrated spirit medium, Dr Henry Slade, was prosecuted by Edwin Ray Lankester, a young zoologist who was later to become the director of the British Museum of Natural History. Slade claimed to be able to demonstrate that his wife's spirit sent him messages on pieces of slate, and Lankester was keen to expose him as a common rogue. Details of the trial were reported at length in *The Times.*

Slade's side was taken by Arthur Conan Doyle, a spiritualist in spite of the remorseless logic displayed by his creation Sherlock Holmes, and also by the great biologist and co-discoverer with Darwin of the theory of evolution, Alfred Henry Wallace. Lankester, in turn, was supported by Charles Darwin, backed up by Thomas Henry Huxley and the great stage magician Maskelyne, who said he could reproduce the alleged phenomena in court. In the end Slade was sentenced to three months' hard labour, but after an appeal the conviction was cancelled on a technicality.

Wallace often championed radical causes, including pacifism, women's rights and conservation. As a student of human evolution he had suggested that since human brains were capable of doing mathematics and composing symphonies, something extra, something spiritual, must have been added to the human species during evolution. Darwin disagreed, totally. As a result of this and other forays Wallace lost the support of many scientists, and when at a later stage Darwin wanted to get him a modest pension he had to write a personal letter to the Prime Minister, Gladstone. After the letter had been passed on to Queen Victoria the pension was granted.

The journey to the afterworld was hazardous, and there were tales of fearsome monsters and sinister ferrymen. Magical formulas were needed. The goddess Ishtar, according to ancient Mesopotamian beliefs, had an unconventional entry to the underworld. She found that the entry was barred by many gates, and at each gate she was deprived of an article of clothing before being allowed past. When she finally arrived she was naked. The Greek and Roman dead were given practical equipment. This included money to pay Charon the ferryman for their passage across the Styx, the mythical river of the lower world, and honey cakes for the fierce dog Cerberus that guarded the entrance to Hades, the world of the Dead.

The Future of Corpse Disposal

As this book began with statistics about the ever-increasing flow of corpses and the question of their disposal, it seems appropriate to end with a few suggestions about how to deal with this deluge in the future.

The history of burial in England and France was summarized in chapter 6. It is a story of filling an available space with bodies, moving to another, filling that, and so on. Having first been lodged in the church itself, corpses were first relegated to graveyards and then, as populations expanded and graveyards closed, to out-of-town cemeteries. These, too, are now becoming crowded, and many have been engulfed by growing cities. As the world population expands remorselessly, larger and larger cemeteries are needed, like the ones in Genoa (Italy), Santos (Brazil), Shenhu (China), and Rookwood (Australia), mentioned in chapter 6. Fortunately, cremation is turning out to be the favoured method of disposal, which means that less space is required: in the space occupied by a single coffin 50–100 urns can be stored.

If, as seems conceivable, religious beliefs wither away rather than undergo rebirth, corpse disposal will eventually be purely a matter for public health authorities. they will be guided by considerations of hygiene, conservation, efficiency and, most importantly, human sentiments and needs. Most people will be destined for the giant crematoria, ending up either in urns or having their ashes scattered. If the 'compostorium' (see chapter 8) should ever turn into a practical, large-scale possibility, corpses would lose their individuality at the point of entry, as they were taken into the recycling process.

The need for simple ceremonies and remembrances will remain, because this is a cornerstone of human nature and is independent of belief in an afterlife. Remembrances would be in the form of small plates or tablets and could probably be discarded after two or three generations. Details of family trees and history, maintained in a central storage system, could be called up at will on the computer screen. The old days of tombs, monuments and religious rituals will seem as antiquated as the horse-drawn omnibus seems to us now.

In 1728 the American statesman, printer, scientist and writer Benjamin Franklin (1706–90) wrote an epitaph for himself which holds out hope for us all:

> The body of
> Benjamin Franklin, printer,
> (Like the cover of an old book,
> Its contents worn out,
> And stripped of its lettering and gilding)
> Lies here, food for worms!
> Yet the work itself shall not be lost,
> For it will, as he believed, appear once more
> In a new
> And more beautiful edition
> Corrected and amended
> By its author!

References

N. Albery, G. Elliot and J. Elliot, *The New Natural Death Handbook*. Rider (Ebury Press), 1997

C. Andrews, *Egyptian Mummies*. British Museum Publications, 1984

P. Aries, *The Hour of Our Death*. Allen Lane, 1981

M. Ashworth and C. Gerada, 'Addiction and Dependence – II. Alcohol'. *British Medical Journal*, 1997, 315: 358–60

M. Barry, 'Metal Residues after Cremation'. *British Medical Journal*, 1994, 308–390.

P.G. Bahn, 'The Making of a Mummy'. *Nature*, 1992, 356: 109.

R.O. Ballow (ed.), *The Pocket World Bible*. Routledge & Kegan Paul, 1948.

J. Bentley, *Restless Bones: The Story of Relics*. Constable, 1985

H.E. Berryman et al., 'Recognition of Cemetery Remains in the Forensic Setting'. *Journal of Forensic Sciences*, 1991, 36: 230–7.

M. Bloch, *Placing the Dead*. Seminar Press, 1971

T.S.R. Boas, *Death in the Middle Ages*. Thames & Hudson, 1972

C. Bradford, *Heart Burial*. Allen & Unwin, 1933

S.G.F. Brandon, *Religions in Ancient History*. Allen & Unwin 1973

T. Browne, *Selected Writings* (ed. G. Keynes). Faber, 1968

Bulletin of the New York Academy of Medicine, special issue, 1996, 62: 5, *Homicide: The Public Health Perspective*.

J. Cairns, *Matters of Life and Death*. Princeton University Press, 1997

J. Carey (ed.), *The Faber Book of Reportage*. Faber, 1987

R. Cecil, *The Masks of Death*. The Book Guild Limited, 1981

R. Chapman, I. Kinnes and K. Randsborg. *The Archeology of Death*. Cambridge University Press, 1981

V.G. Childe, *Prehistoric Communities of the British Isles*. W & R Chambers, 1949

J. Church (ed.), *Social Trends, 1997*. HMSO.

A. and E. Cockburn (eds) *Mummies, Disease, and Ancient Culture*. Cambridge University Press, 1980

C.B. Cohen and A.R. Jonsen, 'The Future of the Fetal Tissue Bank'. *Science*, 1993, 262: 1663

F.J. Cole, *History of Comparative Anatomy*. Macmillan, 1944

G. Cope, *Death, Dying, and Disposal*. SPCK, 1970

D. Cox, *Relics and Shrines*. Allen & Unwin, 1985

J.L.C. Dall et al. (eds), *1992 Sandoz Lectures in Gerontology*. Academic Press, 1993

A.R. David and E. Tapp, *The Mummy's Tale*. Michael O'Mara, 1992.

J.D. Davies, *Cremation Today and Tomorrow*. Grove Books, 1990

W.R. Dawson, 'Contributions to the History of Mummification'. *Proceedings of the Royal Society of Medicine*, 16 Feb. 1927, p. 832

A. Desmond and J. Moore, *Darwin*. Penguin, 1992

J. Diamond, *The Rise and Fall of the Third Chimpanzee*. Vintage, 1992

R. Doll and R. Peto, *The Causes of Cancer*. Oxford University Press, 1981

W.D. Edwards, W.J. Gabel and F.E. Hosmer, 'On the Physical Death of Jesus Christ'. *Journal of the American Medical Association*, 1996, 255: 1455–63

G. Elliot, *The Twentieth-century Book of the Dead*. Allen Lane 1972

D.J. Enright, *The Oxford Book of Death*. Oxford University Press, 1987

R. Etienne, *Pompeii: The Day a City Died*. Thames & Hudson, 1992

R.C.W. Ettinger, *The Prospects of Immortality*. Sidgwick & Jackson, 1965

R.C. Finucane, *Appearances of the Dead: A Cultural History of Ghosts*. Junction Books, 1982

D. Field and N.J. James, 'Where and How People Die', in *The Future for Palliative Care* (ed. D. Clark). Open University Press, 1993

B.A.J. Fisher, *Techniques of Crime Scene Investigation*. CRC Press, 1993

C. Francome, *Abortion Practice in Britain and the United States*. Allen & Unwin, 1986

G. Gallup, *Adventures in Immortality*. Souvenir Press, 1983

F.H. Garrison, 'The Bone Called "Luz"'. *New York Medical Journal*, 1910, 92: 149–51

W. Gaylin, Harvesting the Dead. *Harpers*, 1974, 52: 23

A. Gibbons, 'Geneticists Trace the DNA Trail of the First Americans'. *Science*, 1993, 259: 312

C. Gittings, *Death, Burial, and the Individual in Early Modern England*. Routledge, 1984

P.V. Glob, *The Bog People*. Paladin, 1971

G. Gorer, *Death, Grief and Mourning in Contemporary Britain*. Crescent Press, 1965

A. Green, *Haunted Houses*. Shire Publications, 1994

D. Greenwood, *Who's Buried Where in England*. Constable, 1982

M. Grey, *Return from Death*. Fontana, Penguin, 1986

M. Grey, *Return from Death – An Exploration of Near-Death Experience*. Arkana, Penguin, 1987

Guinness Book of Records, 1992. Guinness Superlatives Ltd, 1992

D. Harding, *The Little Book of Life and Death*. Arkana, Penguin, 1988

P. Harris, *What to Do when Someone Dies. Which?* Consumer Guide, 1995 edn. Consumers Association, 1995

R. Harris, *Murders and Madness*. Clarendon Press, 1989

R. Harris and J. Paxman, *A Higher Form of Killing*. Triad/Paladin, 1983

R. Hertz, 'Contribution à une Etude sur la Representation Collective de la Mort'. *L'Année Sociologique*, 1907, 10: 48–137

J. Hinton, *Dying*. Penguin, 1967

G. Hogg, *Cannibalism and Human Sacrifice*. Pan Books, 1961

R. Hughes, *The Fatal Shore*. Pan Books, 1988

S.C. Humphreys and H. King (eds), *Mortality and Immortality: The Anthropology and Archeology of Death*. Academic Press, 1981

Derek Humphry, *Final Exit: The Practicalities of Self-deliverance and Assisted Suicide for the Dying*. Hemlock Society, Eugene, Oregon, USA, n.d.

A. Huxley, *The Perennial Philosophy*. Chatto & Windus, 1946

A. Huxley, *The Devils of Loudun*. Chatto & Windus, 1952

R. Karsten, 'The Head Trophy of the Jibaro Indians' (1923), in *Primitive Heritage*, edn. M. Mead and N. Calas, Victor Gollancz, 1954

R.J. Kellett, 'Infanticide and Child Destruction – the Historical, Legal and Pathological Aspects'. *Forensic Science International*, 1992, 53: 1–28

Sir Geoffrey Keynes (ed.), *Sir Thomas Browne: Selected Writings*. Faber, 1968

B. Knight, *Simpsons Forensic Medicine*, 10th edn. Arnold, 1991

B. Knight, *Forensic Pathology*, 2nd ed. Arnold, 1996

R. Krogman and M.Y. Iscan, *The Human Skeleton in Forensic Medicine*, 2nd edn. C. Thomas, 1986

E. Kubler-Ross, *Questions and Answers on Death and Dying*. Macmillan, 1974

J. Layard, *Stone Men of Malekula*. Chatto & Windus, 1942

B. Lees et al., 'Differences in Proximal Femur Bone Density over Two Centuries'. *Lancet*, 1993, 341: 673

J.H. Lehman, *The First Boer War*. Cape, 1972

R.J. Lifton, *The Nazi Doctors: Medical Killing and the Psychology of Genocide*. Basic Books, 1986

J. Litten, *The English Way of Death: The Common Funeral Since 1450*. Robert Hale, 1992

N. Llewellyn, *The Art of Death. Visual Culture in the English Death Ritual 1500–1800*. Reaktion Books, 1991

A.D. Lopez, G. Caselli and T. Valkonen (eds), *Adult Mortality in Developed Countries: From Description to Explanation*. Clarendon Press, 1995

C. Malone, A. Bonanno, T. Gouder, S. Stoddart and D. Trump, 'The Death Cults of Prehistoric Malta'. *Scientific American*, Dec. 1993

P. McCullagh, *Brain Dead, Brain Absent, Brain Donors*. John Wiley, 1993

M. MacDonald and T.R. Murphy, *Sleepless Souls*. Clarendon Press, 1990.

H. Mellor, *London Cemeteries: An Illustrated Guide; Gazetteer*, 2nd edn. Gregg International, 1985

D.W. Meyers, *The Human Body and the Law*, 2nd edn. Edinburgh University Press, 1990

R.L. Miller et al., 'Paleoepidemiology of Schistosoma Infection in Mummies'. *British Medical Journal*, 1992, 304: 555

A. Mills, 'Mercury and Crematorium Chimneys'. *Nature*, 1990, 346: 615

J. Mitford, *The American Way of Death*. Hutchinson, 1963

G.H. Mullin, *The Tibetan Tradition*. Arkana, Penguin, 1986

C.J.L. Murray and A.D. Lopez, 'Global Mortality, Disability and the Contribution of Risk Factors: Global Burden of Disease Study'. *Lancet*, 1997, 349: 1436.

B.P. Norfleet, *Looking at Death*. David R. Godline, 1993

F. Parkes-Weber, *Aspects of Death and Correlated Aspects of Life in Art, Epigram, Poetry*, 3rd edn. Fisher Unwin, 1918

L.A. Parry, *Some Famous Medical Trials*. Churchill, 1927

C.J. Polson and T.K. Marshall, *The Disposal of the Dead*, 3rd edn. English Universities Press, 1975

R.W. Purcell and S.J. Gould, *Finders, Keepers: Eight Collectors*. Norton, 1992

S. Rinpoche, *The Tibetan Book of Living and Dying*. Rider/Random House, 1992

C. Roberts and K. Manchester, *The Archeology of Disease*. Cornell University Press, 1995

J. Robertson, A.M. Ross and L.A. Gurgoyne, *DNA in Forensic Science*. Ellis Horwood, 1990

W.C. Rodriguez, 'Decomposition of Buried Bodies and Methods that may Aid their Location'. *Journal of Forensic Sciences*, 1985, 30: 836–52

W.C. Rodriguez, Insect Activity and its Relation to Decay Rates of Human Cadavers in East Tennesse. *Journal of Forensic Sciences*, 1983, 28: 423

R. Richardson and B. Hurwitz, 'Jeremy Bentham's Self-image: An Exemplary Bequest for Dissection'. *British Medical Journal*, 1987, 295: 195–8

R. Richardson, *Death, Dissection and the Destitute*. Penguin Books, 1989

C.P. Richter, 'Rats, Man, and the Welfare State'. *American Psychologist*, 1959, 14: 18–28

R. Schafer, 'Dead Faces'. *Granta*, 1989, no. 27. Granta Publications, Penguin

E.L. Schneider and J.W. Rowe (eds), *Handbook of the Biology of Aging*. Academic Press, 1996

A.W.B. Simpson, *Cannibalism and the Common Law*. University of Chicago Press, 1984

K. Simpson, *Forty Years of Murder*. Harrap, 1978

K.G.V. Smith and A.M. Easton, 'The Entomology of the Cadaver'. *Medical Science and Law*, 1970, 10: 208–15

J. Soustelle, *The Daily Life of the Aztecs*. Pelican Books, 1964

A.J. Spencer, *Death in Ancient Egypt*. Pelican, 1982

C. Stringer, 'Secrets of the Pit of the Bones'. *Nature*, 1993, 362: 501

J.H. Taylor, *Unwrapping a Mummy*. British Museum Press, 1995

J.M.C. Toynbee, *Death and Burial in the Roman World*. Thames & Hudson, 1971

G.M. Trevelyan, *English Social History*. Penguin, 1967

G.C. Vaillant, *Aztecs of Mexico*. Pelican Books, 1950

P.L. Walker et al., 'Age and Sex Bias in the Preservation of Human Skeletal Remains'. *American Journal of Physical Anthropology*, 1988, 76: 183–8

D. Wallechinsky, I. Wallace and A. Wallace, *The Book of Lists*. Corgi Books, 1978

D. Wisniewski (ed.), *Annual Abstract of Statistics, 1997*. HMSO.

WHO. *The World Health Report, 1997*. World Health Organization

R. Wilkins, *The Fireside Book of Death*. Robert Hale,1990

A. Wilson and H. Levy, *Burial Reform and Funeral Costs*. Oxford University Press, 1938

J.W. Wilson, *Funerals Without God*. British Humanist Association, 1989

P. Ziegler, *The Black Death*. Collins, 1969

Index

Aberfan 23
Aboriginals 49, 143, 145, 170, 239
abortion 52, 62–6, 67
acceptance of death 323, 332
accidents and natural disasters 22–3, 25, 26–8
acid bath 188
adipocere 124
advice and support for the bereaved 337
afterlife xii-xiii, 126, 127, 171, 179, 190, 318, 326–8, 338–57
  ghosts, spectres and hauntings 353–5
  journey to the 355–6
  near death experiences (NDEs) 349–53
  reincarnation 349
  religious beliefs 341–9
ageing 81–107
  ageing process 89–93
  in animals 83–4
  causes 93–102
  longevity 86–7, 93–4, 96, 98, 105–7
  in the test tube 87–9
agelessness 5–7
AID (artificial insemination by donor) 277
AIDS see HIV/AIDS
Alaska 34, 208
Albigensians 42, 139
alcohol 16, 31, 78
  alcohol-related murder 76–7
  as a preservative 217–18
  premature deaths and 15, 17, 78, 306–7
Aleutian Islands 208
Alexander the Great 190
Alzheimer's disease 50, 90–1, 100, 104, 230
amphorae 142

anencephaly 117, 268–9
animals
  ageing 83–4
  animal attacks 27
  cremation 183
  longevity 95–6, 98, 99, 106
  mummification 206
  predation on corpses 122, 170–1
  sleep requirements 325
  transplantation from 274–5
antidepressant drugs 307
antimony 307
antioxidants 97–8
Antony and Cleopatra 44
Aristotle 59
arsenic 75, 76, 149, 307
Askew, Lucy 86
attitudes to death xii, 48, 317–20
Aum Shinrikyo 80
Australia 165, 173, 184, 207, 239
Austria 67
auto-erotic practices 305
autolysis 120
autopsies see post-mortem examinations
Aztecs 71, 291, 293

Babylonia 133, 156
Bacon, Francis 318
Bahrain 167–8
Balfour's test 115
barrows 141, 157
Baume, Pierre 248–9
beards 256
Bede 111
Beethoven, Ludwig van 167
beheading 56
Bentham, Jeremy 218–19
Berlioz, Hector 249
biblical attitudes to death 332

biological clock 112
birth control 64
birth rates 10
birth records 93
Black Death 22, 138
Black Prince 127
Blair, Robert xi
blood donors 275–6
bloodstains 298
body bags 139
body composition 111–12
body snatchers 243–6
body temperature 96, 112, 301
Boer War 138
bog men 123, 215–17, 237, 292
bone marrow transplants 276
bones 93, 123, 232–3, 252–4, 297–8, 321
Borgia, Cesare 75
Borneo 145
Bosnia 71
brain 85, 93, 113, 192–3, 197–8, 233, 313–14, 340
brain death 46, 115, 116–18, 260
brain-stem death 117, 118, 340
brass commemorative plaques 167
Brazil 26, 72, 166, 263–4
'broken heart' phenomenon 337
Brooke, Rupert 342–3
Brookwood 164
Brown-Séquard, Charles 103
Browne, Sir Thomas 43, 171
bubonic plague 138
Buddhism 40, 44, 53, 133, 135, 136, 143, 146, 170, 173, 266, 291, 347–8, 349
burial 132–68, 338–9
  abhorrence of 136, 169–70

in ancient
    civilizations   133
at sea   153, 154–5
Buddhist customs   135,
    136
burial position   133
cavalry burials   154
cemeteries *see* cemeteries
Christian customs   134
deep burial   148–9
delayed burial   145–6
earliest known   132
foetal position   133, 143
funerals *see* funerals
garden burials   160, 168
Hindus and Sikhs   135, 136
intramural burial   159
Islamic customs   134–5,
    141
Jewish customs   133–4
mass graves   137–8, 140,
    149
paupers' burials   137, 148
perpetual burial rights   165
pit burials   136–7, 141
plague victims   138
premature burial   144–5,
    260
reburial   146–9
separate burial of parts of
    the body   311
ship burials   153–4
tree burial   145, 170
vertical burial   133, 136
'walk away' burial   169
woodland burials   152,
    168
burial clubs and
    societies   151, 247
Burke and Hare   245
Burma   26
burning
    suttee   36–7, 44, 174–5
    war dead   176–8
    of witches, heretics and
        martyrs   175–6
    see also cremation
burns   304
Byron, Lord   178, 252–3, 260

Caesar, Julius   173
cairns   157
Calvin, John   284
Canada   180
Canary Islands   207
cancer   28–9, 50, 51, 87, 88,
    330–2
cannibalism   77, 184–8, 288–
    9, 291
    gourmet cannibalism
        185–6

ritualistic cannibalism
    184–5
survival cannibalism
    186–8
canopic jars   202
capital punishment   53–9, 60
car crash tests   250
carbon monoxide   32, 308
cardiovascular disease   13, 28
Carrel, Alexis   87
Castaing, Dr Edme   76
catacombs   159, 161, 208,
    282
catalepsy   144
causes of death
    accidents and natural
        disasters   22–8
    cancer   28–9
    cardiovascular disease   13,
        28
    cigarette smoking   13–14
    environmental causes   15–
        16
    infectious and parasitic
        diseases   13, 19–22
    main causes (global)   12,
        14–15
    socio-economic factors   13
    war, massacre,
        starvation   16–19
    see also euthanasia;
        homicide; suicide
cells
    apoptosis   4, 101
    cell biology   87–9
    cell death   3–4, 100–2, 113
    cell division   88, 99–100
    cell suicide   101
    germ cells   5, 88
    Hela cells   88, 308
    nerve cells   100, 113, 114
    property rights   308–9
    T-cells   100–1, 258
cemeteries   159–66, 171,
    323, 356
    lawn cemeteries   164
    modern cemeteries   163–6
    multi-storey
        cemeteries   165–6
    municipal
        cemeteries   160–1, 162
cenotaphs   167
centenarians   93–4, 99, 103
chamber-tombs   136–7
Charlemagne   178, 311–12
Charles I of England   56,
    193–4
charnel houses   137
Charon   129
Chatterton, Thomas   39
chemical execution   59

chemical weapons   18–19
children
    attitude to death   329
    child suicide   34–5
    human sacrifice   211, 290
    infant and child
        mortality   10, 322–3
    infanticide   59–62
    stillborn babies   61, 134,
        269, 335
Chile   208, 210–11
China   14, 31, 34, 35, 40, 62,
    140, 180, 184, 268
    burial   146, 158, 159, 164
    capital punishment   54, 58
    eugenics law   67
    infanticide   61
chlorine   18
choking to death   27
Christianity   132, 134, 178
Christie, John   78–9
Churchill, Winston   319
churchyards   43, 152, 159–
    60, 161–2, 356
Cicero   287–8
cloning   7, 215, 221–3
coffin flies   122
coffins   141–3, 203
    anthropoid coffins   143,
        199
    biodegradable
        coffins   131, 143, 189
    common coffins   137
    cubic shape   143
    log-coffins   143
    opening of   149
    rent-a-coffin   137, 143
    sealing   131
    wood and reed
        coffins   141, 142
    zinc and lead-lined   131,
        142
columbarium   145, 163, 181
common graves   137, 148
compostorium   188–9, 356
computer imaging   220, 249,
    250–1
Confucianism   44
contagion   125
corneal transplants   255–6,
    259
coroners   114–15, 149
corpse
    abuse of   286–94
    commercial
        exploitation   292, 294
    corpse-dressing   141
    decomposition   24, 111,
        119, 125–6
    exposure to the
        elements   169–71

identifying 296–300, 302
laying out 128–9, 152
medicinal uses of 280–1
odour 121, 125, 130
property rights in 227–8, 263–4
punishing 286–8
removal of parts of 310–14
unidentified corpses 295
uses for 227–85
viewing 129–30, 152, 163, 320
watching vigils 152
corpse disposal see burial; cremation
corticol death 114
Cotton, Mary Ann 75
Creech, Thomas 33
cremation 133, 134, 135, 145, 163, 169, 171–83, 248, 356
  animals 183
  arguments for and against 171–2
  ashes, disposal of 152, 164, 172, 181–3, 322
  changing attitudes to 178, 179–80
  Hindu cremation 173–4
  modern crematoria 180–1
crimes passionels 54
criminals, executed 53–9, 60, 242–3, 268
Crippen, Dr 298
Cromwell, Oliver 287
crucifixion 56–7
cryptopreservation 213–15
cult of the dead 156–7
Cummings, Frederick 327–8
cutting, stabbing and chopping 304
  suicide by 33
cyanide 59, 307–8
Cyprus 149

danse macabre 322
Darlington, C. D. 66
Darwin, Charles 146, 147, 355
de Gaulle, Charles 150
death agony 113
death certificates 114
death masks see masks and effigies
death rates 10, 13
death rattle 112
death, signs of 115
decomposition 111, 122–4, 125–6, 154, 295
Defoe, Daniel 22

Delacroix, Eugène 104
Deng Xiaoping 262
Denmark 34, 157, 216
Descartes, René 119, 340, 342
diarrhoeal infections 20, 100
diatom test 123–4
Dickinson, Emily 32, 334
Dido 39
dietary restriction 95–6, 103, 104
disability adjusted life years (DALYs) 14, 15
dismemberment 287
dissection 218, 227–49, 286
  alternatives to 249
  body snatchers 243–6
  corpses, sources of 242
  divination 231–2
  donated bodies 248–9
  ethical and legal problems 238–40, 246
  executed criminals 242–3
  history of 241–2
  imported bodies 248
  mummies 234–6
  paupers' corpses 246–8
  post-mortem examinations 228–31, 302
divination, using corpses for 231–2
DNA 5, 6, 98, 100, 101, 106, 215, 222, 296
  fingerprinting 298–9
  mitochondrial DNA 237–8, 300
  Neanderthal DNA 234
  preservation of 220–1
  property rights 309–10
  recovery 237–8
Douce, Francis 145
Downs syndrome 94
Doyle, Arthur Conan 355
dreams 324, 325, 326
Dresden 139
drowning 76, 123–4, 305–6
  suicide by 32
drugs 16, 32, 307
Dryden, John 327
duelling 72–3
Dunblane 79
Durkheim, Emile 40
dying process 112–13, 317–18

East Africa 170–1
Easter Island 234, 238
Egypt 133, 136–7, 141, 149–50, 155, 156, 159, 166, 173, 192–3, 195–207, 234–6, 343, 355

Egyptian Book of the Dead 196
Einstein 314
electric chair 58–9
electrocution 304–5
electroplating the dead 219
elixir of life 102–3
Elizabeth I of England 136
Elliot, Gil 17
embalming 129, 130, 191–5, 207
embryo development 4, 270
embryo use 257, 279–80
Epicurus 332
ethical and legal problems 115–16, 238–41, 263–4, 270–4
ethnic cleansing 71
eugenics 66–71
eunuchs 311
euthanasia 45–52, 330
  alternatives to 48–52
  compulsory euthanasia 52, 70
  doctrine of double effect 51
  physician-assisted suicide 45–6, 47–51
  the 'right to die' 46–8
Evans, Timothy 79
evolution 4, 5–6, 7, 68, 84
exercise, life expectancy and 103
exhumation 148, 149
extinction 7–8

felo de se 33, 43
fertility treatment 277–80
fetal tissue, use of 269, 271, 272, 274
fingerprints 296–7
Finland 10, 34, 67
firing squads 58
'first-degree' murder 80
First World War 138
flies 121, 122
foreign lands, death in 146
forensic pathology 122, 230
formaldehyde 130, 192, 217, 218, 227
fossils 124–5, 233
foundling hospitals 60
France 58, 72–3, 125, 129, 139, 157, 160–1, 163
Frankenstein 260
Franklin, Benjamin 217–18, 357
free radicals 97–8
freezing 89, 210–15
  cryptopreservation 213–15

natural freezing   210–13
Freud, Sigmund   324, 326,
    329
funeral directors   150
funeral processions   150,
    160–1
funeral supermarkets   xii
funerals   xii, 149–53, 335
    burial clubs   151, 247
    current costs   151–2
    Egyptian   203
    funeral wakes   152, 336
    nocturnal funerals   151
    non-religious
        funerals   152–3
    paupers' funerals   137, 148
    therapeutic effect   152

Galen   241, 340
Galileo   312
gallows   55–6
Galton, Sir Francis   66
Gandhi, Mahatma   38–9
Ganges   171, 174
garden burials   160, 168
garden cemeteries   160
garrotte   57–8
gas chambers   59, 69, 140
Gay, John   328
genes   5, 6, 258
    genetic defects   66, 67
    genetic mutation   98
    and life-span   98–9
genocide   17, 70, 139–40
George II of England   143
Germany   48, 52, 58, 67–9,
    73, 144–5, 180, 263, 280
gerontology   86, 102
Ghana   143
ghosts, spectres and
        hauntings   353–5
gibbets   55, 242
GIFT (gamete intra-fallopian
        transfer)   277
Giza   156
Gladstone, William   217
Goethe   39, 324
gold, preservation in   219
Gompertz, Benjamin   81
Gordon, General Charles   167
Gozo   158
Grabaulle Man   216
Grace, Princess   130
grave goods   132, 339
graveyards see churchyards
Greece   44, 62, 72, 129, 142,
    146, 173, 177–8
Greek Orthodox Church   145
Greenland   34, 44, 208
growth hormones   280–1
guillotine   58

guns   74, 79
    gun wounds   303–4
    suicide by gunshot   31, 32,
        35
gypsies   69, 140

Haarmann, Fritz   288–9
Hades   356
Haigh, John George   188
hair   90, 113, 255, 256, 275
Haldane, J.B.S.   114, 190, 252
hanging   54–6, 305
    suicide by   31–2
Hannibal   38
hara-kiri   38
Hardy, Thomas   31, 313
Harvey, William   241
Hayflick, Leonard   88
Hazlitt, William   319
health and happiness   11–12
heart burial   312–13
heart transplants   257
Heimlich manoeuvre   27
hell   172, 343, 344
Hemlock Society   47
Henry V of England   146
Henry VIII of England   55,
    193
hepatitis B   21, 29
heretics   176
hero-stones   167
Herod   59
Herodotus   197, 199, 252
Highgate Cemetery   161
Hinduism   44, 135, 136, 171,
    173–4, 266, 275, 346, 349
Hippocratic oath   50, 273
Hiroshima   139, 182
Hitler, Adolf   18, 67–8, 69,
    71, 296, 297
HIV/AIDS   15, 20, 21, 50, 80,
    125, 259
Ho Chi Minh   195
Holmes, Oliver Wendell   94
homicide   52–80
    abortion   62–6
    eugenics   66–71
    execution of
        criminals   53–9
    extenuating
        circumstances   53
    infanticide   59–62
    manslaughter   80
    murder   71–80
    sacrifice   71
Horemkenesi   200–2
hospices   330, 331
hospital deaths   xii, 129, 323
Huguenots   139
Human Genome Project   71,
    220, 221, 309

human remains, attitudes
    to   238–41
human sacrifice   71, 217,
    289–90, 293
    child sacrifice   211, 290
    motives and occasions
        for   140, 289–90
Humane Society   44–5
Hume, David   43, 119
Hungary   34
Hussein, Saddam   215
hypnotism   73–4
hypothermia   306

illness brought on by
    grief   336–7
immortality   5–7, 327
immune system, changes in
    the   98
impact of death   14–16
Incas   207, 210–11
India   26, 73, 92, 167, 171,
    247, 265–6, 291
    burial   136, 145, 170
    cremation   173–4
    suttee   36–7, 44, 174–5
infant and child
    mortality   10, 322–3
infanticide   59–62
infectious and parasitic
    diseases   13, 19–22,
    87
inquests   114
Inquisition   176
insect invasion of a
    corpse   121–2
insulin   308
intensive care   115, 116
Inuit   37, 240
Iran   56
Ireland   34, 173, 184, 216
Islam   44, 63, 129, 132, 133,
    134–5, 141, 173, 197,
    266, 311, 336, 345–6
Israel   166
Italy   34, 58, 163, 166, 180,
    280
    see also Romans
IVF (in vitro
    fertilization)   277

Jack the Ripper   77
Jainism   52–3
James, Henry   348
Japan   35, 37, 48, 62–3, 67,
    72, 208, 266
    burial   165–6
    life expectancy   82, 87
Jenner, Edward   24
Jesus Christ   40, 42, 57, 284,
    344

Jews and Judaism 42, 69, 133–4, 140, 159, 178, 267, 335, 345
Joan of Arc 176
Johnson, Dr Samuel 104, 111
jumping to one's death 32–3, 306
Jung, Carl Gustav 348
Jutland 167

kamikaze pilots 38–9
Kant, Immanuel 119
Keats, John 162
Kevorkian, Dr Jack 49
kidney transplants 257, 258, 261, 262, 265, 266
Kim Il Sung 195

Lafarge, Marie 75
Lawrence, D.H. 183
Leary, Timothy 183
Lee, John 56
Lenin, Vladimir Ilyich 130, 195, 281, 345
Leningrad 17, 139, 164
Leonardo da Vinci xiii
Les Innocents 125, 161
lethal injection 54, 59
life expectancy 10–11, 81–2, 86–7
    increasing 102–4
life-span 11, 82–3, 84–5, 95, 99, 102, 103, 106
life-support systems 115
lightning 26–7
Lindow Man 216
living, rate of 95–6
living wills 46–7
Livingstone, David 313
Locke, John 119
longevity 86–7, 93–4, 96, 98, 105–7
Lucretia 38
Luther, Martin 284
Luz 133
lynching 56

Madagascar 158
maggots 121–2
malaria 20
Malta 158
manslaughter 60, 80
Maoris 185–6, 239, 240
Marcos, Ferdinand 166, 215
Maria Theresa, Empress 48
Martensson, Bertil 107
Martinez, Miguel 119
martyrs 40, 42, 176, 283
Masada 41

masks and effigies 130, 166–7, 320
mass graves 137–8, 140, 149
mass murder 79–80
massacres 16, 138, 139–40
mastabas 155, 204, 205
measles 21
medicinal uses of corpses 280–1
megalithic tombs 157, 158
memento mori 282, 318, 322
memorial sculptures 167
Mengele, Dr Josef 70, 300
mercury salts 218
Metchnikov, Ilya 103
Mexico 27, 156, 208, 291
miscarriages 63, 269–70
missing persons 72, 295
Mohammed 346
Monroe, Marilyn 40
Montaigne, Michel Eyquem de 4–5
morgues/mortuaries 129, 152
morphia 51, 76
Mortensen, Christian 86
Mount Everest 210
Mount Li tomb 158
mourning customs 335–6
mourning and grieving 333–7
Mozart, Wolfgang Amadeus 237, 282, 320
Mudgett, H. W. 79
mugging 75
mummification 136, 193, 195–209, 320, 343
    animals 206
    communal burials 207
    dating mummies 234–5
    Egyptian process 197–9
    internal organs 202–3
    natural mummification 123, 196, 208
    reconstructed faces 236–7
    'self-mummification' 208
    shrunken heads 209
    smoke-curing 207
    unwrapping a mummy 200–2, 234
murder 54, 71–80
murder investigations 300–1
murder weapons 74–5
Muslims see Islam
mustard gas 18, 19
mutilation 77
myocardial degeneration 89

Napoleon 311
narcolepsy 326

Native Americans 44, 123, 133, 145, 240
'natural causes' 82
natural disasters 23, 25–6, 138
Navohi 189
Nazis 67, 68, 69, 70, 71, 140, 308
Neanderthals 84, 126, 233–4
near death experiences (NDEs) 119, 349–53
necessity of death 4–5, 106
Necker, Suzanne 217
necrophilia 294
necropolis 157
Nehru, Jawaharlal 174
Nelson, Lord 139, 217
neolithic burials 146–7, 157
nerve gases 18–19
Netherlands 47–8, 164, 180, 263
New Guinea 145, 147, 184, 207
New Zealand 10
Newton, Isaac 312, 319–20
Nitschke, Dr Phillip 49
Norsemen 153
Norway 34, 67, 153

Oates, Captain Lawrence 37
obituaries 337
O'Brien, Charles 238–9
occult practices 77
Okinawa 139
old people and death 37, 52, 330–2
    see also ageing
Omar Khayyam 353
organ and tissue transplants 115–16, 117, 135, 255–80
    anencephalic infant donors 268–9
    animal organs 274–5
    cadavers, use of 261–3
    commercial trafficking 264, 265–8
    donors 259–61
    earliest transplants 255, 259
    ethical issues 263–4, 270–4
    executed prisoner, organs from 268
    fertility treatment 277–80
    from foetuses 269–70
    live related donors 265
    medical problems 258, 259
    organ rejection 258
    religious attitudes to 135, 266

using dead or renewable
  human material 275–
  80
ossuaries 137, 148
osteoarthritis 232–3
ova donation 278–9
overlaying of infants 60

Pacific islands 153
Pakistan 170
Papua New Guinea 184–5
paracetamol 307
Parsee 170
Pascal, Blaise 3
passage graves 173
paupers' burials 137, 148
Pavlova, Anna 240
Pepys, Samuel 193, 256, 336
Peron, Eva 192
persistent vegetative
  state 114, 117, 264–5
personhood 119
Peru 137, 141, 207, 211, 291
Peter the Great 254
petrification 124
philosophical aspects of
  death 118–19
phosgene 18
Pierrepoint, Albert 56
pit burials 136–7, 141
placental burial 254, 311
placentas 254–5
plague 22, 125, 138
plague pits 138
Plato 59, 340
Pliny 177, 178
Plunkett, Sir Oliver 287
Poe, Edgar Allen 144
poisoning 75–6, 306
  suicide by 32, 35
Poland 48
poliomyelitis 114
Polynesia 184, 238
Pompeii 138, 177
population growth xi, xii, 9
Port Arthur 79
Portugal 34
post-mortem
  examinations 228–31,
  302
Potter, Beatrix 182
premature death 11
preservation
  alcohol 217–18
  eccentric methods 218–20
  formaldehyde 130, 192,
  217, 218
  mercury salts 218
  silicone 218
  see also embalming;
  freezing; mummification

privation 17
progeria 94, 95
protein abnormalities 97
Protestant Cemetery,
  Rome 162, 179
psychical phenomena 354
Purgatory 345
putrefaction 120–1, 122–4,
  125, 129, 130
pyramids 150, 155–6, 167

Quakers 160
quality of life 11–12

radioactivity 27–8
Raleigh, Sir Walter 194
reincarnation 349
relics 281–5, 311–12, 322
remembering the dead 126–
  8
Renan, Joseph 339
respiratory infections 20
resurrection 134, 171, 172,
  178, 190, 344–5
resurrectionists see body
  snatchers
Richard I of England 312
Richter, C. P. 68
rigor mortis 120, 129, 301
road deaths 23, 25, 33
rock tombs 155
Roman Catholicism 34, 134,
  172, 176, 179, 336, 345
Romans 44, 57, 62, 129, 133,
  142, 150, 151, 156–7,
  159, 166, 173, 231, 312
Romeo and Juliet 39
Rookwood Necropolis 165
Roose, Richard 75
Rosenberg, Julian and
  Ethel 59
Rossetti, Dante Gabriel 149
Rousseau, Jean-Jacques 43,
  341
rune-stones 167
Russia 16, 272
Russian royal family 299–
  300
Ruysch, Frederik 253–4

'sageing' 85
St Francis Xavier 285
Salem 175–6
Samaritans 45
sarcophagus 141
sarin 19
Sartre, Jean-Paul 119
sati-stones 167, 174
Saudi Arabia 33, 56
scalps 312
scavengers 170–1

schists 157
sea burials 153, 154–5
Second World War 139
self-sacrifice 37
semen donation 277–8, 309
Seneca 30–1, 44
serial murderers 77–9, 80
sexual life 91
Shakespeare, William 39, 85,
  125–6, 132, 324, 326
sharks 27, 170, 303
Shaw, George Bernard 100
Shelley, Mary
  Wollstonecraft 260
Shelley, Percy Bysshe 127,
  160, 162, 178–9, 260,
  313, 323–4, 340
Shintoism 44, 143, 173, 286
ship burials 153–4
shiva 335
shrouds 140–1, 142
shrunken heads 209
Siberia 212
Sicily 208
Siddal, Elizabeth 149
Sidney, Sir Philip 150–1
signs of death 115
Sikhs 129, 135, 173, 182,
  349
Silbury Hill 157
silicone 218
Singapore 67
skeletons 123, 228, 322
skin grafts 113, 276–7
skulls 147, 233, 252, 282,
  321, 322
  building faces from 236–7
sleep, comparison of death
  with 323–4
sleeping sickness 21
smallpox 24, 221
smoking, deaths attributable
  to 13–14, 15, 29
snakebite 26
Socrates 317
Solomon Islands 170
somnambulistic crimes 74
soul/spirit 119, 134, 215,
  339, 340, 342, 349
South Sea Bubble 38
Spain 34, 57, 180
spiritualism 354, 355
spontaneous human
  combustion 304
Stalin, Joseph 195
starvation 37
stillborn babies 61, 134, 269,
  335
stings 10, 27
Stopes, Marie 64
strangling 75, 305

strychnine 308
Sudan 155
Sudden Infant Death
     Syndrome (SIDS) 60
suffragettes 40
suicide 30–45, 286, 295,
     304, 305, 307
   burials 134
   chronic suicidal
     behaviour 33
   decriminalization 43
   fake suicides 33
   imitative behaviour 40
   mass suicides 40–2
   mental condition and 35,
     37
   methods 31–3
   national rates of 34, 44
   physician-assisted
     suicide 45–6, 47–51
   popular
     misconceptions 36
   prevention 44–5
   reasons for 36–42
   religious and cultural
     attitudes to 42–4
   ritual suicide 38
   romantic suicides 39
   significant factors 34–6
   suicide pacts 39
   threats of 39–40
   time, chosen 42
   unsuccessful attempts 31
Sumatra 184
Sumeria 132, 339
surrogate motherhood 277,
   280
suttee 36–7, 44, 174–5
Sutton Hoo 154
sweating sickness 23
Sweden 10, 34, 67, 72, 93,
   157
Swift, Jonathan 86, 107,
   188, 288
Swinburne, A. C. 328
Switzerland 48, 67, 180

tabun 19
Taiwan 268
taphephobia 144
Tasmania 238, 280
teeth 124–5, 235–6, 244,
   254, 297
Tennyson, Alfred, Lord 148–
   9, 291, 327
terminal illness 48, 50, 51,
   330
terrorism 19, 79
tetanus 21
theosophy 349

Thomas à Becket 283, 313
Thuggees 73
Tibet 169, 170, 183, 208, 291
Tibetan Book of the
     Dead 136, 323, 348
Tibetan Buddhism 136, 347–8
time of death 112, 120, 301
Titanic 154
Tollund Man 192, 216
Tolstoy, Leo 341
tomb-cult 162–3
tomb robbers ·155, 203–6
Tomb of the Unknown
     Warrior 167
tombs and monuments 127,
     155–8, 167–8
tombstones and
     headstones 155, 159,
     162
Torres Strait Islands 207
torture 77
Towers of Silence 170
toxin theory of disease 97
transplantation see organ and
     tissue transplants
trepanning 233
tuberculosis 20, 21, 235
tumuli 157
Tutankhamun 203–4, 311
Twain, Mark 85–6

Uganda 82
undertakers 150, 336
United Kingdom 10, 12, 23,
     26, 27, 29, 31, 34, 43, 51,
     60, 61, 72, 74, 77–9, 80,
     93, 134, 138, 180, 184
   abortions 63–5
   blood donors 276
   burial 137, 141, 142, 154,
     157, 159–60, 161, 162,
     164
   capital punishment 53–4,
     55–6, 58
   centenarians 94
   cremation 182
   embalming 193–4
   fertility treatment 279
   funerals 150–3
   grave-robbing 243–6
   life expectancy 81, 86–7
   organ transplants 261–3
United States 10, 23, 26, 28,
     31, 34, 35, 49, 61, 80,
     117, 246
   abortions 65–6
   blood donors 276
   burial 142, 162, 163–4
   capital punishment 54,
     58, 59

centenarians 94
cremation 180, 182
embalming 191
eugenics 67
fertility treatment 279
homicide 72
life expectancy 82, 87,
   104
organ transplants 262

vaccination 24, 26, 29
vampires 148
vaults 159
Veronoff, Serge 103
Vesalius 133–4, 241
viral infections 101
viscera 130, 146, 202–3, 310
Visible Human Project 250–
   1
Voyager Trust 133
vultures 170

wakes 336
Wallace, Alfred Henry 355
war booty 312
war graves 138–9, 154
wars 16, 17, 18, 19
Wellington, Duke of 73, 142
Werners syndrome 94, 95,
   100
Wesley, John 43
West Kennet Long
     Barrow 157
whales 303
Whitman, Walt 319
whooping cough 21
wigs 256
William the Conqueror 312–
   13
Williams, John 54
wills 151, 153
winding sheets 141
witches 175–6
Wither, George 128
Woolf, Virginia 32
Wordsworth, William 120,
   327
World Health Organization
     (WHO) 11, 12, 24
worms 121–2

xenogenic transplants 274–5

Young, Edward xi
Young, Graham 76, 231

Zen Buddhism 347
Zenca 44
ziggurats 156
Zoroastrianism 169–70, 353